Lecture Notes in Computer Science   8571

Commenced Publication in 1973
Founding and Former Series Editors:
Gerhard Goos, Juris Hartmanis, and Jan van Leeuwen

Holger Giese   Barbara König (Eds.)

Graph Transformation

7th International Conference, ICGT 2014
Held as Part of STAF 2014
York, UK, July 22-24, 2014
Proceedings

Springer

Volume Editors

Holger Giese
Hasso-Plattner-Institut an der Universität Potsdam
Prof.-Dr.-Helmert-Straße 2-3, 14482 Potsdam, Germany
E-mail: holger.giese@hpi.uni-potsdam.de

Barbara König
Universität Duisburg-Essen, Fakultät für Ingenieurwissenschaften
Abteilung für Informatik und Angewandte Kognitionswissenschaft
Lotharstraße 65, 47057 Duisburg, Germany
E-mail: barbara_koenig@uni-due.de

ISSN 0302-9743   e-ISSN 1611-3349
ISBN 978-3-319-09107-5   e-ISBN 978-3-319-09108-2
DOI 10.1007/978-3-319-09108-2
Springer Cham Heidelberg New York Dordrecht London

Library of Congress Control Number: 2014942862

LNCS Sublibrary: SL 1 – Theoretical Computer Science and General Issues

Typesetting: Camera-ready by author, data conversion by Scientific Publishing Services, Chennai, India

Printed on acid-free paper

Springer is part of Springer Science+Business Media (www.springer.com)

Foreword

Software Technologies: Applications and Foundations (STAF) is a federation of a number of the leading conferences on software technologies. It was formed after the end of the successful TOOLS federated event (http://tools.ethz.ch) in 2012, aiming to provide a loose umbrella organization for practical software technologies conferences, supported by a Steering Committee that provides continuity. The STAF federated event runs annually; the conferences that participate can vary from year to year, but all focus on practical and foundational advances in software technology. The conferences address all aspects of software technology, from object-oriented design, testing, mathematical approaches to modeling and verification, model transformation, graph transformation, model-driven engineering, aspect-oriented development, and tools.

STAF 2014 was held at the University of York, UK, during July 21–25, 2014, and hosted four conferences (ICMT 2014, ECMFA 2014, ICGT 2014 and TAP 2014), a long-running transformation tools contest (TTC 2014), 8 workshops affiliated with the conferences, and (for the first time) a doctoral symposium. The event featured 6 internationally renowned keynote speakers, and welcomed participants from around the globe.

The STAF Organizing Committee thanks all participants for submitting and attending, the program chairs and Steering Committee members for the individual conferences, the keynote speakers for their thoughtful, insightful, and engaging talks, the University of York and IBM UK for their support, and the many ducks who helped to make the event a memorable one.

July 2014

Richard F. Paige
General Chair
STAF 2014

Preface

ICGT 2014 was the 7th International Conference on Graph Transformation held during July 22–24, 2014 in York. The conference was affiliated with STAF (Software Technologies: Applications and Foundations) and it took place under the auspices of the European Association of Theoretical Computer Science (EATCS), the European Association of Software Science and Technology (EASST), and the IFIP Working Group 1.3, Foundations of Systems Specification.

ICGT 2014 continued the series of conferences previously held in Barcelona (Spain) in 2002, Rome (Italy) in 2004, Natal (Brazil) in 2006, Leicester (UK) in 2008, Enschede (The Netherlands) in 2010, and in Bremen (Germany) in 2012, following a series of 6 International Workshops on Graph Grammars and Their Application to Computer Science from 1978 to 1998.

Dynamic structures are a major cause for complexity when it comes to model and reason about systems. They occur in software architectures, configurations of artefacts such as code or models, pointer structures, databases, networks, etc. As interrelated elements, which may be added, removed, or change state, they form a fundamental modeling paradigm as well as a means to formalize and analyze systems. Applications include architectural reconfigurations, model transformations, refactoring, and evolution of a wide range of artefacts, where change can happen either at design or at run time. Dynamic structures occur also as part of semantic domains or computational model for formal modeling languages.

Based on the observation that all these approaches rely on very similar notions of graphs and graph transformations, theory and applications of graphs, graph grammars and graph transformation systems have been studied in our community for more than 40 years. The conference aims at fostering interaction within this community as well as attracting researchers from other areas to join us, either in contributing to the theory of graph transformation or by applying graph transformations to already known or novel areas, such as self-adaptive systems, overlay structures in cloud or P2P computing, advanced computational models for DNA computing, etc.

The conference program included three joint sessions with ICMT 2014, the 7th International Conference on Model Transformation, where two of these sessions were composed of papers accepted at ICMT 2014 and one of papers accepted for ICGT 2014. The proceedings of ICGT 2014 consist of one invited paper, titled "Parameterized Verification and Model Checking for Distributed Broadcast Protocols" by Giorgio Delzanno, and 17 contributions, which were selected following a thorough reviewing process.

The volume starts with the invited paper. The further papers are divided into the thematic topics verification, meta-modeling and transformations, rewriting and applications in biology, graph languages and graph transformation, and applications.

We are grateful to the University of York and the STAF Conference for hosting ICGT 2014, and would like to thank the authors of all submitted papers, the members of the Program Committee as well as the subreviewers.

Particular thanks go to Andrea Corradini for organizing the Doctoral Symposium as part of the STAF Conference and the organizers of the satellite workshops related to ICGT 2014 and affiliated with the STAF Conference:

- 5th International Workshop on Graph Computation Models (GCM 2014), organized by Rachid Echahed, Annegret Habel, and Mohamed Mosbah
- 8th International Workshop on Graph-Based Tools (GraBaTs 2014), organized by Matthias Tichy and Bernhard Westfechtel

We are also grateful to Leen Lambers for her support as publicity chair. Finally, we would like to acknowledge the excellent support throughout the publishing process by Alfred Hofmann and his team at Springer, and the helpful use of the EasyChair conference management system.

July 2014

Holger Giese
Barbara König
Program Chairs
ICGT 2014

Organization

Program Committee

Paolo Baldan	Università di Padova, Italy
Luciano Baresi	Politecnico di Milano, Italy
Paolo Bottoni	Sapienza – Università di Roma, Italy
Andrea Corradini	Università di Pisa, Italy
Juan de Lara	Universidad Autónoma de Madrid, Spain
Rachid Echahed	CNRS, Laboratoire LIG, France
Gregor Engels	Universität Paderborn, Germany
Claudia Ermel	Technische Universität Berlin, Germany
Holger Giese	Hasso-Plattner-Institut Potsdam, Germany
Reiko Heckel	University of Leicester, UK
Frank Hermann	University of Luxembourg, Luxembourg
Hans-Jörg Kreowski	Universität Bremen, Germany
Barbara König	Universität Duisburg-Essen, Germany
Leen Lambers	Hasso-Plattner-Institut Potsdam, Germany
Tihamer Levendovszky	Vanderbilt University, USA
Fernando Orejas	Universitat Politècnica de Catalunya, Spain
Francesco Parisi-Presicce	Sapienza – Università di Roma, Italy
Detlef Plump	University of York, UK
Arend Rensink	University of Twente, The Netherlands
Leila Ribeiro	Universidade Federal do Rio Grande do Sul, Brazil
Andy Schürr	Technische Universität Darmstadt, Germany
Pawel Sobociński	University of Southampton, UK
Gabriele Taentzer	Philipps-Universität Marburg, Germany
Pieter Van Gorp	Eindhoven University of Technology, The Netherlands
Daniel Varro	Budapest University of Technology and Economics, Hungary
Albert Zündorf	Universität Kassel, Germany

Additional Reviewers

Arifulina, Svetlana	Habel, Annegret
Deckwerth, Frederik	Koch, Andreas
Dyck, Johannes	Kuske, Sabine
Ehrig, Hartmut	Nachtigall, Nico
George, Tobias	Poskitt, Christopher M.

Radke, Hendrik
Raesch, Simon-Lennert

Yan, Hui
Zambon, Eduardo

Table of Contents

Invited Contribution

Verification

Meta-Modelling and Model Transformations

Rewriting and Applications in Biology

Graph Languages and Graph Transformation

Applications

Parameterized Verification and Model Checking for Distributed Broadcast Protocols

Giorgio Delzanno

DIBRIS, Università di Genova, Italy
Giorgio.Delzanno@unige.it

Abstract. We report on recent research lines related to parameterized verification and model checking applied to formal models of distributed algorithms. Both approaches are based on graph rewriting and graph transformation systems. Case-studies include distributed mutual exclusion protocols like Ricart-Agrawala, routing protocols like link reversal, and distributed consensus protocols like Paxos. Verification algorithms for restricted classes of models exploit finite-state abstractions, symbolic representations based on graph orderings, the theory of well-structured transition systems, and reachability algorithms based on labeling procedures.

1 Introduction

In this paper we present our recent work on graph-based languages and tools for the formal specification and validation of distributed systems, protocols, and algorithms.

A first research line is focused on foundational issues related to verification problems of topology-sensitive models of distributed protocols. The goal of this kind of study is to identify models, beyond the finite-state case, for which verification is still possible. In this setting we consider a graph-based representation of networks and a process model based on communicating automata. Automata are located on individual nodes of the graph that represents a network. Broadcast communication can be adopted as a general synchronization mechanism even for networks with unknown topology. The operational semantics is expressed via updates rules for node labels, i.e., via ad hoc graph transformation rules. An interesting research question here is decidability and complexity of verification problems for both synchronous and asynchronous broadcast communication. Parameterized verification can be reduced to reachability problems formulated for initial configurations with arbitrary topology and number of nodes. Our methodology, based on a detailed analysis of models of increasing complexity (synchronous communication, asynchronous, clocks, registers), allowed us to draw decidability boundaries for reachability problems with parametric initial configuration.

A second research line is focused on the application of verification tools working at different level of abstractions like Promela/Spin [43], based on automata theory, and Groove [42], based on graph transformation systems, on examples of

H. Giese and B. König (Eds.): ICGT 2014, LNCS 8571, pp. 1–16, 2014.
© Springer International Publishing Switzerland 2014

complex distributed algorithms (leader election, distributed routing, distributed consensus) again based on broadcast communication mechanisms. The comparison of tools like Spin and Groove on a common set of case-studies provides some general guidelines that can be used to reduce the cost of state-space exploration for asynchronous systems and to collect requirements for an ideal verification tool for validation of the considered class of protocols. In the rest of the paper we will give more details on the above mentioned research lines.

2   Parameterized Verification

In this section we describe our investigations on decidability and complexity issues for verifications of automata-based models of distributed systems situated on graphs.

2.1   Broadcast Protocols

Broadcast protocols have been introduced by Emerson and Namjoshi in [32] as a formal model for cache coherence protocols. A broadcast protocol is a tuple $\mathcal{P} = \langle Q, \Sigma, R, Q_0 \rangle$, where Q is a finite set of control states, Σ is a finite alphabet, $R \subseteq Q \times (\{!!a, ??a, !a, ?a \mid a \in \Sigma\} \cup \{\tau\}) \times Q$ is the transition relation, and $Q_0 \subseteq Q$ is a set of initial control states. The label $!a$ (resp. $?a$) represents a rendez-vous between two processes on message $a \in \Sigma$ (a zero-capacity channel). The label $!!a$ (resp. $??a$) represents the capability of broadcasting (resp. receiving) a message $a \in \Sigma$. Given a process $\mathcal{P} = \langle Q, \Sigma, R, Q_0 \rangle$, a configuration with n-processes is a tuple $\gamma = \langle q_0, \ldots, q_n \rangle$ of control states such that q_i is the current state of the i-the process for $i : 1, \ldots, n$. We use Γ (resp. Γ_0) to denote the set of configurations (resp. initial configurations) associated to $\mathcal{P}$. Note that even if Q_0 is finite, there are infinitely many possible initial configurations (the number of Q_0-graphs). For $q \in Q$ and $a \in \Sigma$, we define the set $R_a(q) = \{q' \in Q \mid \langle q, ??a, q' \rangle \in R\}$ which contains the states that can be reached from the state q when receiving the message a. We assume that $R_a(q)$ is non empty for every a and q, i.e. nodes always react to broadcast messages.

Definition 1. *Given a process $\mathcal{P} = \langle Q, \Sigma, R, Q_0 \rangle$, a BP is defined by the transition system $\langle \Gamma, \rightarrow_{fc}, \Gamma_0 \rangle$ where the transition relation $\rightarrow_{fc} \subseteq \Gamma \times \Gamma$ is such that: for $\gamma, \gamma' \in \Gamma$ with $\gamma = \langle q_0, \ldots, q_n \rangle$, we have $\gamma \rightarrow_{fc} \gamma'$ iff $\gamma' = \langle q'_0, \ldots, q'_n \rangle$ and one of the following condition holds*

- *$\exists i$ s.t. $\langle q_i, !!a, q'_i \rangle \in R$ and $q'_j \in R_a(q_j)$ for every $j \neq i$.*
- *$\exists i \neq j$ s.t. $\langle q_i, !a, q'_i \rangle, \langle q_j, ?a, q'_j \rangle \in R$ and $q'_l = q_l$ for every $l \neq i, j$.*
- *$\exists i$ s.t. $\langle q_i, \tau, q'_i \rangle \in R$ and $q'_j = q_j$ for every $j \neq i$.*

Properties of Broadcast Protocols have been studied e.g. in [34,20,18]. More specifically, by applying a counting abstraction as in [41] infinite set of configurations (with variable number of components) can be represented as Petri net markings in which we count occurrences of states in Q. Using counting abstraction, it is possible to verify parameterized properties of Broadcast Protocols by encoding their transition relation using Petri nets with transfer arcs and solve marking coverability [34]. Constraints over integer variables can then be used to symbolically explore the behavior or a protocol with an arbitrary number of processes (e.g. $x_{q_0} \geq 1$ expresses all configurations with at least one process in state q_0) [20,18].

2.2 Broadcast Protocols with Data

In [17,19] we introduced a register-based model of distributed systems with fully connected topologies that naturally extends the Broadcast Protocols of [32]. These models combine multiset rewriting over atomic formulas with constraints (equality and comparisons of variables ranging over an infinite dense domain like the reals). Atomic formulas can be viewed as nodes in a distributed system containing a single piece of information. Parameterized verification is expressed here in terms of coverability of a given multiset of atoms, a natural extension of the notion of coverability adopted in Petri nets. Coverability turns out to be decidable for monadic predicates with equality and comparison constraints. In [1] we extended the result to a class of constraints over integer values called gap-order constraints. In [4,24] we studied the relative expressive power of different infinite-state models including the multiset rewriting models in [17,19].

In the rest of this section we focus our attention on graph-based models with broadcast communication, a topology-sensitive extension of Broadcast Protocols.

3   Distributed Broadcast Protocols

In [27] we introduced an extension of Broadcast Protocols, called Ad Hoc Networks (AHN), in which processes are distributed on a graph. To formalize this idea, let us first define a Q-graph as a labeled undirected graph $\gamma = \langle V, E, L \rangle$, where V is a finite set of *nodes*, $E \subseteq V \times V \setminus \{\langle v, v \rangle \mid v \in V\}$ is a finite set of *edges*, and L is a labeling function from V to a set of labels Q. We use $L(\gamma)$ to represent all the labels present in γ. The nodes belonging to an edge are called the *endpoints* of the edge. A process is a tuple $\mathcal{P} = \langle Q, \Sigma, R, Q_0 \rangle$, where Q is a finite set of control states, Σ is a finite alphabet, $R \subseteq Q \times (\{!!a, ??a \mid a \in \Sigma\} \cup \{\tau\}) \times Q$ is the transition relation, and $Q_0 \subseteq Q$ is a set of initial control states. The label $!!a$ (resp. $??a$) represents the capability of broadcasting (resp. receiving) a message $a \in \Sigma$. As for Broadcast Protocols, we define $R_a(q) = \{q' \in Q \mid \langle q, ??a, q' \rangle \in R\}$ as the set of states that can be reached from the state q when receiving the message a and assume that $R_a(q)$ is non empty for every a and q. We also consider local transitions of the form $\langle q, \tau, q' \rangle$. We do no consider here rendez-vous communication. Given a process $\mathcal{P} = \langle Q, \Sigma, R, Q_0 \rangle$, a configuration is a Q-graph and an initial configuration is a Q_0-graph. We use Γ (resp. Γ_0) to denote the set of configurations (resp. initial configurations) associated to $\mathcal{P}$. Note that even if Q_0 is finite, there are infinitely many possible initial configurations (the number of Q_0-graphs). We assume that each node of the graph is a process that runs a common predefined protocol defined by a communicating automaton with a finite set Q of control states. Communication is achieved via selective broadcast, which means that a broadcasted message is received by the nodes which are adjacent to the sender. Non-determinism in reception is modeled by means of graph reconfigurations. We next formalize this intuition.

Definition 2. *Given a process* $\mathcal{P} = \langle Q, \Sigma, R, Q_0 \rangle$, *a AHN is defined by the transition system* $\langle \Gamma, \rightarrow, \Gamma_0 \rangle$ *where the transition relation* $\rightarrow \subseteq \Gamma \times \Gamma$ *is such that: for* $\gamma, \gamma' \in \Gamma$ *with* $\gamma = \langle V, E, L \rangle$, *we have* $\gamma \rightarrow \gamma'$ *iff* $\gamma' = \langle V, E, L' \rangle$ *and one of the following condition holds*

- $\exists v \in V$ *s.t.* $\langle L(v), !!a, L'(v) \rangle \in R$ *and* $L'(u) \in R_a(L(u))$ *for every* $\langle u, v \rangle \in E$, *and* $L(w) = L'(w)$ *for any other node* w.
- $\exists v \in V$ *s.t.* $\langle L(v), \tau, L'(v) \rangle \in R$, *and* $L(w) = L'(w)$ *for any other node* w.

The model is inspired to graph-based models of distributed systems presented in [46,52,54,55]. Related models have been proposed in [33,12,51].

Dynamic network reconfigurations can be modeled by adding transitions in which the set of edges is non-deterministically changed we extend the transition relation in order to include the following case: $\gamma, \gamma' \in \Gamma$ with $\gamma = \langle V, E, L \rangle$, we have $\gamma \rightarrow \gamma'$ if $\gamma' = \langle V, E', L \rangle$ for some $E' \subseteq V \times V \setminus \{\langle v, v \rangle \mid v \in V\}$.

We use $\rightarrow^*$ to denote the reflexive and transitive closure of $\rightarrow$.

Parameterized verification problems for our model can be defined by considering the following type of reachability queries.

Definition 3 (Coverability). *Given a process* $\mathcal{P}$, *a transition system* $\langle \Gamma, \rightarrow, \Gamma_0 \rangle$, *and a control state* q, *the coverability problem consists in checking whether or not there exists* $\gamma_0 \in \Gamma_0$ *and* $\gamma_1 \in \Gamma$ *s.t.* $\gamma_0 \rightarrow^* \gamma_1$ *and* $q \in L(\gamma_1)$.

We use the term parameterized to remark that the initial configuration is not fixed a priori. In fact, the only constraint that we put on the initial configuration is that the nodes have labels taken from Q_0 without any information on their number or connection links. Similar problems can be studied for the variations of the basic model with the following features: node crashes, asynchronous communication, messages with data fields and nodes with local memory.

According to our semantics, the number of nodes stays constant in each execution starting from the same initial configuration. As a consequence, when fixing the initial configuration γ_0, we obtain finitely many possible reachable configurations. Thus, checking if there exists γ_1 reachable from a given γ_0 s.t. $q \in L(\gamma_1)$ is a decidable problem. The problem here is state explosion and, for a fixed number of nodes, the combinatorial explosion of the number of distinct initial configurations. We will come back to this point (finite-state verification) in the second part of the paper describing graph-based methods for controlling state explosion.

Checking the parameterized version of the reachability problem is in general much more difficult. The problem easily gets undecidable since for parametric initial configurations we have to deal with an infinite family of transition systems (one for each initial graph). Despite of it, we can still find interesting restrictions to the model or to the set of considered configurations for which coverability is decidable.

3.1   Synchronous Broadcast

Let us first consider synchronous broadcast communication. For this model, in [27] we have proved that coverability is undecidable without dynamic reconfiguration, i.e., when the topology never changes during execution. The proof is based on a discovery protocol implemented by running the same process on each node of the network (whose shape is unknown). The discovery protocol controls interferences by forcing states receiving more copies of the same message into special dead states. This strategy can then be used to navigate into unknown networks and to select one by one nodes that belong to a subgraph with a given shape. A simple shape like a list of fixed but arbitrary length is enough to run a simulation of a Two Counter machine (the list models the maximal aggregate value of the counters). Interestingly, the problem becomes decidabile when considering non-deterministic reconfiguration steps. Non-deterministic reconfiguration steps destroy the influence of the topology on the behavior of individual nodes. The results can then be proved via a reduction to coverability for Petri nets. A broadcast is simulated via a rendez vous with an arbitrary subset of nodes in the network. It is interesting to remark that, as in other models like channel systems, see e.g. [9], the loss of information has the effect of simplifying the verification task (the state space becomes more regular) while it complicates the design of protocols (programming in the model is harder).

To emphasize this point, in [26] we have shown that, in presence of non-deterministic reconfigurations, the decision procedure for coverability has polynomial time complexity in the size of the input protocol. The proof is based on a labeling algorithm that exploits monotonicity properties of the semantics with reconfigurations steps. The algorithm can be viewed as a saturation process that adds to the visited set every state that can be generated by via synchronization step assuming that we have an arbitrary number of processes with states taken from the current visited set. The algorithm requires at worst as many step as the number of control states in the protocol. More complex parametric reachability properties in which the target configurations are generated by constraints on the number of occurrences of control states can still be decided but with increasing complexity. For instance, the problem becomes exponential when target states are described by conjunctions of interval constrains defined over occurrences of states [26].

3.2   Restricted Topologies

In [27,28] we have introduced a restricted form of coverability in which configurations are required to belong to a fixed a priori subclass of graphs (e.g. stars, fully connected graphs, etc.). A quite interesting example of non trivial class consists of all undirected graphs in which the length of simple paths is bounded by the same constant k (k-bounded path graphs). For $k \geq 1$, we still have an infinite set of graphs (e.g. $k = 2$ contains all stars with diameter two). Notice that fully connected graphs are not bounded path.

For synchronous broadcast communication without reconfiguration, parameterized verification is still decidable for bounded path graphs. The results follows from a non trivial application of the theory of well-structured transition systems [3,38]. Well-structured transition systems require a well-quasi ordering (wqo) on the set of configurations and transition systems that are monotone with respect to the ordering.

Definition 4. *A quasi order* $(A, \sqsubseteq)$ *is a wqo if for every infinite sequence of elements* $a_1, a_2, \ldots, a_i, \ldots$ *in* A, *there exist indexes* $i < j$ *s.t.* $a_i \sqsubseteq a_j$.

Our model is well-structured on the class of bounded path graphs with respect to the induced subgraph relation. Let us define the ordering starting from the usual subgraph relation.

Definition 5. *Given two graphs* $G = \langle V, E, L \rangle$ *and* $G' = \langle V', E', L' \rangle$, G *is in the subgraph relation with* G', *written* $G \sqsubseteq_s G'$, *whenever there exists an injection* $f : V \to V'$ *such that, for every* $v, v' \in V$, *if* $\langle v, v' \rangle \in E$, *then* $\langle f(v), f(v') \rangle \in E'$.

The induced subgraph ordering has the following stronger requirements.

Definition 6. *Given two graphs* $G = \langle V, E, L \rangle$ *and* $G' = \langle V', E', L' \rangle$, G *is in the* induced subgraph *relation with* G', *written* $G \sqsubseteq_i G'$, *whenever there exists an injection* $f : V \to V'$ *such that, for every* $v, v' \in V$, $\langle v, v' \rangle \in E$ *if and only if* $\langle f(v), f(v') \rangle \in E'$.

The two orderings are not equivalent. As an example, a path with three nodes is a subgraph, but not an induced subgraph, of a ring of the same order. Subgraph and induced subgraph are well-quasi ordering for bounded path graphs a result due to Ding [31]. Monotonicity of $\to$ w.r.t. an order $\sqsubseteq$ is defined as follows.

Definition 7. *Given* G_1, G_2 *and* G_3 $\to$ *is monotone if, whenever* $G_1 \to G_2$ *and* $G_1 \sqsubseteq G_3$, *there exists* G_4 *s.t.* $G_3 \to G_4$ *and* $G_2 \sqsubseteq G4$.

Broadcast communication is monotone with respect to induced subgraph but not with respect to subgraph. These two property can be used to obtain a well-structured transition systems for our extended notion of broadcast protocols over the set of bounded path configurations. The additional property needed to obtain a verification algorithm is the existence of an algorithm to compute a finite representation of the set of predecessor states of an upward closed set of configurations. An upward closed set (ideal) of configurations is a set S such that if $G \in S$ and $G \sqsubseteq G'$ then $G' \in S$. The predecessors of a given set of configurations are defined as follows

$$pre(S) = \{G | G \to G', \ G' \in S\}$$

The above mentioned condition requires that if we have a finite representation F of an upward closed set of configurations S, then we can algorithmically compute a finite representation $F' = Pre(F)$ of $pre(S)$. If $\sqsubseteq$ is a wqo, F and F' are typically the finite bases of S and $pre(S)$. The algorithm that decides coverability is based on a backward reachability analysis in which we use bases of upward closed sets to represent predecessors. We maintain intermediate results in a set of bases I. Intermediate sets can be compared by using the $\sqsubseteq$ orderings on elements. We define then the sequence $I_0 I_1 \ldots$ as follows:

- $I_0 = \{U\}$,
- $I_{i+1} = I_i \cup \{pre(B) | B \in I_i\}$ for $i \geq 0$.

Since $\sqsubseteq$ is a wqo, then the sequence necessarily stabilizes, i.e., there exists k s.t. I_{k+1} represents the same set of configurations as I_k (i.e. I_k is a least fixpoint). When a fixpoint has been detected, to decide coverability it remains to check whether the initial states belong to the denotation of I.

The algorithm in [27] operates on finite representations of infinite set of configurations. The decidability result can be extended to a slightly more general class of graphs that includes both stars and cliques [28]. The bounded path restriction is used in [28] on graphs obtained after collapsing cliques into single nodes. More precisely, the maximal clique graph K_G associated to a graph $G = (V, E, L)$ is the bipartite graph $\langle X, W, E', L' \rangle$ in which $X = V$, $W \subseteq 2^V$ is the set of maximal cliques of G, for $v \in V, w \in X$, $\langle v, w \rangle \in E'$ iff $v \in w$; $L'(v) = L(v)$ for $v \in V$, and $L'(w) = \bullet$ for $w \in W$. We reformulate the bounded path condition on the maximal clique graph associated to a configuration (i.e. the length of the simple paths of K_G is at most n). For $n \geq 1$, the class BPN_n consists of the set of configurations whose associate maximal clique graph has n-bounded paths (i.e. the length of the simple paths of K_G is at most n). The ordering we are interested in is defined on maximal clique graphs as follows.

Definition 8. *Assume* $G_1 = \langle V_1, E_1, L_1 \rangle$ *with* $K_{G_1} = \langle X_1, W_1, E_1', L_1' \rangle$, *and* $G_2 = \langle V_2, E_2, L_2 \rangle$ *with* $K_{G_2} = \langle X_2, W_2, E_2', L_2' \rangle$ *with* G_1 *and* G_2 *both connected graphs. Then,* $G_1 \sqsubseteq_m G_2$ *iff there exist an injection* $f : X_1 \rightarrow X_2$ *and* $g : W_1 \rightarrow W_2$, *such that*

1. *for every* $v \in X_1$, *and* $C \in W_1$, $v \in C$ *iff* $f(v) \in g(C)$;
2. *for every* $v_1, v_2 \in X_1$, *and* $C \in W_2$, *if* $f(v_1) \sim_C f(v_2)$, *then there exists* $C' \in W_1$ *s.t.* $f(v_1) \sim_{C'} f(v_2)$;
3. *for every* $v \in X_1$, $L(v) = L(f(v))$;
4. *for every* $C \in W_1$, $L(C) = L(g(C))$.

It holds that $G_1 \sqsubseteq_m G_2$ iff $G_1 \sqsubseteq_i G_2$ (G_1 is an induced subgraph of G_2). Furthermore, the resulting ordering is still a wqo and the transition relation $\rightarrow$ is still monotone.

3.3   Faults and Conflicts

In [29] we have studied the impact of node and communication failures on the coverability problem for our model of ad hoc network protocols. We started our analysis by introducing node failures via an intermittent semantics in which a node can be (de)activated at any time. Coverability is decidable under the intermittent semantics. Decidability derives from the assumption that nodes cannot take decisions that depend on the current activation state (e.g. change state when the node is turned on). We then consider two restricted types of node failure, i.e., node crash (a node can only be deactivated) and node restart (when it is activated, it restarts in a special restart state). Coverability becomes undecidable in these two semantics. We considered then different types of communication failures. We first consider a semantics in which a broadcast is not guaranteed to reach all neighbors of the emitter nodes (message loss). Coverability is again decidable in this case. We then introduce a semantics for selective broadcast specifically designed to capture possible conflicts during a transmission. A transmission of a broadcast message is split into two different phases: a starting and an ending phase. During the starting phase, receivers connected to the emitter move to a transient state. While being in the transient state, a reception from another node generates a conflict. In the ending phase an emitter always moves to the next state whereas connected receivers move to their next state only when no conflicts have been detected. In our model we also allow several emitters to simultaneously start a transmission. Decidability holds only when receivers ignore corrupted messages by remaining in their original state. Moreover, in all cases the above mentioned models in which coverability is decidable the decision procedure can be defined via polynomial time reachability algorithm similar to that used in the case of reconfigurations.

3.4   Time

In [7] we have considered a timed version of AHN in which each node has a finite number of dense/discrete clocks. Time elapsing transitions increase all clocks at the same rate. The resulting model extends the Timed Networks model of [11] with an underlying connection graph. When constraining communication via a complex connection graph, the decidability frontier becomes much more complex. For nodes equipped with a single clock, coverability is already undecidable for graphs in which nodes are connected so as to form stars with diameter five. The undecidability result can be extended to the more general class of graphs with bounded simple path (for some bound $N \geq 5$ on the length of paths). We remark that in the untimed case coverability is decidable for bounded path topologies and stars. Coverability is undecidable for fully connected topologies in which each timed automaton has at least two clocks. Decidability holds for special topologies like stars with diameter three and fully connected graphs if nodes have at most one clock. For discrete time coverability is decidable for nodes with finitely many clocks for fully connected topologies and graphs with bounded path.

3.5   Asynchronous Broadcast

In [30] we have enriched the model in order to consider asynchronous communication implemented via mailboxes attached to individual nodes and consider different policies for handling mailboxes (unordered and fifo) and the potential loss of messages. In this model nodes have an additional local data structure that models the mailbox. Interestingly, even if the model is apparently richer than the synchronous one, coverability is still decidable in some case. More specifically, we consider a mailbox structure $\mathbb{M} = \langle \mathcal{M}, del?, add, del, \flat \rangle$, where $\mathcal{M}$ is a denumerable set of elements denoting possible mailbox contents; $add(a, m)$ denotes the mailbox obtained by adding a to m, $del?(a, m)$ is true if a can be removed from m; $del(a, m)$ denotes the mailbox obtained by removing a from m when possible, undefined otherwise. Finally, $\flat \in \mathcal{M}$ denotes the empty mailbox. We call an element a of m *visible* when $del?(a, m) = true$. Their specific semantics and corresponding properties changes with the type of mailbox considered. A protocol is defined by a process $\mathcal{P} = \langle Q, \Sigma, R, q_0 \rangle$ as in the AHN model with the same notation for broadcast messages.

Definition 9. *Configurations are undirected $Q \times \mathcal{M}$-graphs. A $Q \times \mathcal{M}$-graph γ is a tuple $\langle V, E, L \rangle$, where V is a finite set of nodes, $E \subseteq V \times V$ is a finite set of edges, and $L : V \to Q \times \mathcal{M}$ is a labelling function.*

C_0 is the set of undirected graphs in which every node has the same label $\langle q_0, \flat \rangle$ that denotes the initial state of individual processes. Given the labeling L and the node v s.t. $L(v) = \langle q, m \rangle$, we define $L_s(v) = q$ (state component of $L(v)$) and $L_b(v) = m$ (buffer component of $L(v)$). Furthermore, for $\gamma \in \mathcal{C}$, we use $L_s(\gamma)$ to denote the union of the set of control states of nodes in γ ($L_s(\gamma) = \bigcup_{v \in V} L_s(v)$ for $\gamma = \langle V, E, L \rangle$).

Definition 10. *For $\mathbb{M} = \langle \mathcal{M}, del?, add, del, \flat \rangle$, an Asynchronous Broadcast Network (ABN) is defined by the transition system $\mathcal{T}(\mathcal{P}, \mathbb{M}) = \langle \mathcal{C}, \Rightarrow_\mathbb{M}, C_0 \rangle$, where $\Rightarrow_\mathbb{M} \subseteq \mathcal{C} \times \mathcal{C}$ is the transition relation defined next.*

For $\gamma = \langle V, E, L \rangle$ and $\gamma' = \langle V, E, L' \rangle$, $\gamma \Rightarrow_\mathbb{M} \gamma'$ holds iff one of the following conditions on L and L' holds:

- *(local) there exists $v \in V$ such that $(L_s(v), \tau, L'_s(v)) \in R$, $L_b(v) = L'_b(v)$, and $L(u) = L'(u)$ for each $u \in V \setminus \{v\}$.*
- *(broadcast) there exists $v \in V$ and $a \in \Sigma$ such that $(L_s(v), !!a, L'_s(v)) \in R$, $L_b(v) = L'_b(v)$ and for every $u \in V \setminus \{v\}$*
  - *if $\langle u, v \rangle \in E$ then $L'_b(u) = add(a, L_b(u))$ and $L_s(u) = L'_s(u)$,*
  - *otherwise $L(u) = L'(u)$;*
- *(receive) there exists $v \in V$ and $a \in \Sigma$ such that $(L_s(v), ??a, L'_s(v)) \in R$, $del?(a, L_b(v))$ is satisfied, $L'_b(v) = del(a, L_b(v))$, and $L(u) = L'(u)$ for each $u \in V \setminus \{v\}$.*

Coverability for ABN is defined as in the case of AHN, i.e., we search for an initial configuration that via a finite number of steps can reach a configuration that exposes a given control state $q \in Q$. Decidability of coverability is strictly related to the policy used to handle mailboxes.

Multiset. The mailbox structure *Bag* is defined as follows: $\mathcal{M}$ is the denumerable set of bags over Σ, $add(a, m) = [a] \oplus m$ (where $[a]$ is the singleton bag containing a), $del?(a, m) = true$ iff $m(a) > 0$, $del(a, m) = m \ominus [a]$, and $\flat \in \mathcal{M}$ is the empty bag $[]$.

When local buffers are treated as bags of messages the coverability problem is decidable. The proof is based on two steps. We can first show that, for the purpose of deciding coverability, we can restrict to fully connected topologies only. We can then use a reduction to the PTIME-complete algorithm of [26].

FIFO. The mailbox structure $FIFO$ is defined as follows: $\mathcal{M}$ is defined as Σ^*; $add(a, m) = m \cdot a$ (concatenation of a and m); $del?(a, m) = true$ iff $m = a \cdot m'$; $del(a, m)$ is the bag m' whenever $m = a \cdot m'$, undefined otherwise; finally, $\flat \in \mathcal{M}$ is the empty string ϵ. When mailboxes are ordered buffers, we obtain undecidability already in the case of fully connected topologies. The coverability problem becomes decidable when introducing non-deterministic message losses. In an extended model in which a node can test if its mailbox is empty, we obtain undecidability with unordered bags and both arbitrary or fully-connected topologies.

3.6   Distributed Broadcast Protocols with Data

In [25] we considered a further refinement step by introducing local registers and data fields in message payloads. We model a distributed network using a graph in which the behaviour of each node is described via an automaton with operations over a finite set of registers. A node can transmit part of its current data to adjacent nodes using broadcast messages. A message carries both a type and a finite tuple of data. Receivers can either test, store, or ignore the data contained inside a message. We assume that broadcasts and receptions are executed without delays (i.e. we simultaneously update the state of sender and receiver nodes).

Our analysis shows that, even in presence of register automata, dynamic reconfiguration can still render the coverability problem easier to solve. More precisely, in fully connected topologies coverability is undecidable for nodes with two registers and messages with one field. Coverability remains undecidable with dynamic network reconfigurations if nodes have two registers but messages have two fields. Decidability holds for $k \geq 1$ registers and a single data field per message for arbitrary topologies and dynamic network reconfiguration. The decision algorithm is based on a saturation procedure that operates on a graph-based symbolic representation of sets of configurations in which the data are abstracted away. This is inspired by similar techniques used in the case of classical register automata [47]. The problem is PSPACE-complete in this case. Finally, for fully connected topologies but without dynamic reconfiguration, coverability for nodes with a single register and messages with a single field is decidable with non elementary complexity. The decidability proof exploits the theory of well-structured transition systems [10,38]. The non-elementary lower bound follows from a reduction from coverability in reset nets [53].

3.7   Other Graph-Based Models

In [8] we apply graph-based transformations to model intermediate evaluations of non-atomic mutual exclusion protocols with universally quantified conditions. Parameterized verification is undecidable in the resulting model. Semi-decision procedures can be defined by resorting to upward closed abstractions during backward search (monotonic abstraction as in [6,23]). In [21] we studied decidability of reachability and coverability for a graph-based specification used to model biological systems calles kappa-calculus [16]. Among other results, we proved undecidability for coverability for graph rewrite systems that can only increase the size of a configuration. Reachability problems for graph-based representations of protocols have also been considered in [5] where symbolic representations combining a special graph ordering and constraint-based representation of relations between local data of different nodes have been used to verify parameterized consistency protocols.

To generalize some of the ideas studied in [21,5,23,27,28], in [13] we have studied reachability and coverability (i.e. reachability of graphs containing specific patterns) for Graph Transformation Systems (GTS). Specifically, by transferring in the GTS setting the results in [27], in [13] we have shown that coverability is decidable for GTS for graph with bounded path graphs ordered via subgraph inclusion. The latter result follows from the theory of well-structured transition systems. A model with topologies represented as acyclic directed graphs has been presented in [2]. Finally, networks of probabilistic automata have been studied in [14].

4   Model Checking

Model checking distributed protocols is a quite challenging task for finite-state and time model checking tools, see e.g. [37,36,48,40,35,44,45,57,15]. In this section we report on our experience with standard model checkers like Spin and less usual approaches based on Graph Transformation Systems like Groove. Promela/Spin provides a consolidated framework for the specification (via automata) and verification (via model checking) of concurrent and distributed systems. Promela provides a process-oriented specification language for distributed programs. The C-like specification language is enriched with process templates, channels, and CSP-like guarded commands. Guarded commands simplify the encoding of event-driven loops (e.g. handshaske protocols). Communication primitives are implemented via bounded capacity channels that model rendez-vous and asynchronous communication. Promela processes are internally represented as finite-state automata. The entire system is obtained by composition of the automata associated to individual processes. System properties are themselves expressed as automata working as monitors (e.g. they recognize finite or infinite error traces). In this setting global states are vectors obtained by conjoining the values of shared and local variables. The Spin model checker is a full state exploration algorithm equipped with powerful heuristics to prune and approximate the search space, e.g., partial order reductions and bit-vector hashing. Groove provides a declarative specification language for distributed systems based on an extended form of Graph Transformation Systems. A configuration is represented as a graph in which nodes are either shared data, states of individual processes, or message instances. The behavior of individual processes and synchronization steps are specified via graph transformations, i.e., rewrite rules that can modify nodes, edges, and labels of the current configuration. Groove specifications focus on the structure of a distributed system (e.g. retain the topology of the underlying system) whereas Promela specification are closer to a software implementation of a protocol in which details of the communication layer are abstracted away via the use of channel-based synchronization primitives. The Groove graph production rules extend Graph Transformation Systems with negative application conditions, nested quantification (e.g. used to apply a given rewrite rule to all matching patterns in the current configuration), symbolic guards and updates on node and edge attributes (e.g. to represent updates of local states). The Groove simulator builds a reachability graph (in which each configuration is itself a graph) by exploiting graph isomorphism for compacting the search space. The use of the enriched graph transformation rules provided by Groove has several advantages. First of all, graphs provide a natural level of abstraction to model asynchronous systems. Indeed, similarly to what is done in Petri nets, unordered channels can be simply viewed as collections of nodes (i.e. using the Petri net terminology as tokens in a dedicated place that represents the buffer). Values can be represented as distinct nodes (dynamically generated by a proposer) without need of introducing integer or other finite domain set of values. Furthermore, quorum transitions can be defined naturally by using productions with universal quantification. Safety requirements are specified using special rules whose activation raise an alarm (i.e. they work like assertions). An interesting type of pattern that is quite useful in this context is built on top of a spatial reachability relation defined over configurations (e.g. to check for the existence of a path in the current graph).

In [22], we have given a formal specification in Promela of a distributed consensus protocol called Paxos [49]. The consensus problem requires agreement among a number of distributed agents for a single data value. Some of the agents may fail, so consensus protocols must be fault tolerant. Initially, each agent proposes a value to all the other ones. Agents can then exchange their information. A correct protocol must ensure that when a node takes the final choice, the chosen value is the same for all correct agents. It is assumed here that messages can be delayed arbitrarily. A subset of processes may crash anytime and restore their state (keeping their local information) after an arbitrary delay. Fisher, Lynch and Patterson have shown that, under the above mentioned assumptions, solving consensus is impossible [39]. In [49] Lamport proposed a (possibly non terminating) algorithm, called Paxos, addressing this problem. Paxos is based on the metaphor of a part-time parliament, in which part-time legislators need to keep consistent records of their passing laws. The protocol can be described via three separate agent roles: proposers that can propose values for consensus, acceptors that accept a value among those proposed, and learners that learn the chosen value. In this setting messages can be duplicated and processed in any order. Furthermore, elections involving quorums of processes are used in different steps of the protocol in order to select particular round identifiers and values. Paxos is a challenging case-study for automated verification methods. Indeed, the search space grows very fast with the number of processes. This is due to the fact that messages can be processed in any order (the protocol is designed for asynchronous communication media) and because the protocol requires handshakes with quorums of processes (i.e. subsets of processes) that may change from one phase to another of the protocol.

In a first Promela model we used counters local to individual processes to keep track of the thresholds needed to check majority for a given round or value as in simpler examples given in [45]. Contrary to informal presentations like those given in [50], Promela provides a non-ambiguous executable semantics in terms of automata. Via a formal analysis extracted from the correctness requirements, which are specified using auxiliary variables and assertions, we managed to prove reduction theorems that we used to further simplify the specification. To optimize the model, we extended Promela with a new type of transitions that atomically check if a given type of messages occur in a channel, a feature that is not directly provided by the language. This kind of transitions are based on a special type of guards called *counting guards*, implemented in Promela using the deterministic step built in predicate (a construct that can be used to add new type of transitions to the language). Counting guards are used to specify quorum transitions without the need of auxiliary states introduced by the use of counters.

In [22] we have modelled the same case-study in Groove. In the experiments performed on the Promela model using full state exploration we managed to consider configurations with 2 proposers and up to 5 acceptors. The use of quorum transitions is very effective in reducing the state space (one order of magnitude less than the model with counters updated at each message reception). Spin also provides different ways to underapproximate the state space. The application of underapproximation is very effective in scaling up the analysis to model with more than 5 acceptors. One of the strength of the Spin model checker seems to be its robustness in finding bugs. Even when executed using underapproximated search, the model checker is still capable to detect unsafe states. When used in this modality (e.g. bit-vector hasing) the use of approximated search combined with quorum transitions allowed us to scale up the analysis to more than 10 role instances. Furthermore, via a parameter exploration, we managed to use Spin to validate the assumptions on the correctness of the algorithm. Indeed Spin, with or without underapproximations, automatically generates error traces for quorum sets that are not adequate w.r.t. the assumptions of the algorithm. On the other side, Groove provides a more abstract and compact way to specify complex process behaviors like quorum transitions and spatial reachability patterns. The experiments with the Groove model checker applied to full state exploration show the effectiveness of symmetry reductions based on the graph structure of configurations. The symmetry reduction is implemented in Groove by the isomorphism check that is applied automatically when exploring a state space. By enabling this modality, Groove improves the exploration of the state space of protocols of two order of magnitude with respect to Spin, however with the need of human ingenuity to encode the protocol rules using a graph-based formalism. Finding a way to compile process-oriented specifications in a graph-based format, perhaps passing through process algebraic languages with message passing, could be an interesting direction to further extend the applicability of tools like Groove to a wider class of systems.

Acknowledgments. The author would like to thank Riccardo Traverso for the work in his PhD Thesis [56] and all other co-authors.

References

1. Abdulla, P., Delzanno, G.: Constrained multiset rewriting. In: AVIS 2006 (2006)
2. Abdulla, P.A., Atig, M.F., Rezine, O.: Verification of directed acyclic ad hoc networks. In: Beyer, D., Boreale, M. (eds.) FMOODS/FORTE 2013. LNCS, vol. 7892, pp. 193–208. Springer, Heidelberg (2013)
3. Abdulla, P.A., Cerans, K., Jonsson, B., Tsay, Y.-K.: General decidability theorems for infinite-state systems. In: LICS 1996, pp. 313–321. IEEE Computer Society (1996)
4. Abdulla, P.A., Delzanno, G., Van Begin, L.: A classification of the expressive power of well-structured transition systems. Inf. Comput. 209(3), 248–279 (2011)
5. Abdulla, P.A., Delzanno, G., Rezine, A.: Automatic verification of directory-based consistency protocols with graph constraints. Int. J. Found. Comput. Sci. 22(4) (2011)
6. Abdulla, P.A., Delzanno, G., Rezine, A.: Approximated parameterized verification of infinite-state processes with global conditions. Formal Methods in System Design 34(2), 126–156 (2009)
7. Abdulla, P.A., Delzanno, G., Rezine, O., Sangnier, A., Traverso, R.: On the verification of timed ad hoc networks. In: Fahrenberg, U., Tripakis, S. (eds.) FORMATS 2011. LNCS, vol. 6919, pp. 256–270. Springer, Heidelberg (2011)
8. Abdulla, P.A., Ben Henda, N., Delzanno, G., Rezine, A.: Handling parameterized systems with non-atomic global conditions. In: Logozzo, F., Peled, D.A., Zuck, L.D. (eds.) VMCAI 2008. LNCS, vol. 4905, pp. 22–36. Springer, Heidelberg (2008)
9. Abdulla, P.A., Jonsson, B.: Undecidable verification problems for programs with unreliable channels. Inf. Comput. 130(1), 71–90 (1996)
10. Abdulla, P.A., Jonsson, B.: Ensuring completeness of symbolic verification methods for infinite-state systems. Theor. Comput. Sci. 256(1-2), 145–167 (2001)
11. Abdulla, P.A., Nylén, A.: Better is better than well: On efficient verification of infinite-state systems. In: LICS 2000, pp. 132–140. IEEE Computer Society (2000)
12. Alberti, F., Ghilardi, S., Pagani, E., Ranise, S., Rossi, G.P.: Automated support for the design and validation of fault tolerant parameterized systems: A case study. ECEASST 35 (2010)
13. Bertrand, N., Delzanno, G., König, B., Sangnier, A., Stückrath, J.: On the decidability status of reachability and coverability in graph transformation systems. In: RTA, pp. 101–116 (2012)
14. Bertrand, N., Fournier, P., Sangnier, A.: Playing with probabilities in reconfigurable broadcast networks. In: Muscholl, A. (ed.) FOSSACS 2014 (ETAPS). LNCS, vol. 8412, pp. 134–148. Springer, Heidelberg (2014)
15. Bokor, P., Serafini, M., Suri, N.: On efficient models for model checking message-passing distributed protocols. In: Hatcliff, J., Zucca, E. (eds.) FMOODS 2010, Part II. LNCS, vol. 6117, pp. 216–223. Springer, Heidelberg (2010)
16. Danos, V., Laneve, C.: Formal molecular biology. Theor. Comput. Sci. 325(1), 69–110 (2004)
17. Delzanno, G.: An overview of msr(c): A clp-based framework for the symbolic verification of parameterized concurrent systems. Electr. Notes Theor. Comput. Sci. 76, 65–82 (2002)
18. Delzanno, G.: Constraint-based verification of parameterized cache coherence protocols. FMSD 23(3), 257–301 (2003)
19. Delzanno, G.: Constraint-based automatic verification of abstract models of multithreaded programs. TPLP 7(1-2), 67–91 (2007)

20. Delzanno, G., Esparza, J., Podelski, A.: Constraint-based analysis of broadcast protocols. In: Flum, J., Rodríguez-Artalejo, M. (eds.) CSL 1999. LNCS, vol. 1683, pp. 50–66. Springer, Heidelberg (1999)
21. Delzanno, G., Di Giusto, C., Gabbrielli, M., Laneve, C., Zavattaro, G.: The κ-lattice: Decidability boundaries for qualitative analysis in biological languages. In: Degano, P., Gorrieri, R. (eds.) CMSB 2009. LNCS, vol. 5688, pp. 158–172. Springer, Heidelberg (2009)
22. Delzanno, G., Rensink, A., Traverso, R.: Graph- versus vector-based analysis of a consensus protocol. In: GRAPHITE (2014)
23. Delzanno, G., Rezine, A.: A lightweight regular model checking approach for parameterized systems. STTT 14(2), 207–222 (2012)
24. Delzanno, G., Rosa-Velardo, F.: On the coverability and reachability languages of monotonic extensions of petri nets. Theor. Comput. Sci. 467, 12–29 (2013)
25. Delzanno, G., Sangnier, A., Traverso, R.: Parameterized verification of broadcast networks of register automata. In: Abdulla, P.A., Potapov, I. (eds.) RP 2013. LNCS, vol. 8169, pp. 109–121. Springer, Heidelberg (2013)
26. Delzanno, G., Sangnier, A., Traverso, R., Zavattaro, G.: On the complexity of parameterized reachability in reconfigurable broadcast networks. In: FSTTCS 2012. LIPIcs, vol. 18, pp. 289–300. Schloss Dagstuhl - Leibniz-Zentrum fuer Informatik (2012)
27. Delzanno, G., Sangnier, A., Zavattaro, G.: Parameterized verification of ad hoc networks. In: Gastin, P., Laroussinie, F. (eds.) CONCUR 2010. LNCS, vol. 6269, pp. 313–327. Springer, Heidelberg (2010)
28. Delzanno, G., Sangnier, A., Zavattaro, G.: On the power of cliques in the parameterized verification of ad hoc networks. In: Hofmann, M. (ed.) FOSSACS 2011. LNCS, vol. 6604, pp. 441–455. Springer, Heidelberg (2011)
29. Delzanno, G., Sangnier, A., Zavattaro, G.: Verification of ad hoc networks with node and communication failures. In: Giese, H., Rosu, G. (eds.) FMOODS/FORTE 2012. LNCS, vol. 7273, pp. 235–250. Springer, Heidelberg (2012)
30. Delzanno, G., Traverso, R.: Decidability and complexity results for verification of asynchronous broadcast networks. In: Dediu, A.-H., Martín-Vide, C., Truthe, B. (eds.) LATA 2013. LNCS, vol. 7810, pp. 238–249. Springer, Heidelberg (2013)
31. Ding, G.: Subgraphs and well quasi ordering. J. of Graph Theory 16(5), 489–502 (1992)
32. Emerson, E.A., Namjoshi, K.S.: On model checking for non-deterministic infinite-state systems. In: LICS 1998, pp. 70–80. IEEE Computer Society (1998)
33. Ene, C., Muntean, T.: A broadcast-based calculus for communicating systems. In: IPDPS 2001, p. 149. IEEE Computer Society (2001)
34. Esparza, J., Finkel, A., Mayr, R.: On the verification of broadcast protocols. In: LICS 1999, pp. 352–359. IEEE Computer Society (1999)
35. Fehnker, A., van Glabbeek, R., Höfner, P., McIver, A., Portmann, M., Tan, W.L.: Automated analysis of AODV using UPPAAL. In: Flanagan, C., König, B. (eds.) TACAS 2012. LNCS, vol. 7214, pp. 173–187. Springer, Heidelberg (2012)
36. Fehnker, A., van Glabbeek, R., Höfner, P., McIver, A., Portmann, M., Tan, W.L.: A process algebra for wireless mesh networks. In: Seidl, H. (ed.) ESOP. LNCS, vol. 7211, pp. 295–315. Springer, Heidelberg (2012)
37. Fehnker, A., van Hoesel, L., Mader, A.: Modelling and verification of the lmac protocol for wireless sensor networks. In: Davies, J., Gibbons, J. (eds.) IFM 2007. LNCS, vol. 4591, pp. 253–272. Springer, Heidelberg (2007)
38. Finkel, A., Schnoebelen, P.: Well-structured transition systems everywhere! Theor. Comput. Sci. 256(1-2), 63–92 (2001)

39. Fischer, M.J., Lynch, N.A., Paterson, M.: Impossibility of distributed consensus with one faulty process. J. ACM 32(2), 374–382 (1985)
40. Függer, M., Widder, J.: Efficient checking of link-reversal-based concurrent systems. In: Koutny, M., Ulidowski, I. (eds.) CONCUR 2012. LNCS, vol. 7454, pp. 486–499. Springer, Heidelberg (2012)
41. German, S.M., Sistla, A.P.: Reasoning about systems with many processes. J. ACM 39(3), 675–735 (1992)
42. Ghamarian, A.H., de Mol, M., Rensink, A., Zambon, E., Zimakova, M.: Modelling and analysis using groove. STTT 14(1), 15–40 (2012)
43. Holzmann, G.J.: The SPIN Model Checker - primer and reference manual. Addison-Wesley (2004)
44. John, A., Konnov, I., Schmid, U., Veith, H., Widder, J.: Towards modeling and model checking fault-tolerant distributed algorithms. In: Bartocci, E., Ramakrishnan, C.R. (eds.) SPIN 2013. LNCS, vol. 7976, pp. 209–226. Springer, Heidelberg (2013)
45. John, A., Konnov, I., Schmid, U., Veith, H., Widder, J.: Towards modeling and model checking fault-tolerant distributed algorithms. In: Bartocci, E., Ramakrishnan, C.R. (eds.) SPIN 2013. LNCS, vol. 7976, pp. 209–226. Springer, Heidelberg (2013)
46. Joshi, S., König, B.: Applying the graph minor theorem to the verification of graph transformation systems. In: Gupta, A., Malik, S. (eds.) CAV 2008. LNCS, vol. 5123, pp. 214–226. Springer, Heidelberg (2008)
47. Kaminski, M., Francez, N.: Finite-memory automata. Theor. Comput. Sci. 134(2), 329–363 (1994)
48. Konnov, I., Veith, H., Widder, J.: Who is afraid of model checking distributed algorithms? In: Unpublished Contribution to: CAV Workshop $(EC)^2$ (2012)
49. Lamport, L.: The part-time parliament. ACM Transactions on Computer Systems 16(3), 133–169 (1998)
50. Marzullo, K., Mei, A., Meling, H.: A simpler proof for paxos and fast paxos. Course Notes (2013)
51. Namjoshi, K.S., Trefler, R.J.: Uncovering symmetries in irregular process networks. In: Giacobazzi, R., Berdine, J., Mastroeni, I. (eds.) VMCAI 2013. LNCS, vol. 7737, pp. 496–514. Springer, Heidelberg (2013)
52. Saksena, M., Wibling, O., Jonsson, B.: Graph grammar modeling and verification of ad hoc routing protocols. In: Ramakrishnan, C.R., Rehof, J. (eds.) TACAS 2008. LNCS, vol. TACAS, pp. 18–32. Springer, Heidelberg (2008)
53. Schnoebelen, P.: Revisiting ackermann-hardness for lossy counter machines and reset petri nets. In: Hliněný, P., Kučera, A. (eds.) MFCS 2010. LNCS, vol. 6281, pp. 616–628. Springer, Heidelberg (2010)
54. Singh, A., Ramakrishnan, C.R., Smolka, S.A.: Query-based model checking of ad hoc network protocols. In: Bravetti, M., Zavattaro, G. (eds.) CONCUR 2009. LNCS, vol. 5710, pp. 603–619. Springer, Heidelberg (2009)
55. Singh, A., Ramakrishnan, C.R., Smolka, S.A.: A process calculus for mobile ad hoc networks. Sci. Comput. Program. 75(6), 440–469 (2010)
56. Traverso, R.: Formal verification of ad hoc networks. PhD thesis, University of Genova (2014)
57. Tsuchiya, T., Schiper, A.: Using bounded model checking to verify consensus algorithms. In: Taubenfeld, G. (ed.) DISC 2008. LNCS, vol. 5218, pp. 466–480. Springer, Heidelberg (2008)

Tableau-Based Reasoning for Graph Properties

Leen Lambers[1] and Fernando Orejas[2]

[1] Hasso Plattner Institut, University of Potsdam, Germany
[2] Dpto de L.S.I., Universitat Politècnica de Catalunya, Barcelona, Spain

Abstract. Graphs are ubiquitous in Computer Science. For this reason, in many areas, it is very important to have the means to express and reason about graph properties. A simple way is based on defining an appropriate encoding of graphs in terms of classical logic. This approach has been followed by Courcelle. The alternative is the definition of a specialized logic, as done by Habel and Penne-mann, who defined a logic of nested graph conditions, where graph properties are formulated explicitly making use of graphs and graph morphisms, and which has the expressive power of Courcelle's first order logic of graphs. In particular, in his thesis, Pennemann defined and implemented a sound proof system for reasoning in this logic. Moreover, he showed that his tools outperform some standard provers when working over encoded graph conditions.

Unfortunately, Pennemann did not prove the completeness of his proof system. In this sense, one of the main contributions of this paper is the solution to this open problem. In particular, we prove the (refutational) completeness of a tableau method based on Pennemann's rules that provides a specific theorem-proving procedure for this logic. This procedure can be considered our second contribution. Finally, our tableaux are not standard, but we had to define a new notion of nested tableaux that could be useful for other formalisms where formulas have a hierarchical structure like nested graph conditions.

Keywords: Graph properties, Graph Logic, Automated deduction, Visual modelling, Graph transformation.

1 Introduction

Graphs are ubiquitous in Computer Science. For this reason, in many areas, it is (or it may be) very important to have the means to express and reason about graph properties. Examples may be, model-driven engineering where we may need to express properties of graphical models, or the verification of systems whose states are modelled as graphs, or if we need to express properties about sets of semi-structured documents, especially if they are related by links, or the area of graph databases, to express integrity constraints. We can follow two different ways. The first one is based on defining an appropriate encoding of graphs in terms of some existing logic. The second way is based on directly defining a logic where graphs are first-class citizens. An example of the first approach is the work of Courcelle who, in a series of papers (see, e.g. [3]) studied systematically a graph logic defined in terms of first-order (or monadic second-order) logic. In particular, in that approach, graphs are defined axiomatically by means of predicates $node(n)$, asserting that n is a node, and $edge(n_1, n_2)$ asserting that there is an edge from n_1 to

H. Giese and B. König (Eds.): ICGT 2014, LNCS 8571, pp. 17–32, 2014.
© Springer International Publishing Switzerland 2014

n_2. Graphs can also be defined as terms over a given algebra A as done in the context of Maude [2,1]. The most prominent example of the second approach is the *logic of nested graph conditions*, that we study in this paper, defined by Pennemann and Habel [6], where graph properties are formulated explicitly making use of graphs and graph morphisms. The origins of this approach can be found in the notion of graph constraint [8], introduced in the area of graph transformation, in connection with the notion of (negative) application conditions, as a form to limit the applicability of transformation rules. However, graph constraints have a very limited expressive power, while nested conditions have the same expressive power as Courcelle's first-order graph logic [6]. A similar approach was first introduced by Rensink in [15].

An advantage of encoding graph properties in terms of some existing logic is that we can reason about them by using methods and tools provided by the given logic, while, in the other case, we would need to define specific dedicated methods. In this sense, in [12], we defined a sound and complete proof system for the restrictive case of graph constraints, including the case where the constraints included conditions on graph attributes [11]. Almost simultaneously, in [13,14], Pennemann defined and implemented a sound proof system for reasoning with nested conditions. Unfortunately, in this work the completeness of his approach was not proven. The problem is related with the difficulty to use induction for building a counter-model in the completeness proof. In this sense, one of the main contributions of this paper is the solution to this open problem, since we prove the completeness of a subset of the rules defined by Pennemann. In particular, we prove the (refutational) completeness of a tableau method based on this subset of rules providing a specific theorem-proving procedure for this logic. This procedure can be considered our second contribution. Moreover, our tableaux are not standard, but we had to define a new notion of *nested tableaux*, that solves the difficulties with induction, and that could be useful for other formalisms where formulas have a hierarchical structure like nested graph conditions.

One may question in which sense this kind of work is relevant or interesting, if (apparently) the encoding approach gives the same expressive power in a simpler way. There are two main reasons: generality and efficiency. On the one hand, the logic of nested graph conditions and our results are not restricted to a given kind of graphs. Our graphs can be directed or undirected, typed or untyped, attributed or without attributes, etc. The only condition is that the given category of graphs should be $\mathcal{M}$-adhesive [9,5] satisfying some additional categorical properties that are used in this paper. Actually, this approach applies also to categories of structures that are not graphs, like sets. On the contrary, when using encodings, each kind of graph structure needs a different encoding. On the other hand, a dedicated theorem-prover may be much more efficient than a generic one when used over the given graph encoding. In particular, in [13,14], Pennemann compares the implementation for his proof system with some standard provers, like VAMPIRE, DARWIN and PROVER9, working over encoded graph conditions. The result is that his implementation outperforms the coding approach. Actually, in a considerable amount of examples, the above standard provers were unable to terminate in the same time needed by Pennemann's proof system.

The paper is organized as follows: In Section 2 we present the kind of graph properties that we consider together with some preliminariy results that we need in the rest of the paper. In Section 3 we then present our new tableau-based reasoning method for graph properties and subsequently in Section 4 we present soundness and completeness for this reasoning method. We conclude the paper with Section 5. Due to space limitations, proofs are only sketched. Detailed proofs can be found in [10].

2 Preliminaries

In this section we present the basic notions for this paper. First we reintroduce the kind of graph properties that we consider. Secondly we introduce a (weak) conjunctive normal form for them together with some shifting results.

For simplicity, we will present all our notions and results in terms of plain directed graphs and graph morphisms[1], i.e.: A *graph* $G = (G^V, G^E, s^G, t^G)$ consists of a set G^V of nodes, a set G^E of edges, a source function $s^G : G^E \rightarrow G^V$, and a target function $t^G : G^E \rightarrow G^V$. Given the graphs $G = (G^V, G^E, s^G, t^G)$ and $H = (H^V, H^E, s^H, t^H)$, a *graph morphism* $f : G \rightarrow H$ is a pair of mappings, $f^V : G^V \rightarrow H^V$, $f^E : G^E \rightarrow H^E$ such that $f^V \circ s^G = s^H \circ f^E$ and $f^V \circ t^G = t^H \circ f^E$. A graph morphism $f : G \rightarrow H$ is a *monomorphism* if f^V and f^E are injective mappings. Finally, two graph morphisms $m : H \rightarrow G$ and $m' : H' \rightarrow G$ are *jointly surjective* if $m^V(H^V) \cup m'^V(H'^V) = G^V$ and $m^E(H^E) \cup m'^E(H'^E) = G^E$. Note that all along the paper, unless it is explicitly said, if we say that f is a morphism, we will implicitly assume that f is a monomorphism.

2.1 Graph Properties Expressed in GL

The underlying idea of the graph logic GL, handled in this paper, is that graph properties can be described stating that certain patterns, consisting of graphs and morphisms (actually, inclusions), must be present (or must not be present) in a given graph. For instance, the simplest kind of graph property, $\exists C$, specifies that a given graph G should include (a copy of) C. For instance, the property $\exists C_1$ with C_1 as depicted on the left of Fig. 1 states that a graph should include a node with a loop. More precisely, the fact that a graph G satisfies a graph property α is expressed in term of the existence (or non-existence) of monomorphisms from the graphs included in α to G, such that some conditions are satisfied. For example, a graph G satisfies the former property if there exists a monomorphism $f : C_1 \rightarrow G$. Obviously, graph properties can be combined using the standard connectives $\vee$, $\wedge$, and $\neg$. For example, $\neg \exists C_1$, where C_1 is the graph on the left of Fig. 1, specifies that a graph should not include loops. We can also describe more complex properties by using more complex patterns or diagrams. For instance, we may consider that the property depicted in Fig. 1, where h_1 and h_2 are two inclusions between the graphs involved, states that there must exist a node with a loop, such that

---

[1] In some examples however, for motivation, we deal with typed attributed graphs. Anyhow, following the approach used in [4], it is straightforward to show that our results generalize to a large class of (graphical) structures. The only condition is that the given category of graphs should be $\mathcal{M}$-adhesive [9,5] satisfying the additional property of having a unique $\mathcal{E}' - \mathcal{M}$ pair factorization [5] (needed for Lemma 3) as well as infinite colimits (see Proposition 1 needed for the Completeness Theorem).

for all pairs of edges connected to that node, there exists another pair of edges completing a rectangle. More precisely, a graph G would satisfy this condition if there exists a monomorphism $f : C_1 \to G$ such that for all $f' : C_2 \to G$, with $f = f' \circ h_1$, there exists $f'' : C_2 \to G$, such that $f' = f'' \circ h_2$.

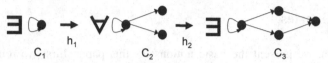

Fig. 1. A graph property

For our convenience, we will express these properties using a nested notation [6] and avoiding the use of universal quantifiers. Moreover, the conditions that we define below are slightly more general than what may seem to be needed. Instead of defining properties about graphs, *nested conditions* define properties of graph monomorphisms.

Definition 1 (condition, nesting level). *Given a finite graph C, a condition over C is defined inductively as follows:*

- *true is a condition over C. We say that true has nesting level 0.*
- *For every monomorphism $a : C \to D$ and condition c_D over a finite graph D with nesting level n such that $n \geq 0$, $\exists(a, c_D)$ is a condition over C with nesting level $n+1$.*
- *Given conditions over C, c_C and c'_C, with nesting level n and n', respectively, $\neg c_C$ and $c_C \wedge c'_C$ are conditions over C with nesting level n and $max(n, n')$, respectively. We restrict ourselves to finite conditions, i.e. each conjunction of conditions is finite.*

We define when a monomorphism $q : C \to G$ satisfies a condition c_C over C inductively:

- *Every morphism q satisfies true.*
- *A morphism q satisfies $\exists(a, c_D)$, denoted $q \models \exists(a, c_D)$, if there exists a monomorphism $q' : D \to G$ such that $q' \circ a = q$ and $q' \models c_D$.*
- *A morphism q satisfies $\neg c_C$ if it does not satisfy c_C and satisfies $\wedge_{i \in I} c_{C,i}$ if it satisfies each $c_{C,i}$ ($i \in I$).*

For example, the property in Fig. 1, would be denoted by the nested condition over the empty graph $\exists(i_{C_1} : \emptyset \to C_1, \neg\exists(h_1 : C_1 \to C_2, \neg\exists(h_2 : C_2 \to C_3, true)))$ with nesting level 3. As said above, nested conditions are more general than needed. The graph properties in our graph logic *GL* are not arbitrary conditions, but conditions over the empty graph. Consequently in this case, the models of a condition are morphisms $\emptyset \to G$ and this is equivalent to say that models of these conditions are graphs G. However, we must notice that if $\exists(a, c)$ is a graph property, in general c is an arbitrary graph condition.

Definition 2 (GL Syntax, GL Semantics). *The language of graph properties GL consists of all conditions over the empty graph $\emptyset$. Given an element $\exists(a, c_D)$ of GL with $a : \emptyset \to D$, we also denote it by $\exists(D, c_D)$. A graph G satisfies a graph property c of GL if the unique morphism $i : \emptyset \to G$ satisfies c.*

2.2   Conjunctive Normal Form and Shifting Results

In this section, we introduce the notion of clause and (weak) conjunctive normal form in GL that is needed in the following section to present tableau reasoning [7] for GL.

Definition 3 ((weak) CNF, subcondition). *A literal ℓ is either a condition of the form $\exists(a,d)$ or $\neg\exists(a,d)$. We say that a literal of the form $\exists(a,d)$ is a* positive literal *and of the form $\neg\exists(a,d)$ is a* negative literal. *Each disjunction of literals is also called a* clause. *A condition is in* weak conjunctive normal form (CNF) *if it is either true, or false, or a conjunction of clauses $\wedge_{j\in J} c_j$, with $c_j = \vee_{k\in K_j} \ell_k$, where for each literal, $\ell_k = \exists(a_k,d_k)$ or $\ell_k = \neg\exists(a_k,d_k)$, d_k is a condition in weak CNF again. Moreover, for each negative literal $\ell_k = \neg\exists(a_k,d_k)$ it holds that a_k is a monomorphism, but not an isomorphism. A condition c is in* CNF *if it is in weak CNF and in addition for each literal $\exists(a,d)$ or $\neg\exists(a,d)$ occurring in c it holds that a is a monomorphism, but not an isomorphism. A* subcondition *of a condition c in (weak) CNF is either a clause or a conjunction of clauses in c.*

The idea of having conditions in weak CNF is to be able to handle them efficiently in our tableau reasoning method. In particular, the conjunction of clauses will allow us to apply the classical tableau rules for CNF formulas. In addition, since negative literals $\neg\exists(a,d)$ where a is not an isomorphism are always satisfiable, we can handle them in a specialized way as will be explained in the following section.

In [13,14], Pennemann describes a procedure, based on some equivalences, for transforming any condition into CNF, which makes it sufficient to formulate our tableau reasoning method based on nested tableaux (see Section 3.2) for graph properties in CNF. It is routine to show, based on the same equivalences, that the transformation to CNF does not increase the nesting level of conditions.

Lemma 1 (transformation to CNF [13,14]). *There exists a transformation for each condition c_C over C according to equivalences listed in [13,14] into an equivalent condition $[c_C]$ in CNF.*

Lemma 2 (transformation to CNF preserves or reduces nesting level). *Given a condition c_C over C with nesting level n, $[c_C]$ has nesting level not greater than n.*

In the following sections we will make use extensively of the following shifting result over morphisms for conditions [5] that allow us to move a condition along a morphism.

Lemma 3 (shift of conditions over morphisms). *There is a transformation Shift such that for each condition c_P over P and each morphism $b : P \to P'$, it returns a condition $Shift(b, c_P)$ over P' such that for each monomorphism $n : P' \to H$ it holds that $n \circ b \models c_P \Leftrightarrow n \models Shift(b, c_P)$.*

Construction 1 (shift of conditions over morphisms). *The transformation Shift is inductively defined as follows:*

$$Shift(b, true) = true.$$
$$Shift(b, \exists(a, c_C)) = \bigvee_{(a',b')\in \mathcal{F}} \exists(a', Shift(b', c_C)) \quad if$$
$$\mathcal{F} = \{(a',b') \text{ jointly surjective} \mid b' \text{ mono and } (1) \text{ commutes}\} \neq \emptyset$$
$$Shift(b, \exists(a, c_C)) = false \text{ if } \mathcal{F} = \emptyset.$$
$$\text{Moreover, } Shift(b, \neg c_P) = \neg Shift(b, c_P) \text{ and}$$
$$Shift(b, \wedge_{i\in I} c_{P,i}) = \wedge_{i\in I} Shift(b, c_{P,i}).$$

(diagram)
$$\begin{array}{ccc} P & \xrightarrow{\ b\ } & P' \\ a\downarrow & (1) & \downarrow a' \\ C & \xrightarrow{\ b'\ } & C' \\ \triangle & & \\ c_C & & \end{array}$$

When shifting conditions over morphisms we need to know if their nesting level is preserved/reduced and if they remain in weak CNF. It is routine to prove the lemma below by induction on the structure of the given condition. But shifting negative literals does not preserve weak CNF. When shifting a negative literal, according to Construction 1 we obtain $Shift(b, \neg\exists(a, c_C)) = \neg Shift(b, \exists(a, c_C)) = \neg\bigvee_{(a',b')\in\mathcal{F}} \exists(a', Shift(b', c_C)) = \bigwedge_{(a',b')\in\mathcal{F}} \neg\exists(a', Shift(b', c_C))$. Notice that we require that a' is a monomorphismIn, which is not required in [5]. We may add this requirement because the models of our conditions are monomorphisms instead of arbitrary morphisms.

Lemma 4 (Shift literal in weak CNF). *Given a positive literal ℓ in weak CNF with nesting level n and a morphism $b : P \to P'$, $Shift(b, \ell)$ is a condition in weak CNF again and it has nesting level less than or equal to n. Given a negative literal ℓ in weak CNF with nesting level n and a morphism $b : P \to P'$, $Shift(b, \ell)$ has nesting level less than or equal to n.*

3 Tableau-Based Reasoning for GL

Analogously to tableaux for plain first-order logic reasoning [7], we introduce tableaux for dedicated automated reasoning for graph properties. We consider *clause tableau reasoning* assuming that the given graph conditions are in weak CNF as introduced in Def. 3. However, we will see that usual tableau reasoning is not sufficient and we will introduce so-called nested tableaux for automated graph property reasoning.

3.1 Tableaux for Graph Conditions

As usual, tableaux are trees whose nodes are literals. The construction of a tableau for a condition c_C in weak CNF can be informally explained as follows. We start with a tableau consisting of the single node *true*. Then, as usual in first-order logic, for every clause $c_1 \vee \ldots \vee c_n$ in c_C we extend all the leaves in the tableau with n branches, one for each condition c_i, as depicted in Fig. 2. The tableau rules that are specific for our logic are the so-called *lift* and *supporting lift* rules, defined by Penneman as part of his proof system [13,14]. In both rules, given two literals $\ell_1 = \exists(a_1, c_1)$ and ℓ_2 in the same branch, we add a new node to that branch with a new literal ℓ_3 that is equivalent to the conjunction of ℓ_1 and ℓ_2. In particular, ℓ_3 is built by pushing ℓ_2 inside the next level of nesting of ℓ_1 by shifting ℓ_2 (see Lemma 3) along a_1. The difference between the lift and supporting lift rules is that, in the former, ℓ_2 is a negative literal while, in the latter, ℓ_2 is a positive literal. Moreover, since at any time in the tableau we want to produce conditions that are in weak CNF and as argued in Lemma 4 shifted negative literals in general do not preserve weak CNF, we apply the transformation rules according to Lemma 1 to each shifted negative literal. The two rules are depicted in Fig. 3. Note that because of Lemma 4 and 1, the tableau rules and in particular the lift and supporting lift rule indeed generate a tableau as given in Def. 4, where each node is a literal in weak CNF. As usual, a closed branch is then a branch where we have found an inconsistency.

Definition 4 (Tableau, branch). *A tableau is a finitely branching tree whose nodes are literals in weak CNF (see Def. 1). A branch in a tableau T is a maximal path in T.*

Definition 5 (Tableau rules). *Given a condition c_C in weak CNF over C, then a tableau for c_C is defined as a tableau constructed with the following rules:*

- *The tree consisting of a single node* true *is a tableau for c_C (initialization rule).*
- *Let T be a tableau for c_C, B a branch of T, and $c_1 \vee \dots \vee c_n$ a clause in c_C, then extend B with n new subtrees and the nodes of the new subtrees labeled with c_i (extension rule).*
- *Let T be a tableau for c_C, B a branch of T, and $\exists(a_1, c_1)$ and $\neg\exists(a_2, c_2)$ literals in B, then extend B with a node equal to $\exists(a_3, c_3)$, where $\exists(a_3, c_3)$ is the condition $\exists(a_1, c_1 \wedge [Shift(a_1, \neg\exists(a_2, c_2))])$ (lift rule).*
- *Let T be a tableau for c_C, B a branch of T, and $\exists(a_1, c_1)$ and $\exists(a_2, c_2)$ nodes in B, then extend B with a node equal to $\exists(a_3, c_3)$, where $\exists(a_3, c_3)$ is the condition $\exists(a_1, c_1 \wedge Shift(a_1, \exists(a_2, c_2)))$ (supporting lift rule).*

Definition 6 (Open/closed branch). *In a tableau T a branch B is* closed *if B contains $\exists(a, false)$ or $false$; otherwise, it is* open.

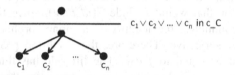

Fig. 2. The extension rule

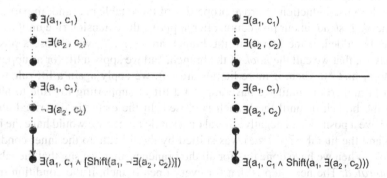

Fig. 3. The lift rule (left) and supporting lift rule (right)

In the completeness proof for our tableau reasoning method (see Theorem 2) we will need to track conditions that, because of the lift and supporting lift rules, move inside other conditions. We do this by means of a successor relation between conditions (and subconditions) for each branch in a given tableau.

Definition 7 (Successor relation for a branch). *For every branch B in a given tableau T, we define the successor relation associated to that branch as the least relation on nested conditions satisfying:*

- *If $\exists(a_1,c_1 \wedge [Shift(a_1,\neg\exists(a_2,c_2))])$ is the literal on a node created using the lift rule from literals $\exists(a_1,c_1)$ and $\neg\exists(a_2,c_2)$, then $[Shift(a_1,\neg\exists(a_2,c_2))]$ is a successor of $\neg\exists(a_2,c_2)$ in B.*
- *If $\exists(a_1,c_1 \wedge Shift(a_1,\exists(a_2,c_2)))$ is the literal on a node created using the supporting lift rule from literals $\exists(a_1,c_1)$ and $\exists(a_2,c_2)$, then $Shift(a_1,\exists(a_2,c_2))$ is a successor of $\exists(a_2,c_2)$ in B.*

Let us see an example that shows what happens if we have a very obvious refutable condition $\exists(a,true) \wedge \neg\exists(a,true)$. Since we have two clauses consisting of one literal, using the extension rule we would get a single branch with the literals $\exists(a,true)$ and $\neg\exists(a,true)$, where a is not an isomorphism. Using the lift rule we would extend the branch with the condition $\exists(a,true \wedge [Shift(a,\neg\exists(a,true))])$, which is equivalent to $\exists(a,true \wedge [\neg Shift(a,\exists(a,true))])$.

According to the construction of $Shift$, $Shift(a,\exists(a,true))$ is a disjunction that would include $\exists(id,true)$, since the diagram on the left commutes, the identity is a monomorphism and (id,id) are jointly surjective. Therefore, the condition $[\neg Shift(a,\exists(a,true))]$ would be equal to $false$. The lift rule has thus created a literal $\exists(a,true \wedge false)$ on the branch manifesting the obvious contradiction between the literals $\exists(a,true)$ and $\neg\exists(a,true)$ in the inner condition of the literal $\exists(a,true \wedge false)$.

Using the lift and supporting lift rules we can thus make explicit the inconsistencies that occur at the outer level of nesting, but we would need additional tableau rules that do something similar at any inner level of nesting. The problem is that, then, it is very difficult to use induction to prove properties of these tableaux (and, in particular, completeness). Instead, in our procedure, after applying the extension rule until no new literals can be added, if the literals in the branch are $\ell_1, \ldots, \ell_n$ we choose a positive literal [2], say ℓ_1, that we call the *hook* of the branch, and we apply a lift (or a supporting lift) rule to ℓ_1 and ℓ_2. Then, if more literals are left, we apply again a lift rule to the result and ℓ_3 and so on, until we have applied a lift or supporting lift rule to all the literals in the branch or until the branch is closed. In the example described above, where we have a positive and negative literal on a single branch we would have the hook $\exists(a,true)$ and the literal $\neg\exists(a,true)$ was shifted by the lift rule to the inner condition of this hook. When we have done this for all the branches, then we say that the tableau is *semi-saturated*. The next step is that for every open branch, if the condition in the leaf is $\exists(a,c)$, then we will proceed to *open* a new tableau for the inner condition c, trying to find a refutation for c. In our example, the leaf of the open branch is equal to $\exists(a,true \wedge [\neg Shift(a,\exists(a,true))]) = \exists(a,true \wedge false)$ and we would therefore open a new tableau for $true \wedge false$. It is obvious that the new tableau will be closed and

---

[2] If all the literals are negative, then no rule can be applied. But in this case, we can conclude that the given condition c_C is satisfiable. The reason is that, if c_C is a condition over C, the identity $id : C \to C$ would be a model for all the literals in the path and, hence, for the condition.

will therefore refute the inner condition of $\exists(a, true \wedge false)$ making also the original condition $\exists(a, true) \wedge \neg\exists(a, true)$ refutable. If in the newly opened tableaux there are still some open branches however, then new tableaux will be created for the conditions in the leaves, and so on. We call such a family of tableaux a *nested tableau* and we will study them in the following section.

Definition 8 (Semi-saturation, hook for a branch). *Given a tableau T for a condition c_C over C, we say that T is semi-saturated if*

- *No new literals can be added to any branch B in T using the extension rule.*
- *And in addition, for each branch B in T, one of the following conditions hold*
  - *B is closed,*
  - *B consists only of negative literals,*
  - *If $E = \{\ell_1, \ldots, \ell_n\}$ is the set of literals added to B using the extension rule, then there is a positive literal $\ell = \exists(a, d)$ in E such that the literal in the leaf of B is $\exists(a, d \wedge_{\ell' \in (E \setminus \{\ell\})} Shift(a, \ell'))$. Then, we say that l is a hook for branch B in T.*

For any condition in weak CNF we can then build a finite semi-saturated tableau.

Lemma 5 (Finiteness and existence of semi-saturated tableau). *Given a condition c_C over C in weak CNF, there exists a semi-saturated and finite tableau T for c_C.*

In the rest of this subsection we relate satisfiability of conditions in weak CNF with satisfiability of their associated tableaux. First, we define in the obvious way tableau satisfiability. Then, we show that if a condition c_C is satisfiable, then any associated tableau cannot have all its branches closed. This means that the tableau rules are sound. The proof is by induction on the structure of the tableau. The base case is trivial. If a node has been added by using the extension rule, then satisfiability of the given condition implies satisfiability of the tableau. Finally, the case of the lift and the supporting lift rules is a consequence of the soundness of these rules as shown by Penneman [13].

Definition 9 (Branch and tableau satisfiability). *A branch B in a tableau T for a condition c_C over C is satisfiable if there exists a morphism $q : C \rightarrow G$ satisfying all the literals in B. In this case we say that $q : C \rightarrow G$ is a model for B and we write $q \models B$. A tableau T is satisfiable if there is a satisfiable branch B in T. If $q \models B$ in T, we also say that q is a model for T and also $q \models T$.*

Lemma 6 (Tableau soundness). *Given a condition c_C in weak CNF and a tableau T for this condition, then if c_C is satisfiable, so is T.*

Now, we show that if a tableau T is semi-saturated and satisfiable, then its associated condition c_C is also satisfiable. Actually, we show something slightly stronger. In particular, if an open branch B in T includes positive literals and the literal in the leaf is satisfiable then c_C is also satisfiable. The first part of the proof of this lemma is trivial, since semi-saturation ensures that each branch includes a literal for each clause in c_C. The second part uses the fact that semi-saturation implies that d includes the shift of all the literals in the branch (except for the hook). Then, satisfaction of this condition implies the satisfaction of all conditions in the branch.

Lemma 7 (Semi-saturation and satisfiability). *If T is a semi-saturated tableau for c_C, then $q \models T$ implies $q \models c_C$. Moreover, if B is an open branch including some positive literals and $\exists(a, d)$ is the literal in the leaf of B, then $q \models \exists(a, d)$ implies $q \models c_C$.*

3.2   Nested Tableaux for Graph Properties

As we have discussed in the previous section, standard tableaux cannot be used in a reasonably simple way as a proof procedure in our logic. So, our approach is based on a notion of *nested tableaux* whose idea is that, for each open branch of a tableau T whose literal in the leaf is $\exists(a,c)$, we open a new tableau T' to try to refute condition c. Then, we say that $\exists(a,c)$ is the *opener* for T' and if a is a morphism from G to G', we say that G' is the context of T', meaning that c is a property of the graph G'.

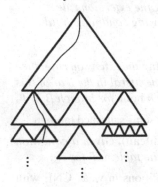

Nested tableaux have nested branches consisting of sequences of branches of a sequence of tableaux in the given nested tableau. We may notice that while our tableaux are assumed to be finite (cf. Lemma 5), nested tableaux and nested branches may be infinite. For simplicity, we will assume that the original condition to (dis)prove is a graph property, i.e. the models of the given condition (if they exist) are intuitively graphs. Formally, this means that the given condition c is a condition over the empty graph, i.e. on the outer level of nesting it consists of literals of the form $\exists(a,d)$ or $\neg\exists(a,d)$, where a is some morphism over the empty graph. Similarly, the models of c are morphisms from the empty graph.

Fig. 4. Nested tableau with nested branch

Definition 10 (Nested tableau, opener, context, nested branch, semi-saturation). *Let $(I, \leq, i_0)$ be a poset with minimal element i_0. A nested tableau NT is a family of triples $\{\langle T_i, j, \exists(a_j, c_j)\rangle\}_{i\in I}$, where T_i is a tableau and $\exists(a_j, c_j)$, called the* opener *of T_i is the literal of an open branch in T_j with $j < i$. Moreover, we assume that there is a unique initial tableau T_{i_1} that is part of the triple $\langle T_{i_1}, i_0, true\rangle$.*

We say that T_{i_1} has the empty context *$\emptyset$ and for any other tableau T_i, with $\langle T_i, j, \exists(a_j, c_j)\rangle \in NT$, if $a_j : A_j \rightarrow A_{j+1}$, we say that T_i has context A_{j+1}.*

A nested branch NB in a nested tableau $NT = \{\langle T_i, j, \exists(a_j, c_j)\rangle\}_{i\in I}$ is a maximal sequence of branches $B_{i_1}, \ldots, B_{i_k}, B_{i_{k+1}}, \ldots$ from tableaux $T_{i_1}, \ldots, T_{i_k}, T_{i_{k+1}}, \ldots$ in NT starting with a branch B_{i_1} in the initial tableau T_{i_1}, such that if B_{i_k} and $B_{i_{k+1}}$ are two consecutive branches in the sequence then the leaf in B_{i_k} is the opener for $T_{i_{k+1}}$.

Finally, NT is semi-saturated if each tableau in NT is semi-saturated.

Definition 11 (Nested tableau rules). *Given a graph property c in CNF, a nested tableau for c is constructed with the following rules:*

- *Let T_{i_1} for c be a tableau constructed according to the rules given in Def. 5, then $\{\langle T_{i_1}, i_0, true\rangle\}$ is a nested tableau for c (initialization rule).*
- *Let $NT = \{\langle T_i, j, \exists(a_j, c_j)\rangle\}_{i\in I}$ be a nested tableau for c and $\exists(a_n, c_n)$ with $a_n : A_n \rightarrow A_{n+1}$ a literal in a leaf of a tableau T_n in NT such that $\exists(a_n, c_n)$ is not the opener for any other tableau in NT, then add to NT a triple $\langle T_j, n, \exists(a_n, c_n)\rangle$, where $j > n$ is an index not previously used in I of NT, and T_j is a tableau for c_n (nesting rule).*

For the proof of the completeness theorem, we need to extend the successor relation defined for the branches of a tableau to a successor relation defined for each nested branch of a nested tableau.

Definition 12 (Successor relation for a nested branch). *For each nested branch NB in a nested tableau NT, we define the successor relation associated to that branch as the least transitive relation on nested conditions satisfying:*

- *If c_1 is a successor of c_2 in a branch B, with context A, in NB, then $\langle A, c_1 \rangle$ is the successor of $\langle A, c_2 \rangle$ in NB*
- *If $\exists(a_n, c_n)$ with $a_n : A_n \to A_{n+1}$ is the opener for a branch B in NB and ℓ is a literal included in B using the extension rule, then $\langle A_{n+1}, \ell \rangle$ is the successor of $\langle A_n, c \rangle$ for each condition c that is a subcondition of c_n.*

If the context of conditions is clear from the context, then we may just say that a condition c is a successor of c' in NB, instead of saying that $\langle A, c \rangle$ is a successor of $\langle A', c' \rangle$ in NB, where A and A' are the contexts of c and c' respectively.

A result that will be needed in the completeness proof is the fact that, if all the successors of a literal in a given branch are satisfiable, then the literal is satisfiable. The proof is by induction on $j - i$, using the definition of the Shift operation and of the successor relation, and the fact that all the tableaux involved are semi-saturated.

Lemma 8 (Satisfiability of successors). *Let $NB = B_0, B_1, \ldots, B_j, \ldots$ be a nested branch in a semi-saturated nested tableau, ℓ_i a literal in B_i with context A_i, $a_{ij} : A_i \to A_j$ the composition $a_{j-1} \circ \cdots \circ a_i$. Then if $\langle A_j, \ell_j \rangle$ is the successor of ℓ_i in B_i and $q_j \models \ell_j$, then $q_j \circ a_{ij} \models \ell_i$.*

As in the case of standard tableaux, a closed nested branch represents an inconsistency detected between the literals in the branch, and an open branch represents, under adequate assumptions, a model of the original condition.

Definition 13 (Open/closed nested branch, nested tableau proof). *A nested branch NB in a nested tableau NT for a graph property c in CNF of GL is closed if NB contains $\exists(a, false)$ or false; otherwise, it is open. A nested tableau is closed if all its nested branches are closed.*

A nested tableau proof for (the unsatisfiability of) c is a closed nested tableau NT for c in CNF of GL according to the rules given in Def. 11.

Example 1. Let us consider another simple example of deduction with our tableau method. Suppose that we want to resolve the condition c:
$$\exists(\emptyset \to Node, true) \land \neg\exists(\emptyset \to Loop, true) \land \neg\exists(\emptyset \to Node, \neg\exists(Node \to Loop, true))$$
Thereby, *Node* is a graph consisting of just one node, and *Loop* is a graph consisting of a node and a loop. That is, c states (i) that there must exist a node, (ii) there cannot be loops and (iii) there is no node that does not include a loop. The tableau associated to c is depicted in Fig. 5. It includes just a branch, since there are no disjunctions in c. The first nodes of this branch include three literals obtained by applying the extension rule. Moreover, the first literal of these three nodes is the hook for the branch, since it is the only positive literal. Then, after semi-saturation, the leaf of the branch would be $\exists(\emptyset \to Node, d)$, where d is depicted at the bottom of the figure.

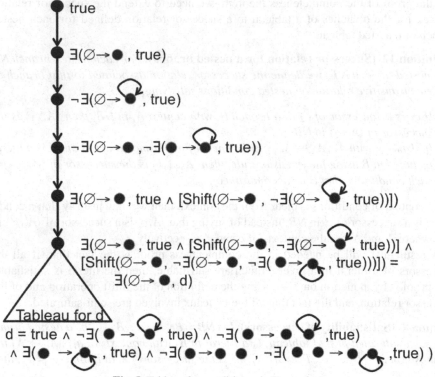

Fig. 5. Tableau for condition c in Example 1

So, the tableau for d with context *Node* is depicted in Fig. 6. Again, it includes a single branch whose hook is the only positive literal in d. In this case, the leaf of the branch would be $\exists(Node \to Loop, d')$, where d' is equivalent to the conjunction of shifting along $Node \to Loop$ with the rest of the literals in d. But in this case, because of shifting the negative literal $\neg\exists(Node \to Loop, true)$ the conjunction d' would include the *false* literal: As a consequence, when opening a tableau for d' at the next level of nesting, the only branch would include the *false* literal, which would close the single branch. Hence, this nested tableau would be a proof of the unsatisfiability of c.

4   Soundness and Completeness

In this section we prove that our tableau method is sound and complete. In particular, soundness means that if we are able to construct a nested tableau where all its branches are closed then we may be sure that our original condition c is unsatisfiable. Completeness means that if a *saturated* tableau includes an open branch, where the notion of saturation is defined below, then the original condition is satisfiable. Actually, the open branch provides the model that satisfies the condition.

Theorem 1 (Soundness). *If there is a nested tableau proof for the graph property c in CNF of GL, then c is unsatisfiable.*

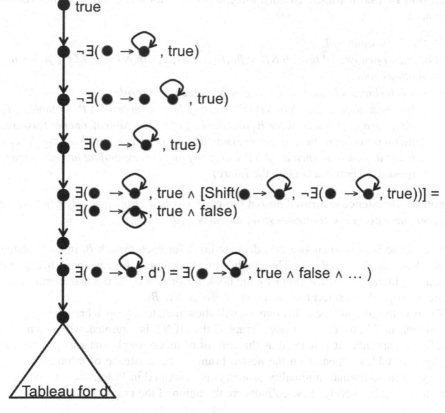

Fig. 6. Tableau for condition d in Example 1

In the proof of this theorem we use Lemma 6 that states the soundness of the rules for constructing (non-nested) tableaux and the fact that if all branches of the nested tableau are closed then it is finite. In particular, we prove by induction on the structure of NT that if c is satisfiable, then it must include an open branch. The base case is a consequence of Lemma 6. For the general case, assuming that the given nested tableau NT_i has an open nested branch NB, we consider two cases, depending on how we extend that NT_i. If the new tableau is not opened at the leaf of NB, then NB is still an open branch of the new tableau. Otherwise, we show that NB can be extended by a branch of the new tableau using Lemma 6.

For the completeness proof, the notion of saturation of nested tableaux is important. As usual, saturation describes some kind of fairness that ensures that we do not postpone indefinitely some inference step. In this case, the main issue concerning fairness is the choice of the hook for each tableau in the given nested tableau. In particular, if a (positive) literal, or its successors are never chosen as hooks we will be unable to make inferences between that literal and other literals, especially, negative literals. This means that we may be unable to find some existing contradictions.

Definition 14 (Saturation). *A nested tableau NT is* saturated *if the following conditions hold:*

1. *NT is semi-saturated.*
2. *For each open nested branch $NB = B_0, B_1, \ldots, B_j, \ldots$ in NT one of the following conditions hold:*
   - *NB includes a branch consisting only of negative literals.*
   - *For each node $\exists(a_j, c_j)$ in NB with nesting level k on branch B_j in tableau T_j either $\exists(a_j, c_j)$ is a hook for B_j and the leaf of B_j is a tableau opener (positive literal is hook), or there is a successor $\exists(a_i, c_i)$ of $\exists(a_j, c_j)$ with nesting level k being a hook in branch B_i of NB with $i > j$ and corresponding tableau opener (positive literal is a hook in the future).*

Lemma 9 (Existence saturated nested tableau). *Given a graph property c in CNF of GL, then there exists a saturated nested tableau NT for c.*

The key issue here is to choose the adequate hook for each branch B_i in each nested branch $NB = B_0, B_1, \ldots, B_j, \ldots$. This can be done by keeping, for each branch, a queue of pending literals. So when choosing the hook for branch B_i, if the first literal in the queue is $\exists(a_j, c_j)$, we select the successor of $\exists(a_j, c_j)$ in B_i.

To prove the completeness theorem we will show that, to any open branch in a given nested tableau NT, we can associate a graph G that, if NT is saturated, will be a model for NT. In particular, it is defined as the colimit of monomorphisms arising from the sequence of tableau openers on the nested branch. The existence of infinite colimits (satisfying an additional minimality property) is described in Proposition 1, which is proven in [12]. Intuitively, these colimits are the union of the graphs in the sequence.

Proposition 1 (Infinite colimits). *Given a sequence of monomorphisms:*

$$G_1 \xrightarrow{\;f_1\;} G_2 \xrightarrow{\;f_2\;} \ldots \xrightarrow{\;f_{i-1}\;} G_i \xrightarrow{\;f_i\;} \ldots$$

there exists a colimit:

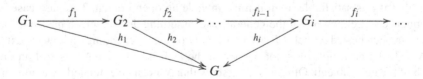

that satisfies that for every monomorphism $g : G' \to G$, such that G' is a finite graph, there is a j and a monomorphism $g_j : G' \to G_j$ such that the diagram below commutes:

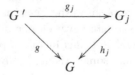

Definition 15 (Canonical model for an open nested branch NB**).** *Given a nested tableau* NT *for c in CNF of GL and given an open nested branch* NB *in* NT, *then the canonical model for* NB *is the empty graph in case that* $NB = B_{i_1}$ *consists of only one branch and otherwise it is defined as the (possibly infinite) colimit*

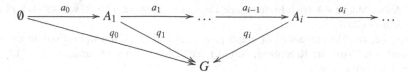

from the sequence of monomorphisms $\emptyset \overset{a_0}{\hookrightarrow} A_1 \overset{a_1}{\hookrightarrow} A_2 \cdots \overset{a_{i-1}}{\hookrightarrow} A_i \overset{a_i}{\hookrightarrow} A_{i+1} \cdots$ *arising from the sequence of tableau openers* $true, \exists(a_0, c_0), \exists(a_1, c_1), \cdots, \exists(a_i, c_i) \cdots$ *on the nested branch* NB *in* NT.

Theorem 2 (Completeness). *If the graph property c in CNF of GL is unsatisfiable, then there is a tableau proof for c.*

In the proof of the Completeness Theorem we show that if there is no tableau proof for c, then c is satisfiable. Because of Lemma 9 we know that there exists a saturated nested tableau NT for c. If NT is not a tableau proof for c, there exists at least one open nested branch NB in NT. Then, we prove by induction on the nesting level of literals that its associated canonical model G (as given in Def. 15) satisfies all the literals in NB. In particular, we show that morphism $q_n : A_n \to G$ satisfies all literals whose context is A_n. This means that each condition on the branch B_{i_1} with empty context within NB is satisfied by $q_0 : \emptyset \to G$ and thus by G. Hence, as a consequence of semi-saturation and Lemma 7, it holds that $G \models c$.

5 Conclusion

In this paper, we have presented a new tableau-based reasoning method for graph properties, where graph properties are formulated explicitly making use of graphs and graph morphisms, having the expressive power of Courcelle's first order logic of graphs. In particular, we proved the soundness and completeness of this method based on Pennemann's [13] rules. Finally, we presented a new notion of nested tableaux that could be useful for other formalisms where formulas have a hierarchical structure like nested graph conditions. With respect to future work, in addition to implementing our approach, we consider that the graph logic GL could be used as the foundations for graph databases, in the same sense as first-order logic is used as foundations for relational databases. In particular there would be two aspects that would need further work. On the one hand, we would need to extend the logic and reasoning method to allow for the possibility to state the existence of paths between nodes in a graph. On the other hand, we would need to characterize a sufficiently expressive fragment of that logic that is not only decidable, but where queries could be computed efficiently.

Acknowledgement. We would like to thank Annegret Habel and Karl-Heinz Pennemann for some interesting discussions on the main ideas of this paper.

References

1. Boronat, A., Meseguer, J.: Automated model synchronization: A case study on uml with maude. ECEASST 41 (2011)
2. Clavel, M., Durán, F., Eker, S., Lincoln, P., Martí-Oliet, N., Meseguer, J., Talcott, C.: All About Maude - A High-Performance Logical Framework. LNCS, vol. 4350. Springer, Heidelberg (2007)
3. Courcelle, B.: The expression of graph properties and graph transformations in monadic second-order logic. In: Rozenberg, G. (ed.) Handbook of Graph Grammars, pp. 313–400. World Scientific (1997)
4. Ehrig, H., Ehrig, K., Prange, U., Taentzer, G.: Fundamentals of algebraic graph transformation. Springer (2006)
5. Ehrig, H., Golas, U., Habel, A., Lambers, L., Orejas, F.: $\mathcal{M}$-adhesive transformation systems with nested application conditions. part 1: Parallelism, concurrency and amalgamation. Math. Struct, in Comp. Sc. (2012) (to appear)
6. Habel, A., Pennemann, K.H.: Correctness of high-level transformation systems relative to nested conditions. Mathematical Structures in Computer Science 19(2), 245–296 (2009)
7. Hähnle, R.: Tableaux and related methods. In: Robinson, J.A., Voronkov, A. (eds.) Handbook of Automated Reasoning, pp. 100–178. Elsevier and MIT Press (2001)
8. Heckel, R., Wagner, A.: Ensuring consistency of conditional graph rewriting - a constructive approach. Electr. Notes Theor. Comput. Sci. 2, 118–126 (1995)
9. Lack, S., Sobocinski, P.: Adhesive and quasiadhesive categories. ITA 39(3), 511–545 (2005)
10. Lambers, L., Orejas, F.: Tableau-based reasoning for graph properties. Tech. rep., Departament de Llenguatges i Sistèmes Informàtics, Universitat Politècnica de Catalunya (2014)
11. Orejas, F.: Symbolic graphs for attributed graph constraints. J. Symb. Comput. 46(3), 294–315 (2011)
12. Orejas, F., Ehrig, H., Prange, U.: Reasoning with graph constraints. Formal Asp. Comput. 22(3-4), 385–422 (2010)
13. Pennemann, K.H.: Resolution-like theorem proving for high-level conditions. In: Ehrig, H., Heckel, R., Rozenberg, G., Taentzer, G. (eds.) ICGT 2008. LNCS, vol. 5214, pp. 289–304. Springer, Heidelberg (2008)
14. Pennemann, K.H.: Development of Correct Graph Transformation Systems, PhD Thesis. Dept. Informatik, Univ. Oldedburg (2009)
15. Rensink, A.: Representing first-order logic using graphs. In: Ehrig, H., Engels, G., Parisi-Presicce, F., Rozenberg, G. (eds.) ICGT 2004. LNCS, vol. 3256, pp. 319–335. Springer, Heidelberg (2004)

Verifying Monadic Second-Order Properties
of Graph Programs

Christopher M. Poskitt[1] and Detlef Plump[2]

[1] Department of Computer Science, ETH Zürich, Switzerland
[2] Department of Computer Science, The University of York, UK

Abstract. The core challenge in a Hoare- or Dijkstra-style proof system for graph programs is in defining a weakest liberal precondition construction with respect to a rule and a postcondition. Previous work addressing this has focused on assertion languages for first-order properties, which are unable to express important global properties of graphs such as acyclicity, connectedness, or existence of paths. In this paper, we extend the nested graph conditions of Habel, Pennemann, and Rensink to make them equivalently expressive to monadic second-order logic on graphs. We present a weakest liberal precondition construction for these assertions, and demonstrate its use in verifying non-local correctness specifications of graph programs in the sense of Habel et al.

1 Introduction

Many problems in computer science and software engineering can be modelled in terms of graphs and graph transformation, including the specification and analysis of pointer structures, object-oriented systems, and model transformations; to name just a few. These applications, amongst others, motivate the development of techniques for verifying the functional correctness of both graph transformation rules and programs constructed over them.

A recent strand of research along these lines has resulted in the development of *proof calculi* for graph programs. These, in general, provide a means of systematically proving that a program is correct relative to a specification. A first approach was considered by Habel, Pennemann, and Rensink [1,2], who contributed weakest precondition calculi – in the style of Dijkstra – for simple rule-based programs, with specifications expressed using *nested conditions* [3]. Subsequently, we developed Hoare logics [4,5] for the graph transformation language GP 2 [6], which additionally allows computation over labels, and employed as a specification language an extension of nested conditions with support for expressions.

Both approaches suffer from a common drawback, in that they are limited to first-order structural properties. In particular, neither of them support proofs about important *non-local* properties of graphs, e.g. acyclicity, connectedness, or the existence of arbitrary-length paths. Part of the difficulty in supporting such assertions is at the core of both approaches: defining an effective construction for

H. Giese and B. König (Eds.): ICGT 2014, LNCS 8571, pp. 33–48, 2014.
© Springer International Publishing Switzerland 2014

the weakest property guaranteeing that an application of a given rule will establish a given postcondition (i.e. the construction of a *weakest liberal precondition* for graph transformation rules).

Our paper addresses exactly this challenge. We define an extension of nested conditions that is equivalently expressive to monadic second-order (MSO) logic on graphs [7]. For this assertion language, and for graph programs similar to those of [1,2], we define a weakest liberal precondition construction that can be integrated into Dijkstra- and Hoare-style proof calculi. Finally we demonstrate its use in verifying non-local correctness specifications (properties including that the graph is bipartite, acyclic) of some simple programs.

The paper is organised as follows. In Section 2 we provide some preliminary definitions and notations. In Section 3 we define an extension of nested conditions for MSO properties. In Section 4 we define graph programs, before presenting our weakest liberal precondition construction in Section 5, and demonstrating in Section 6 its use in Hoare-style correctness proofs. Finally, Section 7 presents some related work before we conclude the paper in Section 8.

Proofs omitted from this paper are available in an extended version [8].

2   Preliminaries

Let $\mathbb{B} = \{\text{true}, \text{false}\}$ denote the set of Boolean values, Vertex, Edge denote (disjoint) sets of node and edge identifiers (which shall be written in lowercase typewriter font, e.g. v, e), and VSetVar, ESetVar denote (disjoint) sets of node- and edge-set variables (which shall be written in uppercase typewriter font, e.g. X, Y).

A *graph* over a label alphabet $\mathcal{C} = \langle \mathcal{C}_V, \mathcal{C}_E \rangle$ is defined as a system $G = (V_G, E_G, s_G, t_G, l_G, m_G)$, where $V_G \subset$ Vertex and $E_G \subset$ Edge are finite sets of *nodes* (or *vertices*) and *edges*, $s_G, t_G \colon E_G \to V_G$ are the *source* and *target* functions for edges, $l_G \colon V_G \to \mathcal{C}_V$ is the node labelling function and $m_G \colon E_G \to \mathcal{C}_E$ is the edge labelling function. The *empty graph*, denoted by $\emptyset$, has empty node and edge sets. For simplicity, we fix the label alphabet throughout this paper as $\mathcal{L} = \langle \{\square\}, \{\square\} \rangle$, where $\square$ denotes the blank label (which we render as ● and ⟶ in pictures). We note that our technical results hold for any fixed finite label alphabet.

Given a graph G, the *(directed) path predicate* $\text{path}_G \colon V_G \times V_G \times 2^{E_G} \to \mathbb{B}$ is defined inductively for nodes $v, w \in V_G$ and sets of edges $E \subseteq E_G$. If $v = w$, then $\text{path}_G(v, w, E)$ holds. If $v \neq w$, then $\text{path}_G(v, w, E)$ holds if there exists an edge $e \in E_G \setminus E$ such that $s_G(e) = v$ and $\text{path}_G(t_G(e), w, E)$.

A *graph morphism* $g \colon G \to H$ between graphs G, H consists of two functions $g_V \colon V_G \to V_H$ and $g_E \colon E_G \to E_H$ that preserve sources, targets and labels; that is, $s_H \circ g_E = g_V \circ s_G$, $t_H \circ g_E = g_V \circ t_G$, $l_H \circ g_V = l_G$, and $m_H \circ g_E = m_G$. We call G, H the *domain* (resp. *codomain*) of g. Morphism g is an *inclusion* if $g(x) = x$ for all nodes and edges x. It is *injective* (*surjective*) if g_V and g_E are injective (surjective). It is an *isomorphism* if it is both injective and surjective. In this case G and H are *isomorphic*, which is denoted by $G \cong H$.

3 Expressing Monadic Second-Order Properties

We extend the nested conditions of [3] to a formalism equivalently expressive to MSO logic on graphs. The idea is to introduce new quantifiers for node- and edge-set variables, and equip morphisms with constraints about set membership. The definition of satisfaction is then extended to require an interpretation of these variables in the graph such that the constraint evaluates to true. Furthermore, constraints can also make use of a predicate for explicitly expressing properties about directed paths. Such properties can of course be expressed in terms of MSO expressions, but the predicate is provided as a more compact alternative.

Definition 1 (Interpretation; interpretation constraint). Given a graph G, an *interpretation* I *in* G is a partial function $I :$ VSetVar $\cup$ ESetVar $\rightarrow$ $2^{V_G} \cup 2^{E_G}$, such that for all variables X on which it is defined, $I(X) \in 2^{V_G}$ if $X \in$ VSetVar (resp. 2^{E_G}, ESetVar). An *(interpretation) constraint* is a Boolean expression that can be derived from the syntactic category Constraint of the following grammar:

> Constraint ::= Vertex '$\in$' VSetVar | Edge '$\in$' ESetVar
> | **path** '(' Vertex ',' Vertex [',' **not** Edge {'|' Edge}] ')'
> | **not** Constraint | Constraint (**and** | **or**) Constraint | **true**

Given a constraint γ, an interpretation I in G, and a morphism q with codomain G, the value of $\gamma^{I,q}$ in $\mathbb{B}$ is defined inductively. If γ contains a set variable for which I is undefined, then $\gamma^{I,q} =$ false. Otherwise, if γ is **true**, then $\gamma^{I,q} =$ true. If γ has the form $x \in X$ with x a node or edge identifier and X a set variable, then $\gamma^{I,q} =$ true if $q(x) \in I(X)$. If γ has the form **path**(v,w) with v,w node identifiers, then $\gamma^{I,q} =$ true if the predicate $\text{path}_G(q(v),q(w),\emptyset)$ holds. If γ has the form **path**$(v,w,\textbf{not}\ e_1|\ldots|e_n)$ with v,w node identifiers and $e_1,\ldots,e_n$ edge identifiers, then $\gamma^{I,q} =$ true if it is the case that the path predicate $\text{path}_G(q(v),q(w),\{q(e_1),\ldots,q(e_n)\})$ holds. If γ has the form **not** γ_1 with γ_1 a constraint, then $\gamma^{I,q} =$ true if $\gamma_1^{I,q} =$ false. If γ has the form γ_1 **and** γ_2 (resp. γ_1 **or** γ_2) with γ_1, γ_2 constraints, then $\gamma^{I,q} =$ true if both (resp. at least one of) $\gamma_1^{I,q}$ and $\gamma_2^{I,q}$ evaluate(s) to true. □

Definition 2 (M-condition; M-constraint). An *MSO condition* (short. *M-condition*) over a graph P is of the form **true**, $\exists_V X [c]$, $\exists_E X [c]$, or $\exists(a \mid \gamma, c')$, where $X \in$ VSetVar (resp. ESetVar), c is an M-condition over P, $a : P \hookrightarrow C$ is an injective morphism (since we consider programs with injective matching), γ is an interpretation constraint over items in C, and c' is an M-condition over C. Furthermore, Boolean formulae over M-conditions over P are also M-conditions over P; that is, $\neg c$, $c_1 \wedge c_2$, and $c_1 \vee c_2$ are M-conditions over P if c, c_1, c_2 are M-conditions over P.

An M-condition over the empty graph $\emptyset$ is called an *M-constraint*. □

For brevity, we write **false** for $\neg$**true**, $c \Rightarrow d$ for $\neg c \vee d$, $c \Leftrightarrow d$ for $c \Rightarrow d \wedge d \Rightarrow c$, $\forall_V X [c]$ for $\neg \exists_V X [\neg c]$, $\forall_E X [c]$ for $\neg \exists_E X [\neg c]$, $\exists_V X_1, \ldots X_n [c]$ for

$\exists_V X_1 [\ldots \exists_V X_n [c] \ldots]$ (analogous for other set quantifiers), $\exists (a \mid \gamma)$ for $\exists (a \mid \gamma, \mathbf{true})$, $\exists (a, c')$ for $\exists (a \mid \mathbf{true}, c')$, and $\forall (a \mid \gamma, c')$ for $\neg \exists (a \mid \gamma, \neg c')$.

In our examples, when the domain of a morphism $a \colon P \hookrightarrow C$ can unambiguously be inferred, we write only the codomain C. For instance, an M-constraint $\exists (\emptyset \hookrightarrow C, \exists (C \hookrightarrow C'))$ can be written as $\exists (C, \exists (C'))$.

Definition 3 (Satisfaction of M-conditions). Let $p \colon P \hookrightarrow G$ denote an injective morphism, c an M-condition over P, and I an interpretation in G. We define inductively the meaning of $p \models^I c$, which denotes that p *satisfies c with respect to I*. If c has the form $\mathbf{true}$, then $p \models^I c$. If c has the form $\exists_V X [c']$ (resp. $\exists_E X [c']$), then $p \models^I c$ if $p \models^{I'} c'$, where $I' = I \cup \{X \mapsto V\}$ for some $V \subseteq V_G$ (resp. $\{X \mapsto E\}$ for some $E \subseteq E_G$). If c has the form $\exists (a \colon P \hookrightarrow C \mid \gamma, c')$, then $p \models^I c$ if there is an injective morphism $q \colon C \hookrightarrow G$ such that $q \circ a = p$, $\gamma^{I,q} = \mathbf{true}$, and $q \models^I c'$.

A graph G *satisfies* an M-constraint c, denoted $G \models c$, if $i_G \colon \emptyset \hookrightarrow G \models^{I_\emptyset} c$, where $I_\emptyset$ is the *empty interpretation in* G, i.e. undefined on all set variables.   □

We remark that model checking for both first-order and monadic second-order logic is known to be PSPACE-complete [9]. However, the model checking problem for monadic second-order logic on graphs of bounded treewidth can be solved in linear time [10].

Example 1. The following M-constraint *col* (translated from the corresponding formula §1.5 of [11]) expresses that a graph is 2-colourable (or bipartite); i.e. every node can be assigned one of two colours such that no two adjacent nodes have the same one. Let γ_{col} denote $\mathbf{not}\ (v \in X\ \mathbf{and}\ w \in X)\ \mathbf{and}\ \mathbf{not}\ (v \in Y\ \mathbf{and}\ w \in Y)$.

$$\exists_V X, Y\ [\ \forall(\ \bullet_v\ ,\ \exists(\ \bullet_v\ \mid\ (v \in X\ \mathbf{or}\ v \in Y)\ \mathbf{and}\ \mathbf{not}\ (v \in X\ \mathbf{and}\ v \in Y)))$$

$$\wedge\ \forall(\ \bullet_v\ \ \bullet_w\ ,\ \exists(\ \bullet_v \!\!\rightarrow\!\! \bullet_w\ ) \Rightarrow \exists(\ \bullet_v\ \ \bullet_w\ \mid \gamma_{\mathsf{col}}))\ ]$$

A graph G will satisfy *col* if there exist two subsets of V_G such that: (1) every node in G belongs to *exactly one* of the two sets; and (2) if there is an edge from one node to another, then those nodes are not in the same set. Intuitively, one can think of the sets X and Y as respectively denoting the nodes of colour one and colour two. If two such sets do not exist, then the graph cannot be assigned a 2-colouring.   □

Theorem 1 (M-constraints are equivalent to MSO formulae). The assertion languages of M-constraints and MSO graph formulae are equivalently expressive: that is, given an M-constraint c, there exists an MSO graph formula φ such that for all graphs G, $G \models c$ if and only if $G \models \varphi$; and vice versa.   □

4   Graph Programs

In this section we define rules, rule application, and graph programs. Whilst the syntax and semantics of the control constructs are based on those of GP 2 [6], the rules themselves follow [1,2], i.e. are labelled over a fixed finite alphabet, and do not support relabelling or expressions. We equip the rules with application conditions (M-conditions over the left- and right-hand graphs), and define *rule application* via the standard double-pushout construction [12].

Definition 4 (Rule; direct derivation). A *plain rule* $r' = \langle L \hookleftarrow K \hookrightarrow R \rangle$ comprises two inclusions $K \hookrightarrow L$, $K \hookrightarrow R$. We call L, R the left- (resp. right-) hand graph and K the interface. An *application condition* ac $= \langle \mathrm{ac}_L, \mathrm{ac}_R \rangle$ for r' consists of two M-conditions over L and R respectively. A *rule* $r = \langle r', \mathrm{ac} \rangle$ is a plain rule r' and an application condition ac for r'.

$$
\begin{array}{ccc}
L & \longleftarrow K \longrightarrow & R \\[2pt]
{\scriptstyle g}\big\downarrow \quad (1) & \big\downarrow \quad (2) & \big\downarrow {\scriptstyle h} \\[2pt]
G & \longleftarrow D \longrightarrow & H
\end{array}
$$

For a plain rule r' and a morphism $K \hookrightarrow D$, a *direct derivation* $G \Rightarrow_{r',g,h} H$ (short. $G \Rightarrow_{r'} H$ or $G \Rightarrow H$) is given by the pushouts (1) and (2). For a rule $r = \langle r', \mathrm{ac} \rangle$, there is a *direct derivation* $G \Rightarrow_{r,g,h} H$ if $G \Rightarrow_{r',g,h} H$, $g \models^{I_\emptyset} \mathrm{ac}_L$, and $h \models^{I_\emptyset} \mathrm{ac}_R$. We call g, h a *match* (resp. *comatch*) for r. Given a set of rules $\mathcal{R}$, we write $G \Rightarrow_{\mathcal{R}} H$ if $G \Rightarrow_{r,g,h} H$ for some $r \in \mathcal{R}$. □

It is known that, given a (plain) rule r, graph G, and morphism g as above, there exists a direct derivation if and only if g satisfies the *dangling condition*, i.e. that no node in $g(L) \setminus g(K)$ is incident to an edge in $G \setminus g(L)$. In this case, D and H are determined uniquely up to isomorphism, constructed from G as follows: first, remove all edges in $g(L) \setminus g(K)$ obtaining D. Then add disjointly all nodes and edges from $R \setminus K$ retaining their labels. For $e \in E_R \setminus E_K$, $s_H(e) = s_R(e)$ if $s_R(e) \in V_R \setminus V_K$, otherwise $s_H(e) = g_V(s_R(e))$, (targets defined analogously) resulting in the graph H.

We will often give rules without the interface, writing just $L \Rightarrow R$. In such cases we number nodes that correspond in L and R, and establish the convention that K comprises exactly these nodes and that $E_K = \emptyset$ (i.e. K can be completely inferred from L, R). Furthermore, if the application condition of a rule is $\langle \mathbf{true}, \mathbf{true} \rangle$, then we will only write the plain rule component.

We consider now the syntax and semantics of graph programs, which provide a mechanism to control the application of rules to some graph provided as input.

Definition 5 (Graph program). *(Graph) programs* are defined inductively. First, every rule (resp. rule set) $r, \mathcal{R}$ and **skip** are programs. Given programs C, P, Q, we have that $P; Q$, $P!$, **if** C **then** P **else** Q, and **try** C **then** P **else** Q are programs. □

Graph programs are *nondeterministic*, and their execution on a particular graph could result in one of several possible outcomes. That outcome could be a graph, or it could be the special state "fail" which occurs when a rule (set) is not *applicable* to the current graph.

A full structural operational semantics is given in [8], but the informal meaning of the constructs is as follows. Let G denote an input graph. Programs $r, \mathcal{R}$

correspond to rule (resp. rule set) application, returning H if there exists some $G \Rightarrow_r H$ (resp. $G \Rightarrow_\mathcal{R} H$); otherwise fail. Program $P; Q$ denotes sequential composition. Program $P!$ denotes as-long-as-possible iteration of P. Finally, the conditional programs execute the first or second branch depending on whether executing C returns a graph or fail, with the distinction that the if construct does not retain any effects of C, whereas the try construct does.

Example 2. Consider the program init; grow! defined by the rules:

where tc is an (unspecified) M-condition over L expressing some termination condition for the iteration (proving termination is not our concern here, see e.g. [13]). The program, if executed on the empty graph, nondeterministically constructs and returns a tree. It applies the rule init exactly once, creating an isolated node. It then iteratively applies the rule grow (each application adding a leaf to the tree) until the termination condition tc holds. An example program run, with $tc = \exists(\,\bullet_1 \bullet\bullet\,)$, is:

$\square$

5   Constructing a Weakest Liberal Precondition

In this section, we present a construction for the *weakest liberal precondition* relative to a rule r and a postcondition c (which is an M-constraint). In our terminology, if a graph satisfies a weakest liberal precondition, then: (1) any graphs resulting from applications of r will satisfy c; and (2) there does not exist another M-constraint with this property that is weaker. (Note that we do not address termination or existence of results in this paper.)

The construction is adapted from the one for nested conditions in [3], and as before, is broken down into a number of stages. First, a translation of postconditions into M-conditions over R (transformation "A"); then, from M-conditions over R into M-conditions over L (transformation "L"); and finally, from M-conditions over L into an M-constraint expressing the weakest liberal precondition (via transformations "App" and "Pre").

First, we consider transformation A, which constructs an M-condition over R from a postcondition (an M-constraint) by computing a disjunction over all the ways that the M-constraint and comatches might "overlap".

Theorem 2 (M-constraints to M-conditions over R). There is a transformation A, such that for all M-constraints c, all rules r with right-hand side R, and all injective morphisms $h: R \hookrightarrow H$,

$$h \models^{I_\emptyset} A(r, c) \quad \text{if and only if} \quad H \models c.$$

Construction. Let c denote an M-constraint, and r a rule with right-hand side R. We define $A(r, c) = A'(\emptyset \hookrightarrow R, c)$ where A' is defined inductively as follows. For injective graph morphisms $p: P \hookrightarrow P'$ and M-conditions over P, define:

$$A'(p, \mathbf{true}) = \mathbf{true},$$
$$A'(p, \exists_V X[c']) = \exists_V X[A'(p, c')],$$
$$A'(p, \exists_E X[c']) = \exists_E X[A'(p, c')],$$
$$A'(p, \exists(a: P \hookrightarrow C \mid \gamma, c')) = \bigvee_{e \in \varepsilon} \exists(b: P' \hookrightarrow E \mid \gamma, A'(s: C \hookrightarrow E, c')).$$

The final equation relies on the following. First, construct the pushout (1) of p and a leading to injective graph morphisms $a': P' \hookrightarrow C'$ and $q: C \hookrightarrow C'$. The disjunction then ranges over the set ε, which we define to contain every surjective graph morphism $e: C' \to E$ such that $b = e \circ a'$ and $s = e \circ q$ are injective graph morphisms (we consider the codomains of each e up to isomorphism, hence the disjunction is finite).

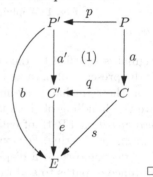

The transformations A, A' are extended for Boolean formulae over M-conditions in the usual way, that is, $A(r, \neg c) = \neg A(r, c)$, $A(r, c_1 \wedge c_2) = A(r, c_1) \wedge A(r, c_2)$, and $A(r, c_1 \vee c_2) = A(r, c_1) \vee A(r, c_2)$ (analogous for A'). □

Example 3. Recall the rule **grow** from Example 2. Let c denote the M-constraint:

$$\exists_V X, Y[\ \forall(\ \bullet_v \bullet_w\ , \exists(\ \bullet_v \bullet_w\ \mid \mathtt{path(v,w)}) \Rightarrow \exists(\ \bullet_v \bullet_w\ \mid \gamma))\ ]$$

for $\gamma = (v \in X \text{ and } w \in Y) \text{ and not } (v \in Y \text{ or } w \in X)$, which expresses that there are two sets of nodes X, Y in the graph, such that if there is a path from some node v to some node w, then v belongs only to X and w only to Y. Applying transformation A:

$$A(\mathbf{grow}, c)$$
$$= A'(\emptyset \hookrightarrow \bullet_1 {\to} \bullet_2\ , c)$$
$$= \exists_V X, Y[\ A'(\emptyset \hookrightarrow \bullet_1 {\to} \bullet_2\ , \forall(\ \bullet_v \bullet_w\ ,$$
$$\exists(\ \bullet_v \bullet_w\ \mid \mathtt{path(v,w)}) \Rightarrow \exists(\ \bullet_v \bullet_w\ \mid \gamma)))\ ]$$
$$= \exists_V X, Y[\ \bigwedge_{i=1}^{7} \forall(\ \bullet_1 {\to} \bullet_2\ \hookrightarrow E_i, \exists(E_i \mid \mathtt{path(v,w)}) \Rightarrow \exists(E_i \mid \gamma))\ ]$$

where the graphs E_i are as given in Figure 1. □

Transformation L, adapted from [3], takes an M-condition over R and constructs an M-condition over L that is satisfied by a match if and only if the

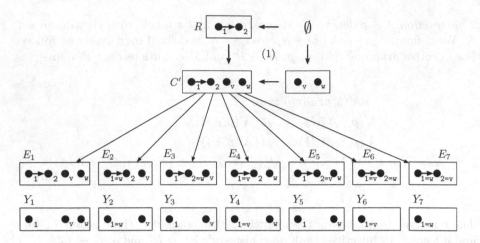

Fig. 1. Applying the construction in Examples 3 and 4

original is satisfied by the comatch. The transformation is made more complex by the presence of path and MSO expressions, because nodes and edges referred to on the right-hand side may no longer exist on the left. For clarity, we separate the handling of these two types of expressions, and in particular, define a *decomposition* LPath of path predicates according to the items that the rule is creating or deleting. For example, if an edge is created by a rule, a path predicate decomposes to a disjunction of path predicates collectively asserting the existence of paths to and from the nodes that will eventually become its source and target; whereas if an edge is to be deleted, the predicate will exclude it.

Proposition 1 (Path decomposition). There is a transformation LPath such that for every rule $r = \langle L \hookleftarrow K \hookrightarrow R \rangle$, direct derivation $G \Rightarrow_{r,g,h} H$, path predicate p over R, and interpretation I,

$$\text{LPath}(r,p)^{I,g} = p^{I,h}.$$

Construction. Let $r = \langle L \hookleftarrow K \hookrightarrow R \rangle$ and $p = \texttt{path}(v, w, \texttt{not } E)$. For simplicity, we will treat the syntactic construct E as a set of edges and identify $\texttt{path}(v, w, \texttt{not } E)$ and $\texttt{path}(v, w)$ when E is empty. Then, define:

$$\text{LPath}(r,p) = \text{LPath}'(r,v,w,E^{\ominus}) \text{ or FuturePaths}(r,p).$$

Here, $E^{\ominus}$ is constructed from E by adding edges $e \in E_L \setminus E_R$, i.e. that the rule will delete. Furthermore, $\text{LPath}'(r, v, w, E^{\ominus})$ decomposes to path predicates according to whether v and w exist in K. If $\text{path}_R(v, w, E^{\ominus})$ holds, then $\text{LPath}'(r, v, w, E^{\ominus})$ returns $\texttt{true}$. Otherwise, if both $v, w \in V_K$, then it returns $\texttt{path}(v, w, \texttt{not } E^{\ominus})$. If $v \notin V_K, w \in V_K$, it returns:

$$\texttt{false or path}(x_1, w, \texttt{not } E^{\ominus}) \texttt{ or path}(x_2, w, \texttt{not } E^{\ominus}) \texttt{ or} \ldots$$

for each $x_i \in V_K$ such that $\text{path}_R(v, x_i, E^\ominus)$. Case $v \in V_K, w \notin V_K$ analogous. If $v, w \notin V_K$, then it returns $\texttt{false}$ or $\texttt{path}(x_i, y_j, \texttt{not } E^\ominus)$ or $\ldots$ for all $x_i, y_j \in V_K$ such that $\text{path}_R(v, x_i, E^\ominus)$ and $\text{path}_R(y_j, w, E^\ominus)$.

Finally, $\text{FuturePaths}(r, p)$ denotes $\texttt{false}$ in disjunction with:

$$(\text{LPath}'(r, v, x_1, E^\ominus) \text{ and } \texttt{path}(y_1, x_2, \texttt{not } E^\ominus) \ldots \text{ and } \texttt{path}(y_i, x_{i+1}, \texttt{not } E^\ominus)$$
$$\ldots \text{ and } \text{LPath}'(r, y_n, w, E^\ominus))$$

over all non-empty sequences of distinct pairs $\langle\langle x_1, y_1 \rangle, \ldots, \langle x_n, y_n \rangle\rangle$ drawn from:

$$\{\langle x, y \rangle \mid x, y \in V_K \wedge \text{path}_R(x, y, E^\ominus) \wedge \neg\text{path}_L(x, y, E^\ominus)\}.$$

□

In addition to paths, transformation L must handle MSO expressions that refer to items present in R but absent in L. To achieve this, it computes a disjunction over all possible "future" (i.e. immediately after the rule application) set memberships of these missing items. The idea being, that if a set membership exists for these missing items that satisfies the interpretation constraints *before* the rule application, then one will still exist once they have been created. The transformation keeps track of such potential memberships via sets of pairs as follows.

Definition 6 (Membership set). A *membership set* M is a set of pairs (x, X) of node or edge identifiers x with set variables of the corresponding type. Intuitively, $(x, \mathsf{X}) \in M$ encodes that $x \in \mathsf{X}$, whereas $(x, \mathsf{X}) \notin M$ encodes that $x \notin \mathsf{X}$.

□

Theorem 3 (From M-conditions over R to L). There is a transformation L such that for every rule $r = \langle\langle L \hookleftarrow K \hookrightarrow R \rangle, \text{ac}\rangle$, every M-condition c over R (with distinct variables for distinct quantifiers), and every direct derivation $G \Rightarrow_{r,g,h} H$,

$$g \models^{I_0} \text{L}(r, c) \text{ if and only if } h \models^{I_0} c.$$

Construction. Let $r = \langle\langle L \hookleftarrow K \hookrightarrow R \rangle, \text{ac}\rangle$ denote a rule and c an M-condition over R. We define $\text{L}(r, c) = \text{L}'(r, c, \emptyset)$. For such an r, c, and membership set M, the transformation L$'$ is defined inductively as follows:

$$\text{L}'(r, \texttt{true}, M) = \texttt{true},$$

$$\text{L}'(r, \exists_\mathsf{V}\mathsf{X}[c'], M) = \exists_\mathsf{V}\mathsf{X}[\bigvee_{M' \in 2^{M_\mathsf{V}}} \text{L}'(r, c', M \cup M')]$$

$$\text{L}'(r, \exists_\mathsf{E}\mathsf{X}[c'], M) = \exists_\mathsf{E}\mathsf{X}[\bigvee_{M' \in 2^{M_\mathsf{E}}} \text{L}'(r, c', M \cup M')]$$

where $M_\mathsf{V} = \{(v, \mathsf{X}) \mid v \in V_R \setminus V_L\}$ and $M_\mathsf{E} = \{(e, \mathsf{X}) \mid e \in E_R \setminus E_L\}$.

For case $c = \exists(a \mid \gamma, c')$, we define:

$$\text{L}'(r, \exists(a \mid \gamma, c'), M) = \texttt{false}$$

if $\langle K \hookrightarrow R, a \rangle$ has no pushout complement; otherwise:

$$\text{L}'(r, \exists(a \mid \gamma, c'), M) = \exists(b \mid \gamma_M, \text{L}'(r^*, c', M))$$

which relies on the following. First, construct the pushout (1), with $r^* = \langle Y \leftarrow Z \hookrightarrow X \rangle$ the "derived" rule obtained by constructing pushout (2). The interpretation constraint γ_M is obtained from γ as follows. First, con-

$$
\begin{array}{ccc}
r: \langle\; L & \xleftarrow{\quad} K \xrightarrow{\quad} & R \;\rangle \\
b \downarrow & (2) \qquad\qquad (1) & \downarrow a \\
r^*: \langle\; Y & \xleftarrow{\quad} Z \xrightarrow{\quad} & X \;\rangle
\end{array}
$$

sider each predicate $x \in \mathsf{X}$ such that $x \notin Y$. If $(y, \mathsf{X}) \in M$ for some $y = x$, replace the predicate with $\mathtt{true}$; otherwise $\mathtt{false}$. Then, replace each path predicate p with $\text{LPath}(r^*, p)$.

The transformation L is extended for Boolean formulae in the usual way, that is, $\text{L}(r, \neg c) = \neg \text{L}(r, c)$, $\text{L}(r, c_1 \wedge c_2) = \text{L}(r, c_1) \wedge \text{L}(r, c_2)$, and $\text{L}(r, c_1 \vee c_2) = \text{L}(r, c_1) \vee \text{L}(r, c_2)$ (analogous for L'). $\qquad\square$

Example 4. Take $\mathtt{grow}, c, \gamma$ and $\text{A}(\mathtt{grow}, c)$ as considered in Example 3. Applying transformation L:

$$\text{L}(\mathtt{grow}, \text{A}(\mathtt{grow}, c)) = \text{L}'(\mathtt{grow}, \text{A}(\mathtt{grow}, c), \emptyset)$$

$$= \exists_{\mathsf{V}} \mathsf{X}, \mathsf{Y} [\; \bigvee_{M' \in 2^{M_{\mathsf{V}}}} \text{L}'(\mathtt{grow}, \bigwedge_{i=1}^{7} \forall(\; \bullet_1 \!\!\rightarrow\!\! \bullet_2 \;\hookrightarrow E_i, \exists(E_i \mid \mathtt{path}(\mathtt{v},\mathtt{w}))$$
$$\Rightarrow \exists(E_i \mid \gamma)), M') \;]$$

$$= \exists_{\mathsf{V}} \mathsf{X}, \mathsf{Y} [\; \bigvee_{M' \in 2^{M_{\mathsf{V}}}} (\bigwedge_{i \in \{1,2,4\}} \forall(\; \bullet_1 \hookrightarrow Y_i, \exists(Y_i \mid \mathtt{path}(\mathtt{v},\mathtt{w})) \Rightarrow \exists(Y_i \mid \gamma))$$
$$\wedge \; \forall(\; \bullet_1\bullet_{\mathtt{v}} \;, \exists(\; \bullet_1\bullet_{\mathtt{v}} \mid \mathtt{path}(\mathtt{v}, 1)) \Rightarrow \exists(\; \bullet_1\bullet_{\mathtt{v}} \mid \gamma_{M'}, \text{L}'(\mathtt{grow}, \mathtt{true}, M')))$$
$$\wedge \; \forall(\; \bullet_1\bullet_{\mathtt{w}} \;, \mathtt{false} \Rightarrow \exists(\; \bullet_1\bullet_{\mathtt{w}} \mid \gamma_{M'}, \text{L}'(\mathtt{grow}, \mathtt{true}, M')))$$
$$\wedge \; \forall(\; \bullet_{1=\mathtt{v}} \;, \mathtt{true} \Rightarrow \exists(\; \bullet_{1=\mathtt{v}} \mid \gamma_{M'}, \text{L}'(\mathtt{grow}, \mathtt{true}, M')))$$
$$\wedge \; \forall(\; \bullet_{1=\mathtt{w}} \;, \mathtt{false} \Rightarrow \exists(\; \bullet_{1=\mathtt{w}} \mid \gamma_{M'}, \text{L}'(\mathtt{grow}, \mathtt{true}, M')))) \;]$$

$$= \exists_{\mathsf{V}} \mathsf{X}, \mathsf{Y} [\; \bigvee_{M' \in 2^{M_{\mathsf{V}}}} (\bigwedge_{i \in \{1,2,4\}} \forall(\; \bullet_1 \hookrightarrow Y_i, \exists(Y_i \mid \mathtt{path}(\mathtt{v},\mathtt{w})) \Rightarrow \exists(Y_i \mid \gamma))$$
$$\wedge \; \forall(\; \bullet_1\bullet_{\mathtt{v}} \;, \exists(\; \bullet_1\bullet_{\mathtt{v}} \mid \mathtt{path}(\mathtt{v}, 1)) \Rightarrow \exists(\; \bullet_1\bullet_{\mathtt{v}} \mid \gamma_{M'}))$$
$$\wedge \; \forall(\; \bullet_{1=\mathtt{v}} \;, \exists(\; \bullet_{1=\mathtt{v}} \mid \gamma_{M'}))) \;]$$

$$= \exists_{\mathsf{V}} \mathsf{X}, \mathsf{Y} [\; \bigwedge_{i \in \{1,2,4\}} \forall(\; \bullet_1 \hookrightarrow Y_i, \exists(Y_i \mid \mathtt{path}(\mathtt{v},\mathtt{w})) \Rightarrow \exists(Y_i \mid \gamma))$$
$$\wedge \; \forall(\; \bullet_1\bullet_{\mathtt{v}} \;, \exists(\; \bullet_1\bullet_{\mathtt{v}} \mid \mathtt{path}(\mathtt{v}, 1)) \Rightarrow \exists(\; \bullet_1\bullet_{\mathtt{v}} \mid \mathtt{v} \in \mathsf{X} \text{ and not } \mathtt{v} \in \mathsf{Y}))$$
$$\wedge \; \forall(\; \bullet_{1=\mathtt{v}} \;, \exists(\; \bullet_{1=\mathtt{v}} \mid \mathtt{v} \in \mathsf{X} \text{ and not } \mathtt{v} \in \mathsf{Y})) \;]$$

where the graphs E_i and Y_i are as given in Figure 1 and $M_{\mathsf{V}} = \{(2, \mathsf{X}), (2, \mathsf{Y})\}$. Here, only one of the subsets ranged over yields a satisfiable disjunct: $M' = \{(2, \mathsf{Y})\}$, i.e. $\gamma_{M'} = (\mathtt{v} \in \mathsf{X} \text{ and } \mathtt{true}) \text{ and not } (\mathtt{v} \in \mathsf{Y} \text{ or } \mathtt{false})$ for $\mathtt{w} = 2$. $\qquad\square$

Transformation App, adapted from Def in [2], takes as input a rule set $\mathcal{R}$ and generates an M-constraint that is satisfied by graphs for which $\mathcal{R}$ is applicable.

Theorem 4 (Applicability of a rule). There is a transformation App such that for every rule set $\mathcal{R}$ and every graph G,

$$G \models \mathrm{App}(\mathcal{R}) \quad \text{if and only if} \quad \exists H.\, G \Rightarrow_{\mathcal{R}} H.$$

Construction. If $\mathcal{R}$ is empty, define $\mathrm{App}(\mathcal{R}) = \mathtt{false}$; otherwise, for $\mathcal{R} = \{r_1, \ldots, r_n\}$, define:

$$\mathrm{App}(\mathcal{R}) = \mathrm{app}(r_1) \vee \cdots \vee \mathrm{app}(r_n).$$

For each rule $r = \langle r', \mathrm{ac} \rangle$ with $r' = \langle L \hookleftarrow K \hookrightarrow R \rangle$, we define $\mathrm{app}(r) = \exists(\emptyset \hookrightarrow L, \mathrm{Dang}(r') \wedge \mathrm{ac}_L \wedge \mathrm{L}(r, \mathrm{ac}_R))$. Here, $\mathrm{Dang}(r') = \bigwedge_{a \in A} \neg \exists a$, where the index set A ranges over all injective graph morphisms $a \colon L \hookrightarrow L^{\oplus}$ (up to isomorphic codomains) such that the pair $\langle K \hookrightarrow L, a \rangle$ has no pushout complement; each $L^{\oplus}$ a graph that can be obtained from L by adding either (1) a loop; (2) a single edge between distinct nodes; or (3) a single node and a non-looping edge incident to that node. □

Finally, transformation Pre (adapted from [1]) combines the other transformations to construct a weakest liberal precondition relative to a rule and postcondition.

Theorem 5 (Postconditions to weakest liberal preconditions). There is a transformation Pre such that for every rule $r = \langle \langle L \hookleftarrow K \hookrightarrow R \rangle, \mathrm{ac} \rangle$, every M-constraint c, and every direct derivation $G \Rightarrow_r H$,

$$G \models \mathrm{Pre}(r, c) \quad \text{if and only if} \quad H \models c.$$

Moreover, $\mathrm{Pre}(r, c) \vee \neg \mathrm{App}(\{r\})$ is the *weakest liberal precondition* relative to r and c.

Construction. Let $r = \langle \langle L \hookleftarrow K \hookrightarrow R \rangle, \mathrm{ac} \rangle$ denote a rule and c denote an M-constraint. Then:

$$\mathrm{Pre}(r, c) = \forall(\emptyset \hookrightarrow L, (\mathrm{Dang}(r) \wedge \mathrm{ac}_L \wedge \mathrm{L}(r, \mathrm{ac}_R)) \Rightarrow \mathrm{L}(r, \mathrm{A}(r, c))).$$

□

Example 5. Take $\mathtt{grow}$, c, γ and $\mathrm{L}(\mathtt{grow}, \mathrm{A}(\mathtt{grow}, c))$ as considered in Example 4. Applying transformation Pre:

$\mathrm{Pre}(\mathtt{grow}, \mathrm{L}(\mathtt{grow}, \mathrm{A}(\mathtt{grow}, c)))$
$= \forall(\; \bullet_1, \mathrm{ac}_L \Rightarrow \exists_v X, Y [\; \bigwedge_{i \in \{1,2,4\}} \forall(\; \bullet_1 \hookrightarrow Y_i, \exists(Y_i \mid \mathtt{path(v,w)}) \Rightarrow \exists(Y_i \mid \gamma))$
$\qquad \wedge\; \forall(\; \bullet_1 \bullet_v \,, \exists(\; \bullet_1 \bullet_v \mid \mathtt{path(v,1)}) \Rightarrow \exists(\; \bullet_1 \bullet_v \mid v \in X \text{ and not } v \in Y))$
$\qquad \wedge\; \forall(\; \bullet_{1=v}\,, \exists(\; \bullet_{1=v} \mid v \in X \text{ and not } v \in Y)) \;])$

where the graphs Y_i are as given in Figure 1. This M-constraint is only satisfied by graphs that do not have any edges between distinct nodes, because of the assertion that every match (i.e. every node) must be in X and not in Y. Were an edge to exist – i.e. a path – then the M-constraint asserts that its target is in Y; a contradiction. □

6   Proving Non-local Specifications

In this section we show how to systematically prove a non-local correctness specification using a Hoare logic adapted from [4,5]. The key difference is the use of M-constraints as assertions, and our extension of Pre in constructing weakest liberal preconditions for rules. (We note that one could just as easily adapt the Dijkstra-style systems of [1,2].)

We will specify the behaviour of programs using *(Hoare) triples*, $\{c\}\ P\ \{d\}$, where P is a program, and c, d are *pre-* and *postconditions* expressed as M-constraints. We say that this specification holds in the sense of *partial correctness*, denoted by $\models \{c\}\ P\ \{d\}$, if for any graph G satisfying c, every graph H resulting from the execution of P on G satisfies d.

For systematically proving a specification, we present a *Hoare logic* in Figure 2, where c, d, e, inv range over M-constraints, P, Q over programs, r over rules, and $\mathcal{R}$ over rule sets. If a triple $\{c\}\ P\ \{d\}$ can be instantiated from an axiom or deduced from an inference rule, then it is *provable* in the Hoare logic and we write $\vdash \{c\}\ P\ \{d\}$. Proofs shall be displayed as trees, with the specification as the root, axiom instances as the leaves, and inference rule instances in-between.

$$[\text{ruleapp}]_{\text{wlp}}\ \ \{\text{Pre}(r, c) \vee \neg\text{App}(\{r\})\}\ r\ \{c\} \qquad [\text{ruleset}]\ \frac{\{c\}\ r\ \{d\}\ \text{for each } r \in \mathcal{R}}{\{c\}\ \mathcal{R}\ \{d\}}$$

$$[\text{comp}]\ \frac{\{c\}\ P\ \{e\} \qquad \{e\}\ Q\ \{d\}}{\{c\}\ P;\ Q\ \{d\}} \qquad\qquad [!]\ \frac{\{inv\}\ \mathcal{R}\ \{inv\}}{\{inv\}\ \mathcal{R}!\ \{inv \wedge \neg\text{App}(\mathcal{R})\}}$$

$$[\text{cons}]\ \frac{c \Rightarrow c'\ \ \{c'\}\ P\ \{d'\}\ \ d' \Rightarrow d}{\{c\}\ P\ \{d\}}$$

Fig. 2. A Hoare logic for partial correctness

For simplicity in proofs we will typically treat $[\text{ruleapp}]_{\text{wlp}}$ as two different axioms (one for each disjunct). Note that we have omitted, due to space, the proof rules for the conditional constructs. Note also the restriction to rule sets in [!], because the applicability of arbitrary programs cannot be expressed in a logic for which the model checking problem is decidable [5].

Theorem 6 (Soundness). Given a program P and M-constraints c, d, we have that $\vdash \{c\}\ P\ \{d\}$ implies $\models \{c\}\ P\ \{d\}$. □

The remainder of this section demonstrates the use of our constructions and Hoare logic in proving non-local specifications of two programs. For the first, we will consider a property expressed in terms of MSO variables and expressions, whereas for the second, we will consider properties expressed in terms of path predicates. Both programs are simple, as our focus here is not on building intricate proofs but rather on illustrating the main novelty of this paper: a Pre construction for MSO properties.

Example 6. Recall the program `init; grow!` of Example 2 that nondeterministically constructs a tree. A known non-local property of trees is that they can be assigned a 2-colouring (i.e. they are bipartite), a property that the M-constraint *col* of Example 1 precisely expresses. Hence we will show that $\vdash \{emp\}\ \mathtt{init;\ grow!}\ \{col\}$, where $emp = \neg\exists(\ \bullet\ )$ expresses that the graph is empty. A proof tree for this specification is given in Figure 3, where the interpretation constraints γ_1 and γ_2 in $\mathrm{Pre}(\mathtt{grow}, col)$ are respectively `(v∈X or v∈Y) and not (v∈X and v∈Y)` and `not (v∈X and w∈X) and not (v∈Y and w∈Y)`.

$$\dfrac{\dfrac{\{\mathrm{Pre}(\mathtt{init}, col)\}\ \mathtt{init}\ \{col\}}{\{emp\}\ \mathtt{init}\ \{col\}} \qquad \dfrac{\dfrac{\{\mathrm{Pre}(\mathtt{grow}, col)\}\ \mathtt{grow}\ \{col\}}{\{col\}\ \mathtt{grow}\ \{col\}}}{\{col\}\ \mathtt{grow!}\ \{col \wedge \neg\mathrm{App}(\{\mathtt{grow}\})\}}}{\vdash \{emp\}\ \mathtt{init;grow!}\ \{col\}}$$

$$\begin{aligned}
\mathrm{Pre}(\mathtt{init}, col) &\equiv col\\
\mathrm{Pre}(\mathtt{grow}, col) &\equiv \forall(\ \bullet_1\ , \neg tc \Rightarrow \exists_{\mathtt{V}}\mathtt{X,Y[}\\
&\qquad \forall(\ \bullet_1 \bullet_{\mathtt{v}}\ , \exists(\ \bullet_1 \bullet_{\mathtt{v}}\ \mid \gamma_1)) \wedge \forall(\ \bullet_{\mathtt{1=v}}\ , \exists(\ \bullet_{\mathtt{1=v}}\ \mid \gamma_1))\\
&\qquad \wedge \forall(\ \bullet_1 \bullet_{\mathtt{v}} \bullet_{\mathtt{w}}\ , \exists(\ \bullet_1 \bullet_{\mathtt{v}} \rightarrow \bullet_{\mathtt{w}}\ ) \Rightarrow \exists(\ \bullet_1 \bullet_{\mathtt{v}} \bullet_{\mathtt{w}}\ \mid \gamma_2))\\
&\qquad \wedge \forall(\ \bullet_{\mathtt{1=v}} \bullet_{\mathtt{w}}\ , \exists(\ \bullet_{\mathtt{1=v}} \rightarrow \bullet_{\mathtt{w}}\ ) \Rightarrow \exists(\ \bullet_{\mathtt{1=v}} \bullet_{\mathtt{w}}\ \mid \gamma_2))\\
&\qquad \wedge \forall(\ \bullet_{\mathtt{1=w}} \bullet_{\mathtt{v}}\ , \exists(\ \bullet_{\mathtt{1=w}} \leftarrow \bullet_{\mathtt{v}}\ ) \Rightarrow \exists(\ \bullet_{\mathtt{1=w}} \bullet_{\mathtt{v}}\ \mid \gamma_2))\\
&\qquad \wedge (\forall(\ \bullet_{\mathtt{1=v}}\ , \exists(\ \bullet_{\mathtt{1=v}}\ \mid \mathtt{not\ v} \in \mathtt{X}))\\
&\qquad\quad \vee \forall(\ \bullet_{\mathtt{1=v}}\ , \exists(\ \bullet_{\mathtt{1=v}}\ \mid \mathtt{not\ v} \in \mathtt{Y})))\ \mathtt{]})
\end{aligned}$$

Fig. 3. Trees are 2-colourable

Observe that $\mathrm{Pre}(\mathtt{grow}, col)$ is essentially an "embedding" of the postcondition *col* within the context of possible matches for `grow`. The second line expresses that every node (whether the node of the match or not) is coloured `X` or `Y`. The following three conjuncts then express that any edges in the various contexts of the match connect nodes that are differently coloured. The final conjunct is of the same form, but is "pre-empting" the creation of a node and edge by `grow`. To ensure that the graph remains 2-colourable, node `1` of the match must not belong to both sets; this, of course, is already established by the first nested conjunct. Hence the first implication arising from instances of [cons], $col \Rightarrow \mathrm{Pre}(\mathtt{grow}, col)$, is valid. The second implication, $emp \Rightarrow \mathrm{Pre}(\mathtt{init}, col)$, is also valid since a graph satisfying *emp* will not have any nodes to quantify over. □

Example 7. An *acyclic graph* is a graph that does not contain any *cycles*, i.e. non-empty paths starting and ending on the same node. One way to test for acyclicity is to apply the rule $\mathtt{delete} = \langle\langle\ \bullet_1 \rightarrow \bullet_2\ \Rightarrow\ \bullet_1 \bullet_2\ \rangle, \mathrm{ac}_L\rangle$ for as long as possible; the resulting graph being edgeless if the input graph was acyclic. Here, ac_L denotes the left application condition $\neg\exists(\ \bullet_1 \rightarrow \bullet_2\ \hookrightarrow\ \bullet \rightarrow \bullet_1 \rightarrow \bullet_2\ ) \vee \neg\exists(\ \bullet_1 \rightarrow \bullet_2\ \hookrightarrow\ \bullet_1 \rightarrow \bullet_2 \rightarrow \bullet\ )$, expressing that in matches, either the source node has indegree 0 or the target node has outdegree 0 (we do not consider the special case of looping edges for simplicity). Note that nodes *within* a cycle would not satisfy this: if a source node has an indegree of 0 for example, there would be no possibility of an outgoing path ever returning to the same node.

We prove two claims about this rule under iteration: first, that it deletes all edges in an acyclic graph; second, that if applied to a graph containing cycles, the resulting graph would not be edgeless. That is, $\vdash \{\neg c\}$ delete! $\{e\}$ and $\vdash \{c\}$ delete! $\{\neg e\}$, for M-constraints c (for cycles), e (for edgeless), $\gamma_c =$ path(v, w, not e) and path(w, v, not e), and proofs as in Figure 4.

$$
\begin{array}{ll}
\dfrac{\{\text{Pre}(\text{delete}, \neg c)\}\ \text{delete}\ \{\neg c\}}{\{\neg c\}\ \text{delete}\ \{\neg c\}} & \dfrac{\{\text{Pre}(\text{delete}, c)\}\ \text{delete}\ \{c\}}{\{c\}\ \text{delete}\ \{c\}} \\[2ex]
\{\neg c\}\ \text{delete!}\ \{\neg c \wedge \neg\text{App}(\{\text{delete}\})\} & \{c\}\ \text{delete!}\ \{c \wedge \neg\text{App}(\{\text{delete}\})\} \\[1ex]
\vdash \{\neg c\}\ \text{delete!}\ \{e\} & \vdash \{c\}\ \text{delete!}\ \{\neg e\}
\end{array}
$$

$$c = \exists(\ \bullet_v \bullet_w\ \mid \text{path}(v, w)\ \text{and}\ \text{path}(w, v))$$

$$e = \neg\exists(\ \bullet_{\overline{v}} \blacktriangleright \bullet_w\ )$$

$$\text{Pre}(\text{delete}, \neg c) = \forall(\ \bullet_1 \xrightarrow{e} \bullet_2\ , \text{ac}_L \Rightarrow$$

$$\wedge \neg\exists(\ \bullet_1 \xrightarrow{e} \bullet_2 \bullet_v \bullet_w\ \mid \gamma_c) \wedge \neg\exists(\ \bullet_1 \xrightarrow{e} \bullet_{2=v} \bullet_w\ \mid \gamma_c)$$

$$\wedge \neg\exists(\ \bullet_1 \xrightarrow{e} \bullet_{2=v} \bullet_w\ \mid \gamma_c) \wedge \neg\exists(\ \bullet_{1=w} \xrightarrow{e} \bullet_2 \bullet_v\ \mid \gamma_c)$$

$$\wedge \neg\exists(\ \bullet_1 \xrightarrow{e} \bullet_{2=w} \bullet_v\ \mid \gamma_c) \wedge \neg\exists(\ \bullet_{1=v} \xrightarrow{e} \bullet_{2=w}\ \mid \gamma_c)$$

$$\wedge \neg\exists(\ \bullet_{1=w} \xrightarrow{e} \bullet_{2=v}\ \mid \gamma_c))$$

$$\text{App}(\{\text{delete}\}) = \exists(\ \bullet_1 \xrightarrow{} \bullet_2\ , \text{ac}_L)$$

Fig. 4. Acyclity (or lack thereof) is invariant

First, observe that Pre(delete, ¬c) is essentially an "embedding" of the post-condition ¬c within the context of possible matches for delete. The path predicates in γ_c now additionally assert (as a result of the L transformation) that paths do not include images of edge e: this is crucially important for establishing the postcondition because the rule deletes the edge. For space reasons we did not specify Pre(delete, c), but this can be constructed from Pre(delete, ¬c) by replacing each ∧ with ∨ and removing each ¬ in the nested part.

The instances of [cons] give rise to implications that we must show to be valid. First, $\neg c \Rightarrow \text{Pre}(\text{delete}, \neg c)$ is valid: a graph satisfying ¬c does not contain any cycles, hence it also does not contain cycles outside of the context of matches for delete. Second, $\neg c \wedge \neg\text{App}(\{\text{delete}\}) \Rightarrow e$ is valid: a graph satisfying the antecedent does not contain any cycles and also no pair of incident nodes for which ac_L holds. If the graph is not edgeless, then there must be some such pair satisfying ac_L; otherwise the edges are within a cycle. Hence the graph must be edgeless, satisfying e.

In the second proof tree, $c \Rightarrow \text{Pre}(\text{delete}, c)$ is valid. A graph satisfying c contains a cycle: clearly, no edge (with its source and target) in this cycle satisfies ac_L; hence the graph satisfies the consequent, since images of edge e cannot be part of the cycle in the graph. Finally, $c \wedge \neg\mathrm{App}(\{\texttt{delete}\}) \Rightarrow \neg e$ is valid: if a graph satisfies the antecedent, then it contains a cycle, the edges of which $\texttt{delete}$ will never be applicable to because of ac_L; hence the graph cannot be edgeless, and satisfies $\neg e$. □

7   Related Work

We point to a few related publications addressing the verification of non-local graph properties through proofs / theorem proving and model checking.

Habel and Radke have considered HR conditions [14], an extension of nested conditions embedding hyperedge replacement grammars via graph variables. The formalism is more expressive than MSO logic on graphs (it is able, for example, to express node-counting MSO properties such as "the graph has an even number of nodes" [15]) but it is not yet clear whether an effective construction for weakest liberal preconditions exists. Percebois et al. [16] demonstrate how one can verify global invariants involving paths, directly at the level of rules. Rules are modelled with (a fragment of) first-order logic on graphs in the interactive theorem prover Isabelle. Inaba et al. [17] address the verification of type-annotated Core UnCAL – a query algebra for graph-structured databases – against input/output graph schemas in MSO. They first reformulate the query algebra itself in MSO, before applying an algorithm that reduces the verification problem to the validity of MSO over trees.

The GROOVE model checker [18] supports rules with paths in the left-hand side, expressed as a regular expression over edge labels. One can specify such rules to match only when some (un)desirable non-local property holds, and then verify automatically that the rule is never applicable. Augur 2 [19] also uses regular expressions, but for expressing forbidden paths that should not occur in any reachable graph.

8   Conclusion

This paper has contributed the means for systematic proofs of graph programs with respect to non-local specifications. In particular, we defined M-conditions, an extension of nested conditions equivalently expressive to MSO logic on graphs, and defined for this assertion language an effective construction for weakest liberal preconditions of rules. We demonstrated the use of this work in some Hoare-style proofs of programs relative to non-local invariants, i.e. the existence of 2-colourings, and the existence of arbitrary-length cycles. Some interesting topics for future work include: extending M-conditions and Pre to support other useful predicates (e.g. an *undirected* path predicate), adding support for attribution (e.g. along the lines of [4,5]), implementing the construction of Pre, and generalising the resolution- and tableau-based reasoning systems for nested conditions [20,21] to M-conditions.

Acknowledgements. The research leading to these results has received funding from the European Research Council under the European Union's Seventh Framework Programme (FP7/2007-2013) / ERC Grant agreement no. 291389.

References

1. Habel, A., Pennemann, K.-H., Rensink, A.: Weakest preconditions for high-level programs. In: Corradini, A., Ehrig, H., Montanari, U., Ribeiro, L., Rozenberg, G. (eds.) ICGT 2006. LNCS, vol. 4178, pp. 445–460. Springer, Heidelberg (2006)
2. Pennemann, K.H.: Development of Correct Graph Transformation Systems. Doctoral dissertation, Universität Oldenburg (2009)
3. Habel, A., Pennemann, K.H.: Correctness of high-level transformation systems relative to nested conditions. Mathematical Structures in Computer Science 19(2), 245–296 (2009)
4. Poskitt, C.M., Plump, D.: Hoare-style verification of graph programs. Fundamenta Informaticae 118(1-2), 135–175 (2012)
5. Poskitt, C.M.: Verification of Graph Programs. PhD thesis, University of York (2013)
6. Plump, D.: The design of GP 2. In: Escobar, S. (ed.) WRS 2011. EPTCS, vol. 82, pp. 1–16 (2012)
7. Courcelle, B., Engelfriet, J.: Graph Structure and Monadic Second-Order Logic: A Language-Theoretic Approach. Cambridge University Press (2012)
8. Poskitt, C.M., Plump, D.: Verifying monadic second-order properties of graph programs: Extended version (2014), http://arxiv.org/abs/1405.5927
9. Flum, J., Grohe, M.: Parameterized Complexity Theory. Springer (2006)
10. Courcelle, B.: The monadic second-order logic of graphs. I. Recognizable sets of finite graphs. Information and Computation 85(1), 12–75 (1990)
11. Courcelle, B.: Graph rewriting: An algebraic and logic approach. In: Handbook of Theoretical Computer Science, vol. B, Elsevier (1990)
12. Ehrig, H., Ehrig, K., Prange, U., Taentzer, G.: Fundamentals of Algebraic Graph Transformation. Springer (2006)
13. Poskitt, C.M., Plump, D.: Verifying total correctness of graph programs. In: Echahed, R., Habel, A., Mosbah, M. (eds.) GCM 2012. Electronic Communications of the EASST, vol. 61 (2013)
14. Habel, A., Radke, H.: Expressiveness of graph conditions with variables. In: Ermel, C., Ehrig, H., Orejas, F., Taentzer, G. (eds.) GraMoT 2010. Electronic Communications of the EASST, vol. 30 (2010)
15. Radke, H.: HR* graph conditions between counting monadic second-order and second-order graph formulas. In: Echahed, R., Habel, A., Mosbah, M. (eds.) GCM 2012. Electronic Communications of the EASST, vol. 61 (2013)
16. Percebois, C., Strecker, M., Tran, H.N.: Rule-level verification of graph transformations for invariants based on edges' transitive closure. In: Hierons, R.M., Merayo, M.G., Bravetti, M. (eds.) SEFM 2013. LNCS, vol. 8137, pp. 106–121. Springer, Heidelberg (2013)
17. Inaba, K., Hidaka, S., Hu, Z., Kato, H., Nakano, K.: Graph-transformation verification using monadic second-order logic. In: Schneider-Kamp, P., Hanus, M. (eds.) PPDP 2011, pp. 17–28. ACM (2011)
18. Ghamarian, A.H., de Mol, M., Rensink, A., Zambon, E., Zimakova, M.: Modelling and analysis using GROOVE. Software Tools for Technology Transfer 14(1), 15–40 (2012)
19. König, B., Kozioura, V.: Augur 2 - a new version of a tool for the analysis of graph transformation systems. In: Bruni, R., Varró, D. (eds.) GT-VMT 2006. ENTCS, vol. 211, pp. 201–210 (2008)
20. Pennemann, K.H.: Resolution-like theorem proving for high-level conditions. In: Ehrig, H., Heckel, R., Rozenberg, G., Taentzer, G. (eds.) ICGT 2008. LNCS, vol. 5214, pp. 289–304. Springer, Heidelberg (2008)
21. Lambers, L., Orejas, F.: Tableau-based reasoning for graph properties. In: Giese, H., König, B. (eds.) ICGT 2014. LNCS, vol. 8571, Springer, Heidelberg (2014)

Generating Abstract Graph-Based Procedure Summaries for Pointer Programs

Christina Jansen* and Thomas Noll

Software Modeling and Verification Group
RWTH Aachen University, Germany
http://moves.rwth-aachen.de/

Abstract. The automated analysis and verification of pointer-manipu-lating programs operating on a heap is a challenging task. It requires abstraction techniques for dealing with complex program behaviour and unbounded state spaces that arise from both dynamic data structures and recursive procedures. In previous work it was shown that hyper-edge replacement grammars provide an intuitive and versatile concept for defining and implementing such abstractions.

Here we extend this approach towards a modular way of reasoning about programs with (possibly recursive) procedures featuring local variables. We propose an interprocedural dataflow analysis to automatically derive procedure contracts, i.e., graph transformations that concisely capture the overall effect of a procedure. Besides its modularity, another advantage of this analysis is that it relieves us from explicitly modelling the call stack on the heap, i.e., heap and control abstraction are clearly separated. The former can now be specified by simple and intuitive hy-peredge replacement grammars describing the data structures only, while the latter is realised by automatically generated procedure contracts.

Keywords: Hypergraphs, Hyperedge Replacement Grammars, Heap Abstraction, Procedure Contracts, Interprocedural Dataflow Analysis.

1 Introduction

Dynamic data structures such as lists and trees, implemented using pointers, are heavily used in e.g. all kinds of application software, but also device drivers, operating systems, and so forth. While pointers offer a flexible concept allowing for very complex program behaviour, pointer programming is error-prone, even being one of the most common sources of bugs in software [4]. Typical problems are dereferencing of null pointers, creation of memory leaks, unsought aliasing effects and the accidental invalidation of data structures through destructive up-dates , i.e., errors which are usually difficult to trace. However, with the flexibility of pointer programs comes the complexity of analysing them, as they generally induce unbounded state spaces. A common approach to tackle this problem is to apply *abstraction techniques* to obtain a finite representation.

* This research is funded by EU FP7 project CARP (http://www.carpproject.eu).

H. Giese and B. König (Eds.): ICGT 2014, LNCS 8571, pp. 49–64, 2014.

Hyperedge replacement grammars (HRGs) have proven to be an intuitive concept for defining and implementing such abstractions [10]. The key idea is to represent heaps as hypergraphs that contain placeholders indicating abstracted fragments of a data structure. Thus they combine concrete and abstract heap parts. Placeholders are realised by deploying *hyperedges*, i.e., edges that connect an arbitrary number of nodes, labelled with nonterminals. The semantics of the nonterminals, i.e. the kind of data structure they represent, is specified by HRGs. Abstraction and its reverse operation, concretisation, are implemented by applying grammar rules in forward and backward direction, respectively.

In [9] we have shown how the HRG formalism can be employed for handling Java bytecode with (recursive) methods and local variables. The approach is based on the explicit modelling of the runtime stack on the heap, where unbounded recursion is dealt with by abstraction using HRGs. However, this requires the development of appropriate HRGs that capture not only the data structures arising during program execution, but additionally their interdependency with the runtime stack, leading to large and intricate grammar specifications. The situation becomes even more complex when tackling the extension to concurrent Java or dynamic thread generation in general, where the list of threads has to be added to the heap representation as a third component.

To overcome this increasing complexity in HRG specification, we advocate an alternative approach, which clearly separates heap abstraction from control abstraction. The former still employs HRGs to only handle the data structures occurring during program execution. For dealing with the program code, modular reasoning on the level of procedures is used. More exactly, the goal is to automatically derive a summary for each procedure in the given program, which abstractly and comprehensively represents the possible effects of its execution. As our method is inspired by work on rely-guarantee reasoning, these summaries are called *contracts*. In essence, a contract specifies a graph transformation in the form of by pre- and postconditions. The former are given by hypergraphs and describe the heap fragments reachable from the procedure parameters upon procedure entry. Postconditions are represented as sets of hypergraphs associated with the respective precondition that describe the possible shapes of this fragment after procedure execution. Restricting the precondition to the reachable fragment is sufficient for capturing the procedure effect as the rest of the heap is non-accessible and therefore immutable. The proposed approach is based on an *interprocedural dataflow analysis* (IPA), which handles the runtime stack and local variables. More concretely, we utilise a demand-driven IPA based on fixpoint iteration, which forms an instance of the general IPA framework introduced in [14]. That is, provided the abstract state space of the program is finite, termination of the program analysis is guaranteed and the results coincide with the meet-over-all-paths solution according to the Interprocedural Coincidence Theorem [14].

We show that graph transformation techniques are particularly suited for dealing with such contracts, for the following reasons. First, they provide a very natural and intuitive way of representing a procedure's behaviour. Second, the analysis is fully automated, discharging the user from the complex challenge of developing procedure contracts by hand (which is, e.g., the case in similar approaches based on separation logic). Third, our framework also nicely deals with *cutpoints*, i.e., situations where the heap fragment that is reachable from the actual procedure parameters is additionally accessible from local variables in the calling context. It thus solves the "important, and still open, problem of handling an unbounded number of live cutpoints under abstraction" [15].

Related Work. In [19] the runtime stack is explicitly represented and abstracted as a linked list, using shape analysis. This is similar to our previous work in [9]. The alternative interprocedural heap abstraction approach developed in the present paper is based on the general IPA framework as described, e.g., in [14,22]. Specific instances have been proposed for the finite, distributive subset case (IDFS; [18]) and the distributive environments case (IDE; [21]). A generalisation of these is presented in [23] where a class of abstract domains and associated transformations is defined which allows to obtain precise and concise procedure summaries. However, these instances are not applicable in our setting due to the combination of recursion and local variables. A framework for interprocedural heap shape analysis in the cutpoint-free case is first proposed in [20] and later generalised in [15] by admitting non-live cutpoints. Moreover, [5] describe a modular interprocedural shape analysis that can handle a bounded number of cutpoints (which are interpreted as additional procedure parameters). However, the analysis is restricted to the setting of singly-linked lists, while the approach proposed in the following deals with all data structures of bounded tree-width.

2   Pointer-Manipulating Programs

In this section we will introduce a simple programming language that supports pointer manipulations and (possibly recursive) procedures with local variables. It is kept minimal to reduce the formal effort in the specification of its syntax and semantics, e.g., global variables and return parameters are omitted. Nevertheless it is sufficient to model most standard concepts that are present in pointer programs or can be extended in a straightforward way to do so.

We consider programs over variables Var and selectors Sel that consist of procedures $p_0, \ldots, p_k$. Each procedure p is defined by its signature $p(x_1, \ldots, x_n)$, where $x_1, \ldots, x_n$ are the formal parameters, followed by a statement block encapsulated by curly brackets. We assume w.l.o.g. that each procedure has a unique entry and exit, in our case provided by the header and the bracket closing the procedure body, respectively. The statement block consists of a sequence of statements, i.e., procedure calls $p(y_1, \ldots, y_n)$, standard branching and looping statements, and pointer assignments. The latter comprise the statements $x = P$, $x.s = P$, and $\mathbf{new}(x)$ where $P ::= \mathbf{null} \mid x \mid x.s$ for $x \in Var$ and $s \in Sel$. As branching and looping conditions we allow equality checks on elements of P, their negation and the connection of those by conjunction and disjunction.

Example 1. As a running example we consider the program given in Fig. 1. It contains two procedures main and trav with parameters head/ tail and cur/ tail.

Note that for simplicity pointer expressions P do not support arbitrary dereferencing depths. This entails no semantic restriction since this

```
main(head, tail){
    if(head != tail){
        trav(head, tail);
    }
}
```

```
trav(cur, tail){
    tmp := cur.next;
    if(tmp != tail){
        trav(tmp, tail);
    }
}
```

Fig. 1. Recursively traversing a list

feature can be emulated by multiple assignments. Object deletion is omitted since a **null**-assignment with subsequent garbage collection has the same effect.

3   Heap Abstraction by Hyperedge Replacement

In the heap abstraction approach based on graph grammars [10], (abstract) heaps are represented by hypergraphs. Hypergraphs are graphs with edges as proper objects, which can connect arbitrarily many nodes. We do not distinguish between isomorphic HGs, i.e., those that are identical modulo renaming of nodes and hyperedges.

Definition 1 (Hypergraph). *Let Σ be a finite alphabet with ranking function $rk : \Sigma \to \mathbb{N}$. A (labelled) hypergraph (HG) over Σ is a tuple $H = (V, E, att, lab, ext)$ where V is a set of nodes and E a set of hyperedges, att : $E \to V^*$ maps each hyperedge to a sequence of attached nodes, lab : $E \to \Sigma$ is a hyperedge-labelling function, and ext $\in V^*$ a (possibly empty) sequence of pairwise distinct external nodes. For $e \in E$, we require $|att(e)| = rk(lab(e))$ and let $rk(e) = rk(lab(e))$. The set of all hypergraphs over Σ is denoted by HG_Σ.*

To set up an intuitive representation of heaps, we use the following notation. We model (concrete) heaps as HGs over the alphabet $\Sigma := Sel \cup Var$ without external nodes. (The latter will be required later for defining the replacement of hyperedges.) Objects are depicted by nodes, variables by edges of rank one and selectors by edges of rank two connecting the corresponding object(s), where selector edges are understood as pointers from the first attached object to the second one. We introduce a node $v_{\mathbf{null}}$ representing **null**, which is unique to every HG. To depict abstract parts of the heap we use nonterminal edges, which carry labels from an additional set of *nonterminals (NTs) N* (and let $\Sigma_N = \Sigma \uplus N$).

Example 2. A typical implementation of a doubly-linked list consists of a sequence of list elements connected by next (n) and previous (p) pointers. Fig. 2 depicts a HG representation of a such a list. The three nodes (circles) represent objects on the heap. A shaded circle indicates an external node, its ordinal (i.e., position in *ext*) is given by the label. The L-labelled box represents an NT edge of rank two indicating an abstracted doubly-linked list between the first and second attached node. Later we will see how these abstract structures are defined. The connections between NT edges and nodes are labelled with their ordinal. For the sake of readability, selectors (n and p) are depicted as directed edges.

Note that not every HG over Σ_N represents a heap. It is necessary that, for each selector $s \in Sel$, every object defines at most one s-pointer, that each variable refers to at most one node, and that exter- **Fig. 2.** Heap representation
nal nodes are absent. HGs that satisfy this requirement are called *heap configurations (HC)* and are collected in the set HC_{Σ_N}. Thus HC_Σ represents all concrete HCs, i.e., those without NT edges. Abstract and concrete HCs with external nodes are called *extended* and are represented by the sets HCE_{Σ_N} and HCE_Σ, respectively. An NT edge of an HC acts as a placeholder for a heap part of a particular shape. We use HRGs to describe its structure.

Definition 2 (Hyperedge replacement grammar). *A hyperedge replacement grammar (HRG) G over an alphabet Σ_N is a finite set of production rules of the form $X \to H$ where $X \in N$, $H \in \mathrm{HG}_{\Sigma_N}$, and $|ext_H| = rk(X)$. The set of all HRGs over Σ_N is denoted by HRG_{Σ_N}.*

Example 3. Fig. 3 specifies an HRG for doubly-linked lists. It employs one NT L of rank two and two production rules. The right one recursively adds one list element, whereas the left one terminates a derivation.

HRG derivations are defined through *hyperedge replacement*, i.e. the substitution of an NT edge

$$L \to$$

Fig. 3. A grammar for doubly-linked lists

by the right-hand side of a corresponding production rule. A sample replacement is illustrated by Fig. 4(a), where the second rule of the grammar in Fig. 3 is applied to the NT edge L in the upper graph. To this aim, the external nodes of the rule's right-hand side are mapped to the nodes of the upper graph as indicated by the dashed arrows, and the L-edge is replaced by the rule's right-hand side. The resulting HC is given in Fig. 4(b).

Note that it is not guaranteed that an HRG derives HCs only. Moreover for abstraction we will apply HRG production rules, besides their

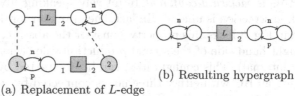

(a) Replacement of L-edge

(b) Resulting hypergraph

Fig. 4. Hyperedge replacement

conventional usage, also in backward direction. HRGs that always yield HCs in both cases are called *heap abstraction grammars* (HAGs). For details, we refer to [13].

In the heap abstraction approach based on HAGs, program states are described by HCs which may contain edges labelled with nonterminals. The structures represented by the NT edges are given by HAGs. The program's state space is obtained by executing its statements on these configurations.

Whenever statements have to be applied which are operating on selectors that are hidden in abstracted parts of the heap, these selectors are exposed beforehand. This operation, denoted by *conc*, is referred to as *concretisation*. In the following we see that both abstraction and concretisation rules are directly given by HAGs.

For HAGs, the forward application of a production rule yields a more concrete HC as certain abstract parts (viz., the NT that is replaced) are concretised. In case more than one rule is applicable, concretisation yields several heaps. For an example concretisation see Fig. 5, where the occurrence of L is either replaced by an empty list (at the bottom, conforming to the first rule), or by a list that is extended by one element (at the top, conforming to the second rule). Thus applying concretisation to the heap depicted on the left yields two successor heaps, one for each possible concretisation. The resulting transition system representing the program behaviour thus over-approximates the transition system in which all heaps are concrete.

The reverse application, i.e., the replacement of an embedding of a rule's right-hand side by its left-hand side, yields an abstraction of a heap. As (forward) rule application is monotonically

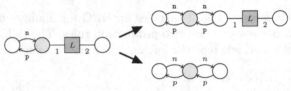

Fig. 5. Concretisation at the shaded node

decreasing with respect to language inclusion, abstraction of a heap configuration yields an over-approximation of the current set of concrete heap configurations. We abstract HCs as far as possible to achieve a compact heap representation, and we refer to the set of all maximal abstractions of an HC H by *fullAbstr*(H). In general, this mapping yields a finite but not a singleton set as typically HCs can be abstracted in different ways. If the order in which abstractions are successively applied does not matter, i.e., all finally yield the same abstract HC, the HAG is *backward confluent*. Intuitively speaking, given a backward confluent HAG, abstraction is unique[1]. In fact, the grammar for doubly-linked lists that is given in Fig. 3 lacks this property: consider the abstract HC that corresponds to the right-hand side of the second production rule (where all nodes are assumed to be internal). This configuration can be abstracted by either of the two rules, yielding an HC with either three nodes connected by two L-edges or two nodes connected by one L-edge. Both cannot be abstracted further by means of the grammar, implying that it is not backward confluent. In this concrete example, the problem can easily be fixed by adding another rule that allows to abstract two adjacent L-edges by a single one. In general, the decidability of this property follows as a direct consequence of the results in [17]. However, it is unknown whether backward confluence completion is always possible in our setting. Nevertheless we will assume backward confluence in the following, as it guarantees that preconditions of procedure summaries do no overlap.

---

[1] In the literature, this is often referred to as *canonical abstraction*.

4   Deriving Procedure Contracts by Interprocedural Analysis

As shown in [9], heap abstraction by HAGs is able to cope with local variables of (recursive) procedures in a naive way by directly including the runtime stack and thus the local variables into the HC and then expanding the procedure code during execution. This bears the disadvantage that in the presence of recursive procedure calls, HAG rules integrating both data structures and the runtime stack must be provided to obtain a finite state-space representation. This is a very complex task. Therefore we consider an instance of an interprocedural analysis (IPA) for recursive programs with local variables as studied in [14]. This IPA calculates procedure contracts, i.e. mappings from input HCs to the set of resulting HCs after the procedure has finished. The runtime stack is handled by extending the analysis information to stacks of these. The results of [14] guarantee that (even in the presence of recursion) stacks with at most two entries have to be considered in the analysis as this suffices to cover all possible calling contexts.

As in our heap abstraction setting the domain of analysis information is generally infinite, we perform a demand-driven analysis, i.e., only reachable HCs are considered as starting configurations of the procedures. Thus the procedure contracts offer a way to reason modularly about the program behaviour.

4.1   IPA Preliminaries

The IPA operates on a system of control flow graphs (SCFG) $S = (F_0, \ldots, F_n)$, representing procedures $p_0 \ldots, p_n$. The nodes N of the flow graphs correspond to the program statements, while the edges E represent the branching structure of the code. We refer to the entry (exit) nodes of the procedures $p_0, \ldots, p_n$ by $N_{entry} := \{s_0, \ldots, s_n\}$ ($N_{exit} := \{r_0, \ldots, r_n\}$), where we assume the entry and exit nodes to correspond to the procedure headers and closing brackets, respectively. Additionally, all nodes representing procedure calls are collected in the set $N_C \subseteq N$. We introduce the following auxiliary functions on SCFGs:

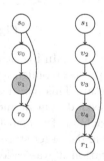

Fig. 6. SCFG

- *entry* : $N_C \to N_{entry}$ (*exit*(n) : $N_C \to N_{exit}$) maps a call node to the corresponding entry (exit) node of the called procedure,
- *caller* : $N_{entry} \to 2^{N_C}$ maps an entry node to the set of nodes calling the corresponding procedure, and
- *pred* : $N \to 2^N$ maps a node to its (direct) predecessors in the SCFG.

Example 4. Fig. 6 provides the system of flow graphs of the program given in Fig. 1. It shows two flow graphs representing procedures main and trav, with the set of procedure entries $N_{entry} = \{s_0, s_1\}$ and exits $N_{exit} = \{r_0, r_1\}$. The nodes

v_0 to v_4 correspond to the statements of the two procedures where procedure calls are indicated by shading, while the edges provide the information on the branching structure of the code. For instance, the node v_2 of trav's flow graph corresponds to the statement tmp := cur.next and is either followed by the if-branch denoted by v_3 or the procedure exit r_1.

Note that the control-flow branching caused by the **if**-statement is shifted to the preceding node. This allows to easily handle the latter as filter operations; for details on the transfer functions we refer to Sect. 4.2.

In our setting, the IPA yields procedure contracts from the domain $D = (\text{HCE}_{\Sigma_N} \to 2^{\text{HCE}_{\Sigma_N}})$ as analysis information. We call each entry of D a *pair*. Thus pair $(S, \{H_1, \dots, H_n\})$ of a contract can be interpreted as a set of graph transformation rules $S \to H_1 \mid \dots \mid H_n$, where the only persistent elements are the external nodes of the graphs. The domain of a mapping in D contains all heap configurations a procedure can be run on. We call these configurations *preconfigurations*. The image of D then provides all heap configurations the execution of the procedure's code may end up in, i.e. the set of all *postconfigurations* reachable from some preconfiguration. Thus, intuitively, the analysis information at a node n of the flow graph G for procedure p corresponds to the aggregated effect of the execution of all statements on any path from the entry node of p to n when started on any of the preconfigurations.

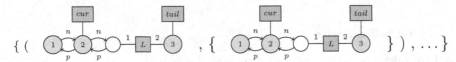

Fig. 7. Contract of procedure trav (excerpt)

Example 5. In Fig. 7, the contract for procedure trav is given. It consists of several pre-postconfigurations-pairs as indicated by the dots, where one is provided in detail. This pair describes the situation where trav is entered with cur set to the second element and tail to the end of a doubly-linked list. Additionally, tail is reachable from cur through at least two next-pointers. The postconfiguration is a singleton set stating that the procedure returns with the same abstract configuration it was called with. This is to be expected as trav traverses the list without modifying it.

To guarantee the uniqueness of the procedure contract's preconfigurations, we assume that HAG G is backward confluent (as discussed at the end of Sect. 3). Note that D together with the operation $\sqsubseteq$, where $\sqsubseteq$ is defined as pointwise application of $\subseteq$ to the images of D, forms a partial order with least element $f_\emptyset : H \mapsto \emptyset$ (for all $H \in \text{HCE}_{\Sigma_N}$).

As already mentioned, the IPA supports recursive procedures with local variables without explicitly representing the runtime stack in the heap configurations. We introduce stacks of analysis information modelling the call hierarchy and utilise the results from [14] stating that it is sufficient to consider stacks of height at most two. Moreover their Interprocedural Coincidence Theorem will allow us to perform a fixpoint interation to obtain the IPA results later on. Stack manipulations are performed by the usual stack operations such as $newstack : D \to Stack$, $push : Stack \times D \to Stack$, $pop : Stack \to Stack$, $top : Stack \to D$.

In the following we start by introducing a local transformer that represents effects of single statements on the analysis information, and a global transformer that aggregates the effects of a sequence of statements. Thus the global transformer calculated for the exit node of a procedure corresponds to the analysis information, i.e., the procedure effect/contract we are finally interested in.

4.2   Local Transformers

The local transformer $[\![.]\!]^* : N \to (Stack \to Stack)$ corresponds to the effect of the execution of a single statement on the stacks of contracts. The local transformer helper $[\![.]\!] : N \to (D \to D)$ corresponds to the application of a single statement to a stack entry. We will see later on that $[\![.]\!]^*$ takes care of fetching the right stack entry to operate on and also handles procedure calls and returns.

Remember that ultimately the procedure contracts should be available at the exit nodes of procedures. Thus the preconfigurations are compiled at procedure entry by $[\![.]\!]^*$. The execution of (non-procedural) statements affect the postconditions, that summarise the procedure effect from entry up to the statement under consideration.

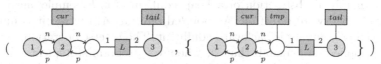

Fig. 8. Effect of pointer assignment

Example 6. Consider the SCFG from Fig. 6 and assume we want to determine the effect of v_2, i.e. the statement tmp := cur.next, on the pair provided in Fig. 7. As described above we apply a local transformer reflecting the statement's effect to the postconditions element-wise resulting in the pair given in Fig. 8.

Formally the local transformer helper $[\![.]\!] : N \to (D \to D)$ is defined as:

$$[\![n]\!](d) := \bigcup_{(S,\{H_1,\ldots,H_m\}) \in d} \{(S, t_{stmt}(H_1) \cup \ldots \cup t_{stmt}(H_m))\}$$

where $stmt$ is the statement corresponding to node n and the transfer functions $t_{stmt} : \text{HCE}_{\Sigma_N} \to 2^{\text{HCE}_{\Sigma_N}} \cup \{error\}$ are defined as follows:

– transformer $t_{x := null}$ sets a variable or pointer x to the **null** node. If the pointer is undefined, a corresponding edge is created before. Thus, $t_{x := null}(H)$ is given by

$$\begin{cases} \{(V_H, E_H, att_H[e_x \mapsto v_{null}], lab_H, ext_H)\} & \text{if } \exists e_x \in E_H.lab(e_x) = x \\ \{(V_H, E_H \cup \{e_x\}, att_H[e_x \to v_{null}], lab_H, ext_H)\} & \text{otherwise} \end{cases}$$

- transformer $t_{x.s := y}$ resets the s-pointer of the node identified by variable x to the node that variable y refers to. This statement is only applicable if s is defined for the x-node, i.e., already refers to an existing node (which is possibly v_{null}). Otherwise, an error is reported. Formally, $t_{x.s := y}(H)$ is defined by

$$\begin{cases} \{(V_H, E_H, att_H[e_s \mapsto & \text{if } \exists e_x, e_y, e_s \in E_H, \forall z \in \{x, y, s\}. \\ att(e_x)(1)att(e_y)(1)], lab_H, ext_H)\} & lab(e_z) = z \wedge att(e_s)(1) = att(e_x)(1) \\ error & \text{otherwise} \end{cases}$$

- x := y, x := y.t, x.s := **null**, x.s := y.t, **new**(x): analogous
- transformer $t_{\text{if } C}$ with $t_{\text{if } C}(H) := \begin{cases} \{H\} & \text{if } C = true \text{ for } H \\ \emptyset & \text{otherwise} \end{cases}$
- transformer t_{else} where **if** C is the corresponding if-branch with

$$t_{\text{else}}(H) := \begin{cases} \{H\} & \text{if } C = false \text{ for } H \\ \emptyset & \text{otherwise} \end{cases}$$

So far, we did not consider abstraction and concretisation during this IPA. Its integration into the analysis is crucial, as our domain is infinite in general. As explained at the end of Sect. 3, the abstraction and concretisation operations are are implemented by the functions *fullAbstr* and *conc*, respectively. The former yields the maximal abstraction of a heap configuration by employing production rules in backward direction. As pointed out earlier, the result is unique if the underlying grammar is backward confluent. The reverse operation, concretisation, is applied on demand, that is, whenever program statements are to be executed which involve selectors hidden in abstracted parts of the heap. Note that the dereferencing depth of pointer assignments in our simple programming language is limited to one such that, provided a normalised HAG [13], one concretisation step is always sufficient, ensuring the finiteness of the resulting set of heap configurations. We assume an extension of *fullAbstr* and *conc* to contracts by pointwise application to each heap configuration in the pairs' second components. Then the transformer for non-procedural statements $[\![.]\!]$ is re-defined in the following way:

$$[\![n]\!] := fullAbstr \circ [\![n]\!] \circ conc$$

(where the definition of function $[\![n]\!]$ on the right-hand side is given at the beginning of this section).

Now that we defined the effect of non-procedural statements on the analysis information, we can specify the local transformer $[\![.]\!]^*$ employing $[\![.]\!]$ to handle call statements. As in the interprocedural setting our analysis information consists of stacks of contracts, the following transformer will additionally manage the stack entries and pass the appropriate one to the local transformer helper that processes non-procedural statements then. This is specified in the first case of the following definition. However, the main contribution of $[\![.]\!]^*$ is the handling of procedure calls, as described in the second case. It proceeds in three steps:

1. A new contract is generated, according to function *new* as given below, and placed on top of the stack. It (only) covers the fragment of the heap that is reachable after the procedure call.
2. The procedure effect is applied to this contract. It is determined by the global transformer (cf. Sect. 4.3) of the corresponding exit node.
3. The modified heap fragment is integrated into the heap configuration of the calling site, specified by function *apply* as given on the next page.

Given $n \in N$ and $stk \in Stack$, $[\![n]\!]^*(stk)$ is formally defined as follows:

$$[\![n]\!]^*(stk) := \begin{cases} push(pop(stk), [\![n]\!](top(stk))) & \text{if } n \in N \setminus N_C \\ return_{(exit(n),n)} \circ [\![exit(n)]\!] \circ call_{(n,entry(n))}(stk) & \text{if } n \in N_C \end{cases}$$

with $call : (N \times N) \to (Stack \to Stack)$

$$call_{(m,n)}(stk) := push(stk, new(m, n, top(stk)))$$

and $return : (N \times N) \to (Stack \to Stack)$

$$return_{(m,n)}(stk) := push(pop(pop(stk)), apply(m, n, top(pop(stk)), top(stk)))$$

Here $new : N \times N \times D \to D$ generates a new contract based on the reachable fragment of the heap configuration, and $apply : N \times N \times$ 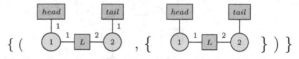 $D \times D \to D$ modifies the caller context according to the procedure contract.

Fig. 9. Contract of procedure main at node v_1.

Example 7. Assume a call to *new* with input $m = v_1$, $n = s_1$ and the contract of Fig. 9. To calculate the heap fragment reachable from the called procedure trav, we first identify the variables passed to the procedure call (head and tail) with the procedure's formal parameters (cur and tail). Thus trav operates on the heap fragment reachable from cur and tail. To be able to identify this reachable fragment on procedure return, we remember all nodes that are shared between this fragment and the remaining heap at the calling site by marking them external. In this case, these cutpoints are simply the heap locations referenced by the variables passed to the procedure call. The such generated reachable fragment forms the first element of the pair. Now remember that a contract should capture the procedure's behaviour up to the node under consideration, i.e. the procedure start s_1 in this case. Thus the set of resulting configurations is the singleton set containing the reachable fragment only, which is equal to the pair given in the main contract of Fig. 9 if head is renamed to cur.

In general, we define $new(m, n, d) := \{(scope(m, n, H_i), \{scope(m, n, H_i)\}) \mid (S, \{H_1, \ldots, H_j\}) \in d, i \in \{1, \ldots, j\})\}$ where $scope(m, n, H)$ performs the following steps:

1. generates a copy H' of H that becomes the precondition of the new contract after modification as depicted in the following steps,

2. renames in H' all parameters in $param(m)$ to $param(n)$, where $param(m)$ corresponds to the variables passed to the called procedure and $param(n)$ to the formal procedure parameters,
3. removes all variables except for $param(n)$ from H',
4. cuts H' to the fragment reachable from $param(n)$, and
5. defines the external nodes of H' as the cutpoints, that is, the heap locations that are referenced by variables passed to the procedure call or those on the border of the heap fragment that is reachable from the calling context.

Function *apply* applies a contract upon procedure return. It takes four arguments: the nodes corresponding to the exit of the called procedure and the calling site, the contract of the calling site where the result of the call is going to be reflected in, and the contract of the called procedure.

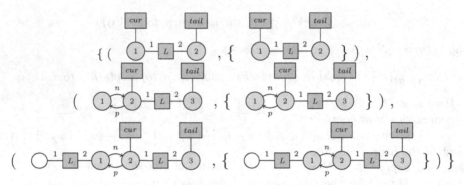

Fig. 10. trav-contract at r_1

Example 8. Assume a call to *apply* with exit node r_1, calling site v_1, the calling site's contract given in Fig. 9 and the contract of trav provided in Fig. 10. Note that this contract slightly differs from the one provided in Fig. 7 as now abstraction is involved. Now *apply* first figures out which of the tuples in its contract to apply. To this aim, it checks if at the calling site the heap fragment reachable from the variables passed to the procedure is equal to one of the contract's preconditions (after parameter renaming). In the example this is indeed the case for the first pair t of Fig. 10 (note that the preconditions of the other pairs do not match because of the preceding list). Then in the current heap configuration at the calling site (i.e. here the set of the contract's postconditions at v_1), the identified subgraph is replaced by all graphs in t's postcondition. The resulting contract states that main actually returns the same abstract HC and thus is equal to the initial contract as specified in Fig. 9.

In general, *apply* : $N \times N \times D \times D \to D$ is given by $apply(m, n, d, d') :=$ $\{(S, \bigcup_{j \in \{1,\ldots,i\}} transform(m, n, H_j, d')) \mid (S, \{H_1, \ldots, H_i\}) \in d\}$. Here, function $transform(m, n, H, d')$ does the following steps for each $(S', \{H'_1, \ldots, H'_j\}) \in d'$:

1. generates T from S' by renaming all parameters $param(m)$ to $param(n)$,
2. if there exists an isomorphism between T and the fragment of H reachable from $param(n)$, replaces this embedding of T in H element-wise by the graphs $\{H'_1, \ldots, H'_j\}$ resulting in the set $\{\bar{H}'_1, \ldots, \bar{H}'_j\}$; due to backward confluence there exists at most one S', stop and return the resulting set, and
3. if there does not exist such an embedding, continues.

4.3 Global Transformer

The global transformer $[\![.]\!] : N \to (Stack \to Stack)$ corresponds to the effect of a sequence of statements, i.e., it captures the effect from procedure entry up to the given node. Upon finishing the analysis, the global transformer of the exit nodes of each procedure assembles the procedure contract as follows (for all $n \in N, stk \in Stack$).

$$[\![n]\!]\,(stk) = \begin{cases} stk & \text{if } n \in S_{entry} \\ cleanup(n, stk) & \text{if } n \in S_{exit} \\ \bigcup\{[\![m]\!] \circ [\![m]\!]\,(stk) \mid m \in pred(n)\} & \text{otherwise} \end{cases}$$

where $cleanup : N \times Stack \to Stack$ removes all local variables in the top-most stack entry to reflect the heap configuration after exiting the current procedure given by the exit node provided to $cleanup$. This is necessary for all exit nodes as their analysis information is inserted at the call sites as the complete procedure effect on the heap. At this point the local variables of the procedure are irrelevant. In detail, $cleanup(n, stk)$ performs the following steps for all $(S, \{H_1, \ldots, H_k\}) \in top(stk)$:

1. Removes all variables except for those in $param(n)$ from $H_1, \ldots, H_k$, and
2. subsequently abstracts all $H_1, \ldots, H_k$.

4.4 Equation System

Using the local and global transformers, we set up the following equation system. To obtain the least solution of the equation system and, thus, the IPA result, fixpoint iteration is performed.

When setting up the equation system we distinguish between the analysis information $pre(n)$ before executing the statement corresponding to node n and $post(n)$ after. The former merges the information from nodes preceding n, while the latter applies the local transformer associated to n.

$$pre(n) := \begin{cases} newstack(\{(init, \{init\})\}) & \text{if } n = s_0 \\ \bigcap\{call_{(m, entry(m))}(pre(m)) \mid m \in caller(n)\} & \text{if } n = N_{entry} \setminus \{s_0\} \\ \bigcap\{post(m) \mid m \in pred(n)\} & \text{otherwise} \end{cases}$$

where $\{(init, \{init\})\}$ is the initial contract with $init$ being the initial heap configuration the main procedure (denoted by s_0) is run on. Moreover $\bigcap$ is defined as the union of the top-most stack components. We perform a demand-driven

IPA, i.e. we do not consider all possible contracts when analysing a procedure but only those that already occurred during analysis. Moreover,

$$post(n) := [\![n]\!]^*(pre(n))$$

Note that $[\![.]\!]^*$ is monotonic, as the join-operator is applied pointwise to each element of the input set and the analysis information is ordered via $\subseteq$, and distributive, as the meet-operator reduces the stack information to the top-most entry and then is applied pointwise to each pair of the input set. Thus the Interprocedural Coincidence Theorem [14] states that the least fixpoint coincides with the interprocedural meet-over-all-paths solution, if the fixpoint-iteration terminates (i.e. abstraction ensures the existence of a least upper bound of D).

Example 9. Consider again the example program for a recursive doubly-linked-list traversal as given in Fig. 1. Upon completion of the fixpoint iteration, the exit nodes r_0 and r_1 of procedures main and trav hold the final contracts the analysis provides. These contracts were already given in Fig. 9 and 10, respectively.

This IPA can be seen as an alternative to the classical approach generating the (abstract) state space of a program as presented in [9]. While in general the IPA yields coarser results than the state space approach, it also features many advantages over the latter. For one, the IPA handles recursive procedures with local variables without the need to model the runtime stack explicitly on the heap. This allows the abstraction of recursive programs without providing HAG rules that consolidate the procedure stack and the program's data structures. Moreover the contracts provided by the IPA offer a way to modularly reason about procedures. That is, whenever an HC appears upon procedure call for which a contract is available, the latter can be applied without re-computation.

5   Conclusions

The contribution of this paper is the development of a novel IPA for automatically deriving procedure contracts for pointer programs. This IPA builds upon an abstraction framework based on HRGs [10]. While in this work recursive procedures with local variables are represented by explicitly modelling and abstracting the call stack [9], here we follow an approach that separates heap from control abstraction. This on the one hand allows a modular reasoning at procedure level, while one the other hand simplifies the user's specification effort on HRGs as the runtime stack abstraction is carried out automatically by the IPA.

The proposed analysis summarises procedure effects with so-called contracts. The latter are given as graph transformation rules and capture all possible procedure effects on the reachable part of the heap in the form of pre-/postconfiguration-pairs. In particular, the approach supports recursive procedures with local variables and cutpoints, that is, heap objects that are shared between the heap fragment a procedure call operates on and the calling context. It turns out that the HRG approach is particularly suited for determining and applying such contracts as it offers an intuitive formalism for describing heap transformations, which moreover can be automatically derived.

The application we are mainly targeting with the proposed IPA is formal verification of software. For instance, the abstract contract representation of a program's behaviour can be exploited to show that it does not accidentally invalidate data structures, e.g., by introducing sharing nodes in a tree [10]. However, a further potential application is testing of software. More concretely, "visual contracts" given as pre-/postcondition specifications have been used in model-based testing to transform use cases into test cases [7]. Our contracts could be applied in a similar fashion to classify test cases, that is, procedure parameters, according to the possible combinations of pre- and postconditions, thus allowing to evaluate coverage criteria.

We are planning to investigate the practical feasability of the IPA approach by comparing its implementation with [9], where the runtime stack is modelled explicitly. Moreover we are extending our techniques to concurrent programs with threads. In this setting, we expect the contract approach to be even more beneficial than in the interprocedural case. While the latter can alternatively be handled by providing HRG rules that allow to abstract the runtime stack in order to get a finite-state representation [9], it seems hopeless to develop similar (efficient) abstraction techniques for the concurrent case [16]. Our idea to overcome the efficiency problem is to develop a thread-modular analysis that, similarly to [6], avoids the enumeration of all possible interleavings between threads. Like in the interprocedural case, this analysis will yield thread contracts that capture all possible effects of the (isolated) execution of the respective thread on (the reachable part of) the heap. To ensure the soundness of this approach, the access to heap objects will be guarded by permissions [1,2,8] in order to detect possible race conditions between threads running concurrently.

Furthermore we are planning to investigate a fragment of HRGs where language inclusion is decidable, which would lift this partially modular reasoning approach to true modularity, because preconfigurations can be checked for language inclusion instead of isomorphism upon contract application. A starting point for this investigation is current research about decidability of entailment in a fragment of SL [11] which is shown to be equivalent to HRGs [3,12].

References

1. Bornat, R., Calcagno, C., O'Hearn, P., Parkinson, M.: Permission accounting in separation logic. In: POPL 2005, pp. 259–270. ACM (2005)
2. Boyland, J.: Checking interference with fractional permissions. In: Cousot, R. (ed.) SAS 2003. LNCS, vol. 2694, pp. 55–72. Springer, Heidelberg (2003)
3. Dodds, M., Plump, D.: From hyperedge replacement to separation logic and back. In: Proc. Doctoral Symp. at the Int. Conf. on Graph Transformation, ICGT 2008. Electronic Communications of the EASST, vol. 16 (2009)
4. Fradet, P., Caugne, R., Métayer, D.L.: Static detection of pointer errors: An axiomatisation and a checking algorithm. In: Riis Nielson, H. (ed.) ESOP 1996. LNCS, vol. 1058, pp. 125–140. Springer, Heidelberg (1996)
5. Gotsman, A., Berdine, J., Cook, B.: Interprocedural shape analysis with separated heap abstractions. In: Yi, K. (ed.) SAS 2006. LNCS, vol. 4134, pp. 240–260. Springer, Heidelberg (2006)

6. Gotsman, A., Berdine, J., Cook, B., Sagiv, M.: Thread-modular shape analysis. In: Proc. ACM SIGPLAN Conf. on Programming Language Design and Implementation, PLDI 2007, pp. 266–277. ACM Press (2007)
7. Güldali, B., Mlynarski, M., Wübbeke, A., Engels, G.: Model-based system testing using visual contracts. In: 35th Euromicro Conf. on Software Engineering and Advanced Applications (SEAA 2009), pp. 121–124 (August 2009)
8. Haack, C., Huisman, M., Hurlin, C.: Permission-based separation logic for multithreaded Java programs. Nieuwsbrief van de Nederlandse Vereniging voor Theoretische Informatica 15, 13–23 (2011)
9. Heinen, J., Barthels, H., Jansen, C.: Juggrnaut – an abstract JVM. In: Beckert, B., Damiani, F., Gurov, D. (eds.) FoVeOOS 2011. LNCS, vol. 7421, pp. 142–159. Springer, Heidelberg (2012)
10. Heinen, J., Noll, T., Rieger, S.: Juggrnaut: Graph grammar abstraction for unbounded heap structures. In: Proc. 3rd Int. Workshop on Harnessing Theories for Tool Support in Software. ENTCS, vol. 266, pp. 93–107. Elsevier (2010)
11. Iosif, R., Rogalewicz, A., Simacek, J.: The tree width of separation logic with recursive definitions. In: Bonacina, M.P. (ed.) CADE 2013. LNCS (LNAI), vol. 7898, pp. 21–38. Springer, Heidelberg (2013)
12. Jansen, C., Göbe, F., Noll, T.: Generating inductive predicates for symbolic execution of pointer-manipulating programs (submitted, 2014)
13. Jansen, C., Heinen, J., Katoen, J.-P., Noll, T.: A local Greibach normal form for hyperedge replacement grammars. In: Dediu, A.-H., Inenaga, S., Martín-Vide, C. (eds.) LATA 2011. LNCS, vol. 6638, pp. 323–335. Springer, Heidelberg (2011)
14. Knoop, J., Steffen, B.: The interprocedural coincidence theorem. In: Pfahler, P., Kastens, U. (eds.) CC 1992. LNCS, vol. 641, pp. 125–140. Springer, Heidelberg (1992)
15. Kreiker, J., Reps, T., Rinetzky, N., Sagiv, M., Wilhelm, R., Yahav, E.: Interprocedural shape analysis for effectively cutpoint-free programs. In: Voronkov, A., Weidenbach, C. (eds.) Ganzinger Festschrift. LNCS, vol. 7797, pp. 414–445. Springer, Heidelberg (2013)
16. Noll, T.G., Rieger, S.: Verifying dynamic pointer-manipulating threads. In: Cuellar, J., Sere, K. (eds.) FM 2008. LNCS, vol. 5014, pp. 84–99. Springer, Heidelberg (2008)
17. Plump, D.: Checking graph-transformation systems for confluence. ECEASST 26 (2010)
18. Reps, T., Horwitz, S., Sagiv, M.: Precise interprocedural dataflow analysis via graph reachability. In: Proc. 22nd ACM SIGPLAN-SIGACT Symp. on Principles of Programming Languages, POPL 1995, pp. 49–61. ACM Press (1995)
19. Rinetzky, N., Sagiv, M.: Interprocedural shape analysis for recursive programs. In: Wilhelm, R. (ed.) CC 2001. LNCS, vol. 2027, pp. 133–149. Springer, Heidelberg (2001)
20. Rinetzky, N., Sagiv, M., Yahav, E.: Interprocedural shape analysis for cutpoint-free programs. In: Hankin, C., Siveroni, I. (eds.) SAS 2005. LNCS, vol. 3672, pp. 284–302. Springer, Heidelberg (2005)
21. Sagiv, S., Reps, T.W., Horwitz, S.: Precise interprocedural dataflow analysis with applications to constant propagation. In: Mosses, P.D., Nielsen, M. (eds.) TAPSOFT 1995. LNCS, vol. 915, pp. 651–665. Springer, Heidelberg (1995)
22. Sharir, M., Pnueli, A.: Two approaches to interprocedural data flow analysis. In: Program Flow Analysis: Theory and Applications, pp. 189–233. Prentice-Hall (1981)
23. Yorsh, G., Yahav, E., Chandra, S.: Generating precise and concise procedure summaries. In: Proc. 35th ACM SIGPLAN-SIGACT Symposium on Principles of Programming Languages, POPL 2008, pp. 221–234. ACM Press (2008)

Generating Inductive Predicates for Symbolic Execution of Pointer-Manipulating Programs

Christina Jansen[1],[*], Florian Göbe[2], and Thomas Noll[1]

[1] Software Modeling and Verification Group
[2] Embedded Software Laboratory
RWTH Aachen University, Germany

Abstract. We study the relationship between two abstraction approaches for pointer programs, Separation Logic and hyperedge replacement grammars. Both employ inductively defined predicates and replacement rules, respectively, for representing (dynamic) data structures, involving abstraction and concretisation operations for symbolic execution. In the Separation Logic case, automatically generating a complete set of such operations requires certain properties of predicates, which are currently implicitly described and manually established. In contrast, the structural properties that guarantee correctness of grammar abstraction are decidable and automatable. Using a property-preserving translation we argue that it is exactly the logic counterparts of those properties that ensure the direct applicability of predicate definitions for symbolic execution.

1 Introduction

Prominent examples of pointer usage are dynamic data structures such as doubly-linked lists, nested lists, trees, and so forth. While pointers are one of the most common sources of bugs in software [8], these errors are often difficult to trace. Thus automated analysis techniques are of great assistance. However, the problem is highly non-trivial as the presence of dynamic data structures generally gives rise to an unbounded state space of the program to be analysed. A common approach is to apply *abstraction techniques* to obtain a finite representation.

In this paper we investigate the relationship between two such techniques. The first, introduced in Sect. 2.2, is based on *Separation Logic (SL)* [18], an extension of Hoare Logic that supports local reasoning in disjoint parts of the heap. It employs inductively defined predicates for specifying the (dynamic) data structures maintained at runtime, such as lists or trees. To support symbolic execution, SL introduces abstraction rules that yield symbolic representations of concrete heap states. Whenever concrete program instructions are to be applied, so-called unrolling rules expose fields of abstracted heap parts on demand to enable the application of the respective operational rule. This yields a (local) *concretisation* of the symbolic heap [1]. More details are provided in Sect. 3.2.

Both abstraction and unrolling rules are dependent on the predicates. This raises the question of the appropriateness of predicate definitions to support

[*] This research is funded by EU FP7 project CARP (http://www.carpproject.eu).

H. Giese and B. König (Eds.): ICGT 2014, LNCS 8571, pp. 65–80, 2014.
© Springer International Publishing Switzerland 2014

the (automated) construction of such rules. Current literature largely ignores this problem; it addresses specific settings such as lists or trees but does not investigate under which conditions a complete set of abstraction and unrolling rules can be generated systematically. Rather, such rules are developed manually.

As it turns out, similar conditions are well-studied for the second approach, which employs a *graph-based* representation of heaps supporting both concrete and abstract parts. The latter is realised by deploying labelled hyperedges, i.e., edges that connect an arbitrary number of nodes. These edges are interpreted as placeholders for data structures that are specified by *hyperedge replacement grammars (HRGs)* (cf. Sect. 2.1). As will be shown in Sect. 3.1, abstraction and concretisation are implemented by applying production rules in backward and forward direction, respectively. To ensure the correctness and practicability of this procedure, the grammars are automatically transformed such that they exhibit certain *structural properties*, which are defined in Sect. 3.3.

The motivation of our work is to exploit these results to support automated predicate generation for symbolic execution in the SL framework. More concretely, predicate definitions should immediately yield sound abstraction and concretisation rules (such as list collapsing and unrolling operations). To this aim, we extend the *semantics-preserving translation* from [6] between SL and HRGs in Sect. 2.3 such that it covers all HRGs describing data structures. Sect. 3.3 then provides logical counterparts of the graph grammar properties and argues why these are necessary for symbolic execution. Finally we show that the translation preserves these structural properties (cf. Sect. 4).

Altogether, these achievements provide the formal basis of an integrated framework for heap abstraction that combines the best of both worlds: first, data structures are specified using the intuitive formalism of HRGs, and the structural properties to ensure correctness of abstraction are automatically established. Then, SL predicate definitions are generated by the property-preserving mapping, and symbolic execution is performed using abstraction and unrolling rules derived from the predicate definitions. The last step can be assisted by a number of tools that support SL reasoning, such as jStar [4], Predator [7], and Smallfoot [2]. The paper concludes with some final remarks in Sect. 5.

Related Work: The research that is closest to ours is described in [5,6]. It establishes the correspondence between a restricted subset of HRGs and a fragment of SL, and shows that SL is actually more expressive than HRGs. Our work generalises this approach by considering heap nodes with more than two selector fields. Moreover, it strengthens the correspondence result by providing characterisations of the structural properties mentioned before for both formalisms, and by showing that they are preserved by the translations in both directions. The work of Dodds and Plump is inspired by [14], which gives a semantics to graph grammars by mapping them to SL formulae. However, these grammars only allow to define tree data structures, and the translation is only one-way, which does not allow to derive any correspondence results.

2   Preliminaries

In the following we introduce the two abstraction formalisms, HRGs and SL, and relate them by extending an existing translation between HRGs and SL [6] in a natural way, such that it covers all HRGs describing data structures. These preliminaries provide the basis to our contribution on the identification and definition of structural properties, that allow for a systematic and automatic generation of abstraction and unrolling rules for symbolic execution based on SL.

To be able to focus on the intuition and formal details regarding these structural properties, we keep this introduction short and refer to [11] for details on definitions and the translation procedure itself.

2.1   Hyperedge Replacement Grammars

In the abstraction approach based on HRGs [10], (abstract) heaps are represented as hypergraphs, i.e., graphs whose edges can connect arbitrarily many nodes.

Definition 1 (Hypergraph). *Let Σ be a alphabet with ranking function rk : $\Sigma \to \mathbb{N}$. A (labelled) hypergraph (HG) over Σ is a tuple $H = (V, E, \text{att}, \text{lab}, \text{ext})$ where V is a set of nodes and E a set of hyperedges,* att : $E \to V^{\star}$ *maps each hyperedge to a sequence of attached nodes such that $|\text{att}(e)| = rk(\text{lab}(e))$,* lab : $E \to \Sigma$ *is a hyperedge-labelling function, and* ext $\in V^{\star}$ *a (possibly empty) sequence of pairwise distinct external nodes. The set of all hypergraphs over Σ is denoted by HG_{Σ}.*

Given a hypergraph H as above, we write V_H, E_H etc. for its components. We do not distinguish between isomorphic HGs (up to renaming). Using the alphabet of selectors Σ introduced in Sect. 2.2, we model (concrete) heaps as HGs over Σ without external nodes. (The latter is required for defining the replacement of hyperedges.) Objects are represented by nodes, and selectors by edges understood as pointers from the first attached object to the second one. We introduce a node v_{null} representing **null**, which is unique to every hypergraph. To represent abstract parts of the heap we use nonterminal edges, which carry labels from an additional set of *nonterminals (NTs) N* (and we let $\Sigma_N = \Sigma \uplus N$).

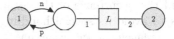

Fig. 1. (Extended) heap as hypergraph

Example 1. A typical implementation of a doubly-linked list consists of a sequence of list elements connected by next (n) and previous (p) pointers. Fig. 1 depicts a HG representation of a such a list. The circles are nodes representing objects on the heap. A shaded circle indicates an external node; its ordinal (i.e., position in ext) is given by the label. The L-labelled box represents an NT edge of rank two indicating an abstracted doubly-linked list between the first and second attached node. Later we will see how these abstract structures are defined. The connections between NT edges and nodes are labelled with their ordinal. For the sake of readability, selectors (**n** and **p**) are depicted as directed edges.

Note that not every HG over Σ represents a valid heap: e.g., it is necessary that, for every selector $s \in \Sigma$, every object has at most one outgoing s-edge, and that external nodes are absent. HGs that satisfy this requirement (cf. [11] for details) are called *heap configurations (HC)*, and are collected in the set HC_{Σ_N}. Thus HC_Σ represents all concrete HCs, i.e., those without nonterminal edges. HCs with external nodes are called *extended* and are represented by the sets HCE_{Σ_N} and HCE_Σ in the abstract and concrete case, respectively.

An NT edge of an HC acts as a placeholder for a heap part of a particular shape. We use *hyperedge replacement grammars* to describe its possible structure.

Fig. 2. A grammar for doubly-linked lists

Definition 2 (Hyperedge Replacement Grammar). *A hyperedge replacement grammar (HRG) over an alphabet Σ_N is a function $G : N \to 2^{HCE_{\Sigma_N}}$ with $|\mathrm{ext}_H| = rk(X)$ for each $X \in N$ and $H \in G(X)$. We call $X \to H$ a production rule. The set of all HRGs over Σ_N is denoted by HRG_{Σ_N}.*

Example 2. Fig. 2 specifies an HRG for doubly-linked lists. It employs one NT L of rank two and two production rules. The right one recursively adds one list element, whereas the left one terminates a derivation.

HRG derivations are defined through *hyperedge replacement*, i.e. the substitution of an NT edge by the right-hand side of a production rule. We forbid the occurrence of the $v_{\mathbf{null}}$ node in grammars to avoid its special handling during hyperedge replacement (as $v_{\mathbf{null}}$ has to be unique in every HC). A sample replacement is illustrated in Fig. 3, where the second rule of the grammar in Fig. 2 is applied to the NT edge L in the upper graph.

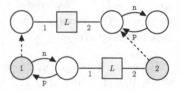

Fig. 3. Replacement of L-edge

To this aim, the external nodes of the rule's right-hand side are mapped to the nodes of the upper graph as indicated by

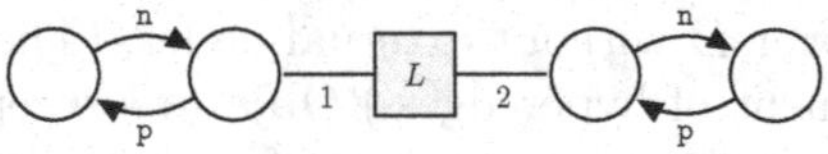

Fig. 4. Resulting hypergraph

the dashed arrows, and the L-edge is replaced by the rule's right-hand side as shown in Fig. 4. We refer to a hyperedge replacement in graph H with result K as a *derivation* from H to K and denote it by $H \Rightarrow K$. $\Rightarrow^\star$ denotes the reflexive-transitive closure of this relation. The formal definition of hyperedge replacement and derivations if given in [11].

The HRG definition does not include a starting graph. Instead, it is introduced as a parameter of the generated language.

Definition 3 (Language of an HRG). *Let $G \in HRG_{\Sigma_N}$ and $H \in HC_{\Sigma_N}$. $L_G(H) = \{K \in HC_\Sigma \mid H \Rightarrow^\star K\}$ is the language generated from H.*

The language of an NT is defined as the language of its handle, a hypergraph consisting of a single hyperedge e attached to $rk(e)$ many distinct nodes only.

Definition 4 (Handle). *Given* $X \in N$ *with* $rk(X) = n$, *an* X-handle *is the hypergraph* $X^{\bullet} = (\{v_1, \ldots, v_n\}, \{e\}, \{e \mapsto v_1 \ldots v_n\}, \{e \mapsto X\}, \varepsilon) \in HC_{\Sigma_N}$, *and* $L(X^{\bullet})$ *is the* language induced by X.

2.2 Separation Logic

SL is an extension of Hoare logic, specifically designed to reason about heaps. It supports the specification of inductively-defined predicates that describe heap shapes. Thus it is particularly interesting for symbolic execution of pointer-manipulating programs, which is studied in e.g. [1].

In SL, a heap is understood as a set of *locations*, given by $Loc := \mathbb{N}$. They are connected via references. To express that a location contains a null reference, we introduce the set $Elem := Loc \cup \{\mathbf{null}\}$. A *heap* is a (partial) mapping $h : Loc \rightharpoonup Elem$. The set of all heaps is denoted by He.

SL features logical variables, denoted by Var, that are interpretated by a (partial) mapping $i : Var \rightharpoonup Elem$. The set of interpretations is denoted by Int. A formula is evaluated over a pair (h, i) $(h \in He, i \in Int)$.

A heap consists of objects that reference other objects. References are represented by a finite set Σ of *selectors*. In the heap representation, we assume these references to reside in successive locations respecting a canonical order and thus reserve $|\Sigma|$ successive locations for each object. We denote the ordinal of selector $s \in \Sigma$ by $cn(s)$, where $0 \le cn(s) < |\Sigma|$, i.e., under $i \in Int$ the location of $x.s$ is $i(x) + cn(s)$. To simplify notation, we assume that a heap containing n objects utilises locations $1, \ldots, n \cdot |\Sigma|$, i.e., we do not distinguish heaps that differ only in their location numbering. Moreover, we assume *safe* interpretations, i.e. interpretations $i \in Int$ where $i(Var) \subseteq \{1, |\Sigma| + 1, 2|\Sigma| + 1, \ldots\} \cup \{\mathbf{null}\}$.

We consider a restricted set of SL formulae where negation $\neg$, **true** and conjunction $\wedge$ in subformulae speaking about the heap is disallowed. This restriction is necessary and due to the fact that HRGs are less expressive than SL [5].

Definition 5 (Syntax of SL). *Let Pred be a set of predicate names. The syntax of SL is given by:*

$$E ::= x \mid \mathbf{null}$$
$$P ::= x = y \mid P \wedge P \qquad\qquad\qquad \textit{pure formulae}$$
$$F ::= \mathbf{emp} \mid x.s \mapsto E \mid F * F \mid \exists x : F \mid \sigma(E, \ldots, E) \qquad \textit{heap formulae}$$
$$S ::= F \mid S \vee S \mid P \wedge S \qquad\qquad\qquad \textit{SL formulae}$$

where $x, y \in Var$, $s \in \Sigma$ *and* $\sigma \in Pred$. *A heap formula of the form* $x.s \mapsto E$ *is called a* pointer assertion. *SLF denotes the set of all SL formulae.*

To simplify notation, we introduce the following abbreviations. Given $\phi \in SLF$, the set $Var(\phi)$ $(FV(\phi))$ collects all (free) variables of ϕ. If $FV(\phi) = \emptyset$, then ϕ is called *closed*. $Atom(\phi)$ denotes the set of all *atomic* subformulae of ϕ, that is, those of the form $x = y$, **emp**, $x.s \mapsto E$ and $\sigma(E, \ldots, E)$. The *predicate calls* in ϕ are given by $pred(\phi) := \{\sigma(x_1, \ldots, x_n) \in Atom(\phi) \mid \sigma \in Pred, x_1, \ldots, x_n \in Var \cup \{\mathbf{null}\}\}$. If $pred(\phi) = \emptyset$, then ϕ is called *primitive*.

Predicate definitions are collected in environments.

Definition 6 (Environment). *A predicate definition for $\sigma \in Pred$ is of the form $\sigma(x_1, \ldots, x_n) := \sigma_1 \vee \ldots \vee \sigma_m$ where $m, n \in \mathbb{N}$, $x_1, \ldots, x_n \in Var$ are pairwise distinct, $\sigma_1, \ldots, \sigma_m$ are heap formulae, **null** does not occur in $\sigma_1 \vee \ldots \vee \sigma_m$, and $FV(\sigma_j) = \{x_1, \ldots, x_n\}$ for each $j \in [1, m]$. We call $\sigma_1 \vee \ldots \vee \sigma_m$ the body of σ. An environment is a set of definitions. All environments are collected in Env.*

Example 3. We define a predicate *dll* specifying doubly-linked lists: $dll(x_1, x_2) := \sigma_1 \vee \sigma_2$ with $\sigma_1 := x_1.n \mapsto x_2 * x_2.p \mapsto x_1$, $\sigma_2 := \exists r : x_1.n \mapsto r * r.p \mapsto x_1 * dll(r, x_2)$. The corresponding environment is $\{dll(x_1, x_2) := \sigma_1 \vee \sigma_2\}$.

The (weak) constraint that **null** is not allowed to appear in the predicate body can again be traced back to HRGs, where a similar restriction was introduced for the sake of simplification. Note that while **null** is disallowed in environments, its usage in SL formulae is admitted. The requirement that each disjunct has to refer to all parameter variables ensures that the HRG resulting from the translation (cf. Sect. 2.3) is ranked. This assumption does not impose a semantic restriction as each SL formula can be translated into an equivalent one by introducing new predicates fulfilling this property. The construction is similar to establishing an equivalent, ranked HRG as shown in [12].

For $\Gamma \in Env$, $Pred_\Gamma$ is the set of all $\sigma \in Pred$ with a definition in Γ. We assume predicate definitions to be in disjunctive normal form, which is no true restriction [11] as every SL formula can be translated into this normal form.

Intuitively, the semantics of a predicate call $\sigma(x_1, \ldots, x_n)$, where σ is defined in environment Γ, is determined by the *predicate interpretation* $\eta_\Gamma(\sigma)$. It is given by the least set of pairs of locations representing the arguments to σ, and heaps that fulfil the unrolling of the predicate's body. Together with the semantics of the remaining SL constructs, this yields the satisfaction relation $\models$. In particular, the separating conjunct is defined by $h, i, \eta_\Gamma \models \phi_1 * \phi_2$ iff there exist $h_1, h_2 \in He$ such that $h_1 \uplus h_2 = h$, $h_1, i, \eta_\Gamma \models \phi_1$, and $h_2, i, \eta_\Gamma \models \phi_2$. Here, $\uplus$ denotes the disjoint union of two functions. We call two SL formulae *equivalent* ($\equiv$) if for any fixed predicate interpretation they are fulfilled by the same set of heap/interpretation pairs.

2.3   Translating HRGs to Environments and Back

The motivation behind this work is to utilise the results on HAG properties for automatic generation of inductively defined predicates suitable for symbolic execution based on SL. To this aim we extend an existing translation between HRGs and SL [6] in a natural way such that it covers all HRGs deriving HCs. In Sect. 3 we provide the properties necessary for sound and automatic abstraction and unrolling using predicate definitions and show that the translation preserves them.

We assume the translation mappings $\mathrm{hrg}[\![.]\!] : Env \to HRG_{\Sigma_N}$ and $\mathrm{env}[\![.]\!] : HRG_{\Sigma_N} \to Env$, and introduce them by an exemplary application of $\mathrm{env}[\![.]\!]$.

Requirements and details are given in [11], where furthermore a criterion for environments, ensuring that hrg[[.]] yields an HRG deriving HCs only, is provided.

Example 4. Consider again the HRG G from Ex. 2. When translating an HRG to an environment, first for each NT $X \in N$ of rank n, a predicate σ_X with parameters $x_1, \ldots, x_n$ is issued (we use $\rightsquigarrow$ to indicate the translation step):
env[[G]]: $N = \{L\}$ and $rk(L) = 2 \rightsquigarrow \sigma_L(x_1, x_2)$
Next each right-hand side of a production rule of X is converted into an SL formula, and the predicate body of σ_X results from disjuncting these.
env[[G]]: $G(X) = \{R_1, R_2\} \rightsquigarrow$ disjuncts σ_1, σ_2 and thus $\sigma_L(x_1, x_2) := \sigma_1 \lor \sigma_2$
For translating the right-hand sides, each node is uniquely identified: external nodes j by x_j, internal nodes v_k by r_k and v_{null} by **null**. The predicate parameters $x_1, \ldots, x_n$ become free variables, $r_1, \ldots, r_m$ will be existentially bound.
env[[G]] for R_1: no internal nodes $\rightsquigarrow$ no quantifiers in σ_1
env[[G]] for R_2: one internal node identified by $r_1 \rightsquigarrow \sigma_2 := \exists r_1 : \ldots$
Then each edge is translated one-by-one using the node identifiers as variables in the corresponding heap formula. Terminal edges become pointer assertions.
env[[G]] for R_1: n- and p-edge $\rightsquigarrow x_1.n \mapsto x_2$ and $x_2.p \mapsto x_1$
env[[G]] for R_2: n- and p-edge $\rightsquigarrow x_1.n \mapsto r_1$ and $r_1.p \mapsto x_1$
NT edges are translated into predicate calls.
env[[G]] for R_2: L-edge $\rightsquigarrow \sigma_L(r_1, x_2)$
The translation is finalised by glueing together the edge translations using $*$.
env[[G]] for R_1: $R_1 \rightsquigarrow \sigma_1 := x_1.n \mapsto x_2 * x_2.p \mapsto x_1$
env[[G]] for R_2: $R_2 \rightsquigarrow \sigma_2 := \exists r_1 : x_1.n \mapsto r_1 * r_1.p \mapsto x_1 * \sigma_L(r_1, x_2)$
Thus G is translated into env[[G]] $= \{\sigma_L(x_1, x_2) := \sigma_1 \lor \sigma_2\}$.

The translation of environments to HRGs, hrg[[.]], shares the same correspondences and thus proceeds analogously, glueing graphs instead of assertions. Table 1 provides an overview of the correspondences between HRGs and SL.

Table 1. Overview: HRG vs. SL

HRG	SL
HC $H \in HC_{\Sigma_N}$	closed formula $\phi \in SLF$
ext. HC $H \in HCE_{\Sigma_N}$	formula $\phi \in SLF$
NT $X \in N$	predicate $\sigma \in Pred$
term. edge e (lab(e) $\in \Sigma$)	pointer assertion $x.s \mapsto y$
NT edge e (lab(e) $\in N$)	call $\sigma(x_1, \ldots, x_n)$
X-rules $G(X)$ ($X \in N$)	def. $\sigma(x_1, \ldots, x_n) = \ldots$
HRG G	environment Γ

3   Generating Predicates for Abstraction and Concretisation

HRGs and SL provide formalisms for analysing pointer-manipulating programs. Due to dynamic allocation and deallocation, the state space of such programs is potentially infinite and so is the necessary number of HCs, or formulae respectively, describing those states. An approach to resolve this issue is abstraction.

In the heap abstraction approach based on HRGs, program states are described by HCs, which may contain edges labelled with nonterminals. HRGs define the structures that are represented by nonterminal edges. The program's state space is obtained by executing its statements on these HCs. Whenever they are operating on fields that are hidden in abstracted parts of the heap, these fields are exposed beforehand. This operation is referred to as *concretisation* and is similar to unrolling of SL predicates. In the following we see that abstraction and concretisation rules are directly given by the HRG.

3.1   Abstraction and Concretisation Using Graph Grammars

For HRGs the forward application of a production rule yields a more concrete heap configuration as certain abstract parts (viz., the NT that is replaced) are concretised. In case more than one rule is applicable, concretisation yields several heaps. For an example concretisation see Fig. 5, where the occurrence of L is either replaced by an empty list (at the bottom, conforming to the first rule), or a list that is extended by one element (at the top, conforming to the second rule). Thus applying concretisation to the heap depicted on the left yields two possible successor heaps, one for each possible concretisation. The resulting transition system representing the program behaviour thus over-approximates the transition system in which all heaps are concrete.

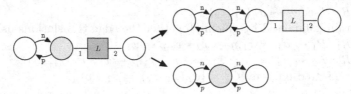

Fig. 5. Concretisation at the shaded node using both L-rules

The reverse application, i.e., the replacement of an embedding of a rule's right-hand side by its left-hand side, yields an abstraction (denoted by $\xrightarrow{abs}$ and $\xrightarrow{abs}{}^*$ for an abstraction sequence, respectively) of a heap. As (forward) rule application preserves language inclusion, abstraction of a heap configuration yields an over-approximation of the current set of concrete heap configurations.

3.2   Abstraction and Unrolling in Separation Logic

Abstraction operations in SL are typically specified by abstraction rules, which define transformations on SL formulae.

Example 5. For example, consider the environment from Ex. 3 defining a doubly-linked list structure. The abstraction rule $abs_1 := \phi * x_1.n \mapsto x_2 * x_2.p \mapsto x_1 \xrightarrow{abs} \phi * dll(x_1, x_2)$, where ϕ is some SL formula, defines an abstraction of a subformula describing a list of length one to a list of arbitrary length.

For the purpose of automated symbolic execution of programs, abstraction rules are usually either pre-defined or given by the user, as e.g. tools like jStar allow [4]. To ensure soundness, these rules must result from valid implications.

Example 6. Consider again the environment and abstraction rule given in Ex. 3. Rule abs_1 is sound as clearly $\phi * x_1.n \mapsto x_2 * x_2.p \mapsto x_1$ implies $\phi * dll(x_1, x_2)$ whereas $\phi * dll(x_1, x_2) \xRightarrow{abs} \phi * x_1.n \mapsto x_2 * x_2.p \mapsto x_1$ yields an underapproximation and is thus not a sound abstraction rule.

The application of abstraction rules yields heap representations that are partially concrete and partially abstract. For example, the heap containing a doubly-linked list of length two represented by the SL formula $\exists x_1, x_2, x_3 : x_1.n \mapsto x_2 * x_2.n \mapsto x_3 * x_2.p \mapsto x_1 * x_3.p \mapsto x_2$ can be abstracted to $\exists x_1, x_2, x_3 : x_1.n \mapsto x_2 * x_2.p \mapsto x_1 * dll(x_2, x_3)$ using the abstraction rule from above.

To generate the abstract state space of a pointer-manipulating program, symbolic execution is performed. The key idea behind symbolic execution using SL is that whenever fields hidden in an abstracted part are to be accessed during execution, they are exposed first. This can e.g. be achieved by performing frame inference using a theorem prover. Thus whenever in a state ϕ, field s of object x is accessed, the prover is requested to provide all formulae $\phi' * x.s \mapsto y$ entailing ϕ. This step is referred to as *unrolling*. Note that for correctness it is necessary to check whether a complete set of such formulae exists. Execution is then resumed and performed on this set, resulting in an over-approximated state space.

Instead of specifying abstraction rules manually and relying on the theorem prover's frame inference procedure during the unrolling step, we propose to generate abstraction and unrolling rules directly from predicate definitions. Similarly to abstraction defined by backward rule application of HRGs, the basic idea is to construct an abstraction rule for each disjunct of the predicate body stating that this disjunct can be abstracted to a call of the predicate. Unrolling is performed by replacing a predicate call by each disjunct of its definition. For instance, concretisation of the formula $\exists r_1, r_2 : dll(r_1, r_2)$ at r_1 yields the following set of formulae (which is a safe over-approximation of the original formula): $\{\exists r_1, r_2 : r_1.n \mapsto r_2 * r_2.p \mapsto r_1, \exists r_1, r_2, r : r_1.n \mapsto r * r.p \mapsto r_1 * dll(r, r_2)\}$, where the two formulae result from unrolling of $dll(r_1, r_2)$ by the first and second disjunct of its body, respectively. To ensure soundness of the generated rules, predicate definitions have to exhibit similar structural properties as HRGs. We show that these impose no real restrictions and how they can be established.

The main benefit of deriving abstraction and unrolling rules from such predicate definitions is the relief of proving soundness of those rules and relying on frame inference. Besides this, we introduce further properties, easily ensured by automated rule generation, improving the automated symbolic execution approach.

3.3 Structural Properties

Applying production rules in forward as well as in backward direction, it is crucial that composition of these operations yield HCs and an (abstract) state space that is a safe over-approximation of the original one. We guarantee this by formulating four properties for HRGs. HRGs exhibiting these properties are suitable for heap abstraction and are thus called *heap abstraction grammars*

(HAGs). It is known that the HAG properties do not restrict the expressivity of the formalism, as every HRG deriving HCs can automatically be transformed into an HAG [12]. Additionally we define the notion of backward confluence for HRGs, which guarantees unique normal forms under abstraction. Backward confluence is beneficial, but not needed for the correctness of the abstraction approach. While there exists a decision procedure, we are not aware of any completion algorithm for establishing backward-confluence of HRGs. For each property we give an equivalent definition on the logic side and show it is preserved under the translations described in Sect. 2.3.

We start by considering the two structural properties, productivity and typedness, that are not immediately necessary for the approach to work correctly, but will pave the way for the more intriguing and mandatory properties, increasingness and local concretisablily, that will be discussed later.

Productivity. An NT is non-productive if it cannot generate any concrete HC. Consider an HRG with non-productive NTs. Then there exist abstract HCs from which no concrete HC can be derived. Thus they represent program states no concrete program execution can end up in. Instead of detecting and discarding such HCs when they appear, we restrict HRGs to productive NTs in advance.

Definition 7 (Productivity). *The set of productive NTs is the least set such that for each element X there exists a production rule $X \to H$ such that either $H \in HC_\Sigma$ or all elements of $\{Z \in N \mid \exists e \in E_H.lab(e) = Z\}$ are productive. A $G \in HRG_{\Sigma_N}$ is productive if all of its NTs are productive.*

The above property implies the common definition of productivity $L(X^\bullet) \neq \emptyset, \forall X \in N$. Productivity can be achieved by removing non-productive NTs and corresponding production rules of a HRG without changing the language [12].

On the logics side, the corresponding property follows along the lines of this definition. Intuitively, a predicate is productive if one of its disjuncts contains no predicate calls (i.e., describes a concrete heap) or if each called predicate is already known to be productive.

Definition 8 (Productivity). *The set of productive predicates is the least subset of Pred such that for each element there exists a disjunct σ_i in its predicate body where either σ_i is primitive or all predicates occurring in $pred(\sigma_i)$ are productive. An environment is productive if each of its predicates is productive.*

Fig. 6. Untyped DSG extension for singly-linked lists

Typedness. Applying production rules backward and forward in an alternating fashion may yield graphs that do not depict a heap, i.e., are no HCs. To illustrate this, consider again the HRG for doubly-linked lists from Fig. 2. Now assume we want to modify this grammar such that it generates partly singly- and partly doubly-linked lists. Thus we extend the HRG with the two rules given in Fig. 6.

Fig. 7 shows an abstraction-concretisation sequence employing this grammar and leading to a graph which is no HC (as there exists a node with two p-successors). *Typedness* ensures that such a case cannot appear. Intuitively we fix the type, i.e., the set of pointers, for each (X, j)-tuple, $X \in N, j \in [1, rk(X)]$ requiring that for each production rule $X \to H$ exactly this set of pointers is derivable at external node $ext_H(j)$.

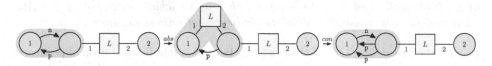

Fig. 7. Abstraction and concretisation yielding an invalid heap configuration

Definition 9 (Typedness). *An NT $X \in N$ with $rk(X) = n$ is typed if, for all $j \in [1, n]$, there exists a set $type(X, j) \subseteq \Sigma$ such that, for every $H \in L(X^\bullet)$ with $V_{X^\bullet} = \{v_1, \ldots, v_n\}$, $type(X, j) = \{lab(e) \mid \exists e \in E_H.att(e)(1) = v_j\}$. An HRG G is typed if each of its NTs is typed.*

Again the counterpart of this property on the logics side is nearly a one-to-one translation of typedness for HRGs: The concrete graphs derivable from an NT X (i.e. $H \in L(X^\bullet)$) translate to heap/interpretation pairs satisfying the predicate body. Note that for a reasonable definition of typedness of SLF formulae it is necessary to assume that the interpretation of parameters $x_1, \ldots, x_n$ is pairwise distinct. This enables us to isolate the selectors derived at $x_j, j \in [1, n]$ (for HRGs this is implicitly given as external nodes are pairwise distinct).

Definition 10 (Typedness). *A predicate definition $\sigma(x_1, \ldots, x_n) := \sigma_1 \vee \ldots \vee \sigma_m \in \Gamma$ is typed if for each x_j ($j \in [1, n]$) there exists a set $type(\sigma, j) \subseteq \Sigma$ such that, for all $h \in He$ and $i \in Int$ with $i(x_j) \neq i(x_k)$ ($j \neq k$) and $h, i, \eta_\Gamma \models \sigma(x_1, \ldots, x_n)$, $type(\sigma, j) = \{s \in \Sigma \mid i(x_j) + cn(s) \in dom(h)\}$. An environment is typed if each of its definitions is.*

Increasingness. Consider a production rule that specifies a replacement of a nonterminal edge with itself. When repeatedly applying this rule in a backward fashion, the abstraction (and thus the heap analysis) does not terminate. Again this behaviour is ruled out beforehand by a syntactic restriction on HRGs, called increasingness. It requires the right-hand side of each production rule to either contain a terminal edge or to be "bigger" (in terms of nodes and edges) than the handle of the left-hand side NT.

Definition 11 (Increasingness). *An NT $X \in N$ is increasing if for all $H \in G(X)$ it holds that either H contains at least one terminal edge, i.e., $lab(E_H) \cap \Sigma \neq \emptyset$, or $|V_H| + |E_H| > rk(X) + 1$. An HRG is increasing if all $X \in N$ are increasing.*

Increasing HRGs guarantee termination of abstraction, as applying rules in backward direction strictly reduces the size of the heap representation.

The analogon of the HRG property for SLF is given below. Here terminal edges correspond to pointer assertions, and the number of nodes and edges in the graph correlates with the number of occurrences of predicate calls and variables.

Definition 12 (Increasingness). *A predicate definition* $\sigma(x_1, \ldots, x_n) := \sigma_1 \vee \ldots \vee \sigma_m$ *is* increasing *if for all* σ_j $(j \in [1, m])$ *it holds that* σ_i *contains a pointer assertion or* $\#pred(\sigma_j) + \#Var(\sigma_j) > n + 1$, *where* $\#pred(\sigma_j)$ *is the number of predicate calls and* $\#Var(\sigma_j)$ *the number of variables occurring in* σ_j. *An environment is* increasing *if all of its predicate definitions are increasing.*

Local Concretisability. One of the key ideas behind symbolic execution of pointer-manipulating programs is that each program statement affects only a local portion rather than the whole heap. Thus if a program statement operates on an abstract heap part, we can perform local concretisation such that the statement can be executed in the concrete part of the heap as if the complete heap was concrete. This necessitates that abstraction can always and everywhere be resolved by concretisation and that this process terminates, i.e., requires only finitely many rule applications. This is generally not ensured by HRGs, as illustrated by the following example.

Consider the HRG G for doubly-linked lists given in Fig. 2 and the HC in Fig. 5 (left). Assume a program that traverses the doubly-linked list from its tail (the rightmost node of the list). Then the p-pointer of this element is abstracted or, put differently, "hidden" in the NT edge L. As seen before, concretisation yields two heap configurations, cf. Fig. 5 (right) where the upper one is locally concrete at the beginning of the list instead of the end. Successive concretisations do not resolve this. Ignoring the second rule yields a terminating concretisation but is unsound as it may under-approximate the state space of the program.

Therefore we formulate a property which guarantees that for each pair (X, j) $(X \in N, j \in [1, rk(X)])$ there exists a subgrammar of HRG G concretising an X-labelled edge at the j-th attached node while preserving the language of G.

Definition 13 (Local Concretisability). *A typed* $G \in HRG_{\Sigma_N}$ *is* locally concretisable *if for all* $X \in N$ *there exist grammars* $G_{(X,1)}, \cdots, G_{(X,rk(X))}$ *with* $dom(G_{(X,j)}) = \{X\}$ *and* $G_{(X,j)} \subseteq G(X), j \in [1, rk(X)]$ *such that* $\forall j$:

1. $L_{G_{(X,j)} \cup (G \backslash G(X))}(X^{\bullet}) = L_G(X^{\bullet})$, *and*
2. $\forall a \in type(X, j), R \in G_{(X,j)}(X) : \exists e \in E_R.lab(e) = a \wedge att_R(e)(1) = ext_R(j)$.

The HRG G from Fig. 2 is locally concretisable at $(X, 1)$ (pick $G_{(X,1)} = G$) but not locally concretisable at $(X, 2)$. Extending G by the rule given

$$L \to \boxed{1} \; {-}\!{-}_{1} \; \boxed{L}_{2} \; \bigcirc \overset{n}{\underset{p}{\longrightarrow}} \bigcirc\, 2$$

Fig. 8. Ensuring local concretisability

in Fig. 8 establishes local concretisability. Thus now we can pick a subgrammar for concretisation depending on where to concretise. That is, we employ all rules in $G_{(X,j)}$ for concretisation at the j-th attached node of an X-labelled edge.

Again there exists a straightforward translation of this property to SLF.

Definition 14 (Local Concretisability). *Let $\Gamma \in Env$ be typed. A predicate definition $\sigma(x_1, \ldots, x_n) := \sigma_1 \vee \ldots \vee \sigma_m \in \Gamma$ is* locally concretisable *if, for all $j \in [1, n]$, there exists a $\sigma'(x_1, \ldots, x_n) := \bigvee \Phi_j$ ($\Phi_j \subseteq \{\sigma_1, \ldots, \sigma_m\}$) such that*

1. *$\forall h \in He, i \in Int$: $h, i, \eta_\Gamma \models \sigma(x_1, ..., x_n) \Longleftrightarrow h, i, \eta_{\Gamma'} \models \sigma'(x_1, ..., x_n)$ where Γ' results from Γ by renaming σ to σ' and*
2. *$\forall \phi \in \Phi_j, s \in type(\sigma, j)$: $x_j.s \mapsto y \in Atom(\phi)$ for some $y \in Var \cup \{\textbf{null}\}$.*

An environment is locally concretisable *if all of its predicate definitions are.*

For an environment satisfying this property we can directly extract the rules for unrolling/concretising a predicate call in a way that guarantees an over-approximation of the program's state space. For instance, consider the environment Γ resulting from the locally concretisable HRG G for doubly-linked lists. As described in Ex. 4, the right-hand sides of the first two rules of G are translated to $\sigma_1 := x_1.n \mapsto x_2 * x_2.p \mapsto x_1$ and $\sigma_2 := \exists r_1.x_1.n \mapsto r_1 * r_1.p \mapsto x_1 * \sigma_L(r_1, x_2)$, while the right-hand side of Fig. 8 translates to $\sigma_3 := \exists r_1.\sigma_L(x_1, r_1) * r_1.n \mapsto x_2 * x_2.p \mapsto r_1$, which yields $\Gamma := \{\sigma_L(x_1, x_2) := \sigma_1 \vee \sigma_2 \vee \sigma_3\}$. Then we can unroll $\sigma_L(x_1, x_2)$ at location x_1 by replacing with both σ_1 and σ_2 as described earlier in this section, while unrolling at x_2 employs the disjuncts σ_2 and σ_3.

Backward Confluence. Being a context-free mechanism, the hyperedge replacement relation is clearly confluent, i.e., the order in which NT edges of a HG are replaced does not matter. However, confluence of heap abstraction is also a desired property as it implies that – under the assumptions that the HRG is increasing and describes the data structures arising in the program under consideration (except from finitely many local deviations) – the resulting abstract state space is finite. Therefore we introduce *backward confluence*: if the order in which abstractions are applied does not matter, i.e., they all finally yield the same abstract HC, these abstractions are referred to as backward confluent. In other words, a backward confluent HRG yields unique abstraction normal forms.

Definition 15 (Backward Confluence). *$G \in HRG_{\Sigma_N}$ is backward confluent if for each $H \in HC_{\Sigma_N}$ and abstractions $H \stackrel{abs}{\Longrightarrow} H_1$ and $H \stackrel{abs}{\Longrightarrow} H_2$ there exists $K \in HC_{\Sigma_N}$ such that $H_1 \stackrel{abs}{\Longrightarrow}^* K$ and $H_2 \stackrel{abs}{\Longrightarrow}^* K$.*

To our knowledge there exists no algorithm that transforms a HRG into an equivalent backward confluent HRG. However, it is decidable whether a HRG is backward confluent ([15], as a consequence of [17]). Notice that we actually consider local confluence. However, for increasing HRGs abstraction is terminating, and thus Newman's Lemma [16] also implies global confluence.

The SLF property is again akin to the HRG's and ensures for abstractions using environments that there always exists a canonical abstraction. To properly reflect the HRG property where we consider abstractions of HCs, i.e., HGs without external nodes, here we consider abstractions of variable-free *SLF*-formulae.

Definition 16 (Confluence under Abstraction). *$\Gamma \in Env$ is confluent under abstraction if, for all closed $\phi \in SLF$ and abstractions $\phi \stackrel{abs}{\Longrightarrow}_\Gamma \phi_1$, $\phi \stackrel{abs}{\Longrightarrow}_\Gamma \phi_2$, there exist $\phi_1 \stackrel{abs}{\Longrightarrow}^*_\Gamma \phi_1'$ and $\phi_2 \stackrel{abs}{\Longrightarrow}^*_\Gamma \phi_2'$ such that $\phi_1' \equiv \phi_2'$.*

4   Property Preservation under Translation

In the preceding section we have motivated the need for additional properties of HRGs, entailing the notion of HAGs. For each of those we gave a (synonymic) characterisation for environments. An HRG or environment featuring all of them is suitable for symbolic execution of pointer-manipulating programs. The following theorem states they are preserved under translation in both directions.

Theorem 1 (Preservation of Properties under Translation). *For a productive, typed, increasing and locally concretisable environment Γ that is confluent under abstraction, $G := \mathrm{hrg}[\![\Gamma]\!]$ is a backward-confluent HAG.*

For a backward confluent HAG G, the environment $\Gamma := \mathrm{env}[\![G]\!]$ is productive, typed, increasing, locally concretisable and confluent under abstraction.

Proof. (sketch) We prove the preservation of each of the five properties separately. As the arguments rely on specifics of the translation between HRGs and SL, we can only give an intuition here; details are given in [11].

Productivity. The preservation of productivity is shown by induction over the number of productive nonterminals or predicate definitions, respectively. The translation of hypergraphs to SL formulae generates exactly one pointer assertion for each terminal and one predicate call for each nonterminal edge, and vice versa. Thus if the hypergraph only contains edges labelled with terminals and nonterminals that are already known to be productive, the resulting formula is productive, too. Furthermore as each rule graph is translated into a disjunct of a predicate definition and each nonterminal into a predicate, the same holds for the generated environment. Similar arguments hold for the only-if-part.

Increasingness. Increasingness follows from two facts: each edge is translated into a pointer assertion or predicate call, and each variable in an SL formula refers to a different location unless stated otherwise. During the translation each node in the HG is assigned a unique identifier used in the translated formula, whereas for the translation from formula to graph two heap locations are merged into the same node only if explicitly stated by a pointer assertion. Thus the number of separating conjuncts/edges and heap locations/nodes is equal.

Typedness. Typedness is proven by contradiction. We assume an untyped nonterminal and typed predicate definition. Then for two HGs where one exhibits pointer s at a fixed external node ext while the other one does not (i.e. representants of the source of untypedness), we show that there exist heap-interpretation models where one interpretation maps the predicate parameter that corresponds to ext to a heap location while the other one does not. This argumentation can directly be lifted to graph grammars and environments, respectively, and can be adopted for the other proof direction as well.

Local Concretisability. For showing the preservation of local concretisability we pick an arbitrary locally concretisable nonterminal X (again the only-if-part works analogously). By the definition of the translation we know that it is translated into a predicate σ_X, where each if its disjuncts directly results from the translation of a rule graph of X. Thus we show that each subgrammar $G_{(X,j)}$ $(j \in [1, rk(X_\sigma)])$ is translated into a predicate definition that fulfils the local concretisability property.

Backward Confluence vs. Confluence under Abstraction. This property is proven by contradiction. We assume an environment confluent under abstraction and the HRG resulting from translation as not backward confluent. Then it is shown that two abstraction sequences without unique normal form can be modelled by similar abstraction operations using the corresponding environment. The resulting formulae are not equal and cannot be further abstracted, which contradicts the assumption that the environment is confluent under abstraction.

The complete proof of this theorem can be found in [11]. For HRGs deriving HCs there exist transformations (productivity, typedness, increasingness, local concretisability) or decision procedures (backward confluence) to handle those properties. Thus they have the potential to serve as the basis for designing suitable environments for symbolic execution of pointer-manipulating programs.

5   Conclusion

In this paper, we established a formal connection between two different approaches for analysing heap-manipulating programs, namely HRGs and inductively defined predicates in SL. We proposed several structural properties of SL formulae, viz., productivity, typedness, increasingness, local concretisability, and backward confluence, that enable the automated generation of a sound set of abstraction and unrolling rules for symbolic execution. Exploiting known results on the HRG counterparts of those properties, we showed that they can be established automatically and impose no restriction on the expressiveness of the SL fragment under consideration. More concretely, we first generalised a previous translation result [5] to HRGs and SL. Subsequently we proved that the properties are preserved by the translation in both directions. These theoretical results pave the way for an integrated approach to heap analysis that combines the intuitive description of data structures using HRGs on the one side and the expressivity and tool support of SL on the other side.

Current research concentrates on algorithms for automated HRG learning on-the-fly. That is, abstraction rules are automatically derived whenever the size of the heap representation reaches a certain threshold [13]. Transferring these learning procedures to SL either directly or indirectly via translation of HRGs would be worthwhile to investigate. Furthermore we are planning to extend the HRG approach to concurrent programs. In particular, we would like to study whether a permission-based SL approach, such as one of those presented in [3,9], could provide the basis for this extension.

References

1. Berdine, J., Calcagno, C., O'Hearn, P.W.: Symbolic execution with separation logic. In: Yi, K. (ed.) APLAS 2005. LNCS, vol. 3780, pp. 52–68. Springer, Heidelberg (2005)
2. Berdine, J., Calcagno, C., O'Hearn, P.W.: Smallfoot: Modular automatic assertion checking with separation logic. In: de Boer, F.S., Bonsangue, M.M., Graf, S., de Roever, W.-P. (eds.) FMCO 2005. LNCS, vol. 4111, pp. 115–137. Springer, Heidelberg (2006)
3. Bornat, R., Calcagno, C., O'Hearn, P., Parkinson, M.: Permission accounting in separation logic. In: POPL, pp. 259–270. ACM (2005)
4. Distefano, D., Parkinson, M.J.: jStar: Towards practical verification for Java. ACM Sigplan Notices 43(10), 213–226 (2008)
5. Dodds, M.: Graph Transformation and Pointer Structures. PhD thesis, The University of York, Department of Computer Science (2008)
6. Dodds, M., Plump, D.: From hyperedge replacement to separation logic and back. ECEASST 16 (2008)
7. Dudka, K., Peringer, P., Vojnar, T.: Predator: A practical tool for checking manipulation of dynamic data structures using logic. In: Gopalakrishnan, G., Qadeer, S. (eds.) CAV 2011. LNCS, vol. 6806, pp. 372–378. Springer, Heidelberg (2011)
8. Fradet, P., Caugne, R., Métayer, D.L.: Static detection of pointer errors: An axiomatisation and a checking algorithm. In: Riis Nielson, H. (ed.) ESOP 1996. LNCS, vol. 1058, pp. 125–140. Springer, Heidelberg (1996)
9. Haack, C., Hurlin, C.: Separation logic contracts for a Java-like language with fork/join. In: Meseguer, J., Roşu, G. (eds.) AMAST 2008. LNCS, vol. 5140, pp. 199–215. Springer, Heidelberg (2008)
10. Heinen, J., Noll, T., Rieger, S.: Juggrnaut: Graph grammar abstraction for unbounded heap structures. In: Proc. 3rd Int. Workshop on Harnessing Theories for Tool Support in Software. ENTCS, vol. 266, pp. 93–107. Elsevier (2010)
11. Jansen, C., Göbe, F., Noll, T.: Generating inductive predicates for symbolic execution of pointer-manipulating programs. Technical Report AIB 2014-08, RWTH Aachen University, Germany (May 2014),
   http://sunsite.informatik.rwth-aachen.de/Publications/AIB/2014/2014-08.pdf
12. Jansen, C., Heinen, J., Katoen, J.-P., Noll, T.: A local Greibach normal form for hyperedge replacement grammars. In: Dediu, A.-H., Inenaga, S., Martín-Vide, C. (eds.) LATA 2011. LNCS, vol. 6638, pp. 323–335. Springer, Heidelberg (2011)
13. Jeltsch, E., Kreowski, H.-J.: Grammatical inference based on hyperedge replacement. In: Ehrig, H., Kreowski, H.-J., Rozenberg, G. (eds.) Graph Grammars 1990. LNCS, vol. 532, pp. 461–474. Springer, Heidelberg (1990)
14. Lee, O., Yang, H., Yi, K.: Automatic verification of pointer programs using grammar-based shape analysis. In: Sagiv, M. (ed.) ESOP 2005. LNCS, vol. 3444, pp. 124–140. Springer, Heidelberg (2005)
15. Nellen, J.: Konfluenzanalyse und Vervollstndigung von Graphersetzungssystemen. Master's thesis, RWTH Aachen University (2010)
16. Newman, M.: On theories with a combinatorial definition of "equivalence". Annals of Mathematics 43(2), 223–243 (1942)
17. Plump, D.: Checking graph-transformation systems for confluence. ECEASST 26 (2010)
18. Reynolds, J.C.: Separation logic: A logic for shared mutable data structures. In: Proc. 17th IEEE Symp. on Logic in Computer Science, pp. 55–74. IEEE (2002)

Attribute Handling for Generating
Preconditions from Graph Constraints

Frederik Deckwerth[*] and Gergely Varró[**]

Technische Universität Darmstadt,
Real-Time Systems Lab,
64283 Merckstraße 25, Darmstadt, Germany
{frederik.deckwerth,gergely.varro}@es.tu-darmstadt.de

Abstract. This paper presents a practical attribute handling approach
for generating rule preconditions from graph constraints. The proposed
technique and the corresponding correctness proof are based on symbolic
graphs, which extend the traditional graph-based structural descriptions
by logic formulas used for attribute handling. Additionally, fully declarative rule preconditions are derived from symbolic graphs, which enable
automated attribute resolution as an integral part of the overall pattern
matching process, which carries out the checking of rule preconditions at
runtime in unidirectional model transformations.

Keywords: static analysis, rule preconditions, attribute handling.

1 Introduction

Graph transformation (GT) [1] as a declarative technique to specify rule-based
manipulation of system models has been successfully employed in many practical,
real-world application scenarios [2] including ones from the security domain [3],
where the formal nature of graph transformation plays an important role.

A recurring important and challenging task is to statically ensure that (global)
negative constraints representing forbidden structures are never allowed to occur
in any system models that are derived by applying graph transformation rules.

A well-known general solution to this challenge was described as a sophisti-
cated constructive algorithm [4], which generates negative application conditions
(NAC) [5] from the negative constraints, and attaches these new NACs to the
left-hand side (LHS) of the graph transformation rules at design time. At runtime, these NAC-enriched left-hand sides block exactly those rule applications
that would lead to a constraint violating model.

This constructive algorithm is perfectly appropriate from a theoretical aspect
for proving the correctness of the approach when system models are graphs without numeric or textual attributes, and negative constraints and graph transformation rules specify only structural restrictions and manipulations, respectively,
but in practical scenarios the handling of attributes cannot be ignored at all.

---

[*] Supported by CASED (www.cased.de).
[**] Supported by the DFG funded CRC 1053 MAKI.

H. Giese and B. König (Eds.): ICGT 2014, LNCS 8571, pp. 81–96, 2014.
© Springer International Publishing Switzerland 2014

A state-of-the-art approach [6] has been recently presented for transforming arbitrary OCL invariants and rule postconditions into preconditions, which implicitly involves the handling of attributes as well. On one hand, the corresponding report lacks formal arguments underpinning the correctness of the suggested algorithm. On the other hand, the proposed transformation manipulates the abstract syntax tree of OCL expressions, consequently, this solution might be negatively affected by the same (performance) issues like any other OCL-based techniques when checking rule preconditions *at runtime*. The main point is that an OCL expression is always evaluated (i) from a single and fix starting point defined explicitly by its context, and (ii) in an imperative manner following exactly the traversal order specified by the user, which is not necessarily suboptimal, but requires algorithmic background from the modeller.

In this paper, we present a practical and provenly correct attribute handling approach for generating preconditions from graph constraints. The proposed technique and the corresponding correctness proof use symbolic graphs [7], which combine graph-based structural descriptions with logic formulas expressing attribute values and restrictions. Additionally, the concept of fully declarative pattern specifications [8,9] is reused in a novel context, namely, as an intermediate language, to which the generated symbolic graph preconditions are converted. Finally, an attribute evaluation order is automatically derived from these declarative pattern specifications together with a search plan for the graph constraints resulting in a new, integrated pattern matching process, which performs the checking of rule preconditions in unidirectional model transformations.

The remainder of the paper is structured as follows: Section 2 introduces basic logic, modeling and graph transformation concepts. The precondition NAC derivation process and the corresponding correctness proof are presented in Sec. 3, while Sec. 4 describes the automated attribute resolution technique. Related work is discussed in Sec. 5, and Sec. 6 concludes our paper.

2   Basic Concepts

2.1   Formal Concepts

Signature and Σ-algebra. A *signature* Σ consists of sort and attribute value predicate symbols, and associates a sort symbol with each argument of each attribute value predicate symbol. A *Σ-algebra* $\mathcal{D}$ defines the symbols in Σ by assigning (i) a carrier set to each sort symbol, and (ii) a relation to each attribute value predicate symbol. The relation is defined on the carrier sets and has to be compatible with respect to the number and sorts of the attribute value predicate arguments. In this paper, we use a signature and a corresponding Σ-algebra that consists of a single sort *Real* that represents the real numbers $\mathbb{R}$ as well as the attribute value predicates symbols *eq*, *gr*, *mult* and *add*. Symbol *eq* is defined by the equality relation on $\mathbb{R}$, symbol *gr* by $gr(x, y) = \{ x, y \in \mathbb{R} \mid x > y \}$, and symbols *mult* and *add* by $mult(x, y, z) = \{ x, y, z \in \mathbb{R} \mid x = y \cdot z \}$ and $add(x, y, z) = \{ x, y, z \in \mathbb{R} \mid x = y + z \}$, respectively.

First-Order Logic Formula. Given a signature Σ and a set of variables X, a *first-order logic formula* is built from the variables in X, the (attribute value) predicate symbols in Σ, the logic operators $\wedge,\vee,\neg,\Rightarrow,\Leftrightarrow$, the constants $\top,\bot$ (meaning true and false) and the quantifiers $\forall$ and $\exists$ in the usual way [10].

Assignment and Evaluation of First-Order Logic Formulas. A *variable assignment* $\sigma : X \to \mathcal{D}$ maps the variables $x \in X$ to a value in the corresponding carrier set of $\mathcal{D}$. A first order logic formula Ψ is evaluated for a given assignment σ in a Σ-algebra $\mathcal{D}$ by first replacing all variables in Ψ according to the assignment σ and evaluating the attribute value predicates according to the algebra and the logic operators in the usual way [10]. We write $\mathcal{D}, \sigma \models \Psi$ iff Ψ evaluates to *true* for the assignment σ; and $\mathcal{D} \models \Psi$, iff Ψ evaluates to *true* for all assignments.

E*-graphs and E*-graph Morphisms. An *E*-graph*[1] is a tuple $G = (V_G, E_G, V_G^{\mathrm{L}}, E_G^{\mathrm{L}}, s_G, t_G, s_G^{\mathrm{L}}, t_G^{\mathrm{L}})$ consisting of a set of graph nodes V_G, graph edges E_G, label nodes V_G^{L}, label edges E_G^{L}, and four functions $s_G, t_G, s_G^{\mathrm{L}}, t_G^{\mathrm{L}}$. The functions $s_G : E_G \to V_G$ and $t_G : E_G \to V_G$ assign source and target graph nodes to the graph edges. The functions $s_G^{\mathrm{L}} : E_G^{\mathrm{L}} \to V_G$ and $t_G^{\mathrm{L}} : E_G^{\mathrm{L}} \to V_G^{\mathrm{L}}$ map the label edges to the (source) graph nodes and (target) label nodes, respectively.

An *E*-graph morphism* $h : G \to H$ *from E*-graph* G *to an E*-graph* H is a tuple of total functions $\langle h_V : V_G \to V_H, h_E : E_G \to E_H, h_{V_{\mathrm{L}}} : V_G^{\mathrm{L}} \to V_H^{\mathrm{L}}, h_{E_{\mathrm{L}}} : E_G^{\mathrm{L}} \to E_H^{\mathrm{L}} \rangle$ such that h commutes with source and target functions, i.e., $h_V \circ s_G = s_H \circ h_E$, $h_V \circ t_G = t_H \circ h_E$, $h_V \circ s_G^{\mathrm{L}} = s_H^{\mathrm{L}} \circ h_{E_{\mathrm{L}}}$, $h_{V_{\mathrm{L}}} \circ t_G^{\mathrm{L}} = t_H^{\mathrm{L}} \circ h_{E_{\mathrm{L}}}$. E*-graphs together with their morphisms form the category **E*-graphs**.

Symbolic Graphs and Symbolic Graph Morphisms. A *symbolic graph* $G^\psi = \langle G, \psi_G \rangle$, which was introduced in [7], consists of an E*-graph part G and a first-order logic formula ψ_G over the Σ-algebra $\mathcal{D}$ using the label nodes in V_G^{L} as variables and elements of the carrier sets of $\mathcal{D}$ as constants.

A *symbolic graph morphism* $h^\psi : \langle G, \psi_G \rangle \to \langle H, \psi_H \rangle$ from symbolic graph $\langle G, \psi_G \rangle$ to $\langle H, \psi_H \rangle$ is an E*-graph morphism $h : G \to H$ such that $\mathcal{D} \models \psi_H \Rightarrow \overline{h}_\psi(\psi_G)$, where $\overline{h}_\psi(\psi_G)$ is the first-order formula obtained when replacing each variable x in formula ψ_G by $h_{V_{\mathrm{L}}}(x)$. Symbolic graphs over a Σ-algebra $\mathcal{D}$ together with their morphisms form the category **SymbGraphs$_{\mathcal{D}}$**.

Pushouts in SymbGraphs$_{\mathcal{D}}$. (1) is a pushout iff it is a pushout in **E*-graphs** and $\mathcal{D} \models \Psi_3 \Leftrightarrow (\overline{g}_1(\Psi_1) \wedge \overline{g}_2(\Psi_2))$.

$$\begin{array}{ccc} \langle G_0, \Psi_0 \rangle & \xrightarrow{h_1^\Psi} & \langle G_1, \Psi_1 \rangle \\ {\scriptstyle h_2^\Psi}\downarrow & (1) & \downarrow{\scriptstyle g_1^\Psi} \\ \langle G_2, \Psi_2 \rangle & \xrightarrow[g_2^\Psi]{} & \langle G_3, \Psi_3 \rangle \end{array}$$

For presentation purposes we consider symbolic graphs G^ϕ to have a conjunction $\phi = p_1(x_{1,1}, \ldots x_{1,n}) \wedge \ldots \wedge p_m(x_{m,1}, \ldots, x_{m,k})$ of attribute value predicates $p_1, \ldots, p_m$ as logic formula.

2.2   Modeling and Transformation Concepts

In this section, metamodels, models and patterns are defined as symbolic graphs.

Metamodels and Models. A *metamodel* is a symbolic graph $MM^\phi = \langle MM, \bot \rangle$, where MM is an E*-graph. The graph nodes $v \in V_{MM}$ and graph edges $e \in E_{MM}$

---

[1] In contrast to E-Graphs [1], E*-Graphs do not provide labels for graph edges.

define *classes* and *associations* in a domain, respectively. The set V_{MM}^L contains one label node for each sort in the given signature. A label edge $e^L \in E_{MM}^L$ from a class $v \in V_{MM}$ to a label node $v^L \in V_{MM}^L$ expresses that class v has an *attribute* e^L of sort v^L.

A *symbolic graph* G^ϕ *conforms to a metamodel* MM^ϕ if all graph nodes V_G and graph edges E_G can be mapped to the classes and associations in the metamodel, and the label edges E_G^L and nodes V_G^L can be mapped to the attributes of corresponding sorts by a symbolic graph morphism $type^\phi : G^\phi \to MM^\phi$.

A *model* M^ϕ *of a metamodel* MM^ϕ is a symbolic graph $M^\phi = \langle M, \phi_M \rangle$ conforming to metamodel MM^ϕ, which has to fulfill the following properties: (i) A model M^ϕ has a label node $x_{val} \in V_M^L$ for each value val in the carrier sets of $\mathcal{D}$. (ii) For each label node x_{val}, the conjunction ϕ_M includes an equality attribute value predicate $eq(x_{val}, val)$ (i.e., $\phi_M = \bigwedge_{val \in \mathcal{D}} eq(x_{val}, val)$). A *model is valid*[2] if each graph node $v_M \in V_M$ has exactly one label edge $e_M^L \in E_M^L$ for each attribute $e_{MM}^L \in E_{MM}^L$ such that $s(e_{MM}^L) = type_V^\phi(v_M)$, $s(e_M^L) = v_M$ and $type_{E_L}^\phi(e_M^L) = e_{MM}^L$.

Graph nodes, graph edges, label nodes and label edges in a model are called *objects*, *links*, *attribute values* and *attribute slots*, respectively.

A *typed symbolic (graph) morphism* $f^\phi : M_1^\phi \to M_2^\phi$ from model M_1^ϕ to M_2^ϕ, both conform to metamodel MM^ϕ, is a symbolic graph morphism that preserves type information, i.e., $type_1^\phi \circ f^\phi = type_2^\phi$.

Example. Figure 1a shows the e-commerce platform metamodel, which consists of the classes Customer, Order, Article and PaymentMethod. A customer has a set of orders (orders) and registered payment methods (paymentMethods) assigned. An order consists of articles and a payment method represented by the associations articles and usedPaymentMethod, respectively. Attributes and their corresponding sorts are represented using the UML class diagram notation. E.g., the class Customer has an attribute reputation of sort double. Additionally, an order has the totalCost attribute that corresponds to the accumulated price of all articles in the articles association. The attribute limit assigns the maximal amount of money admissible in a single transaction to a payment method.

Patterns, Negative Constraints and Model Consistency. A *pattern* P^ϕ is a symbolic graph $P^\phi = \langle P, \phi_P \rangle$ that conforms to a metamodel MM^ϕ. Additionally a pattern has no duplicate attributes, i.e., each graph node $v_P \in V_P$ has at most one label edge $e_P^L \in E_P^L$ for each attribute $e_{MM}^L \in E_{MM}^L$ such that $s(e_{MM}^L) = type_V^\phi(v_P)$, $s(e_P^L) = v_P$ and $type_{E_L}^\phi(e_P^L) = e_{MM}^L$.

A *pattern* P^ϕ *matches a model* M^ϕ if there exists a typed symbolic morphism $m^\phi : \langle P, \phi_P \rangle \to \langle M, \phi_M \rangle$ such that functions $m_V : V_P \to V_M, m_E : E_P \to E_M$ and $m_{E_L} : E_P^L \to E_M^L$ are injective and $\mathcal{D} \models \phi_M \Rightarrow \overline{m}(\phi_P)$. The morphism m^ϕ is called *match*. All such morphisms, denoted as $\mathcal{M}'_\phi$, are called match morphisms.

---

[2] Note that this requirement is only necessary to align the concept of models including attributes with the behaviour of our Eclipse Modeling Framework (EMF) based implementation (Sec. 4). The results of Sec. 3 are not affected by this assumption.

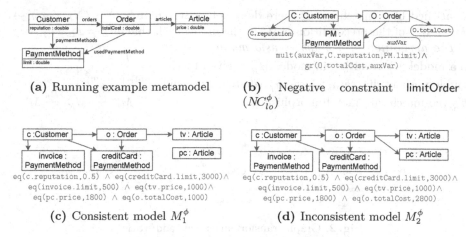

(a) Running example metamodel

(b) Negative constraint limitOrder (NC_{lo}^{ϕ})

(c) Consistent model M_1^{ϕ}

(d) Inconsistent model M_2^{ϕ}

Fig. 1. The e-commerce scenario

A *negative constraint* NC^{ϕ} is a pattern to declaratively define forbidden subgraphs in a model. A model M^{ϕ} is *consistent* with respect to a negative constraint $NC^{\phi} = \langle NC, \phi_{NC} \rangle$, if there does not exist a match $m^{\phi} : NC^{\phi} \to M^{\phi}$.

Example. Figure 1b shows a global negative constraint limitOrder (NC_{lo}^{ϕ}) that prohibits a customer (C) to have an order (O) whose totalCost exceeds the product of its reputation and the limit of the used payment method PM. Figures 1c and 1d show the models M_1^{ϕ} and M_2^{ϕ}, respectively, where label nodes are not explicitly drawn. The model M_1^{ϕ} is consistent w.r.t. constraint limitOrder. Model M_2^{ϕ} is inconsistent w.r.t. constraint limitOrder, since the cost (`o.totalCost`) of the order are greater than the product of the payment method limit (`credit-Card.limit`) and the customer reputation (`c.reputation`). More specifically, we can find a match $m : NC_{lo} \to M_2$ for the graph part of the constraint NC_{lo} in the model M_2 such that $\mathcal{D} \models \phi_{M_2} \Rightarrow \overline{m}(\phi_{NC_{lo}})$ holds for the formula $\phi_{NC_{lo}}$ of the constraint NC_{lo} after label replacement $\overline{m}(\phi_{NC_{lo}})$.

Symbolic Graph Transformation. A *graph transformation rule* $\mathbf{r}^{\phi} = \langle L \xleftarrow{l} K \xrightarrow{r} R, \phi \rangle$ consists of a left hand side (LHS) pattern $\langle L, \phi \rangle$, a gluing pattern $\langle K, \phi \rangle$ and a right hand side (RHS) pattern $\langle R, \phi \rangle$ that share the same logic formula ϕ. Morphisms l^{ϕ}, r^{ϕ} are typed symbolic morphisms that are (i) injective for graph nodes and all kinds of edges, (ii) bijective for label nodes, and (iii) $\mathcal{D} \models \phi \Leftrightarrow \overline{l}(\phi) \Leftrightarrow \overline{r}(\phi)$. These morphisms are denoted by $\mathcal{M}_{\phi}$.

The LHS and RHS of graph transformation rule $\mathbf{r}^{\phi}$ can be augmented with *negative application conditions* (NACs) $n_L^{\phi} : L^{\phi} \to N_L^{\phi} \in NAC_L$ (precondition NAC) and $n_R^{\phi} : R^{\phi} \to N_R^{\phi} \in NAC_R$ (postcondition NAC), where n_L^{ϕ} and n_R^{ϕ} are match morphisms.

A rule $\mathbf{r}^{\phi} = \langle L \xleftarrow{l} K \xrightarrow{r} R, \phi \rangle$ with negative precondition NACs NAC_L is *applicable* to a model M^{ϕ} iff (i) there exists a match $m^{\phi} : \langle L, \phi \rangle \to \langle M, \phi_M \rangle$ of the LHS $\langle L, \phi \rangle$ in $\langle M, \phi_M \rangle$, and (ii) the precondition NACs in NAC_L are satisfied by the current match m^{ϕ}. A precondition NAC $n_L^{\phi} : L^{\phi} \to N_L^{\phi} \in NAC_L$

is satisfied by a match m^ϕ if there does not exist a match $x_L^\phi : \langle N_L, \phi_{N_L} \rangle \rightarrow \langle M_L, \phi_{M_L} \rangle$ of the precondition NAC in the model such that $m_L = x_L \circ n_L$.

The application of a graph transformation rule $\mathbf{r}^\phi$ to a model M_L^ϕ resulting in model M_R^ϕ is given by the double pushout diagram, where m_L^ϕ (match), m_K^ϕ and m_R^ϕ (co-match) are match morphisms.

$$
\begin{array}{ccccc}
L^\phi & \xleftarrow{l^\phi} & K^\phi & \xrightarrow{r^\phi} & R^\phi \\
m_L^\phi \downarrow & & m_K^\phi \downarrow & & m_R^\phi \downarrow \\
M_L^\phi & \xleftarrow{l_{M_L}^\phi} & M_K^\phi & \xrightarrow{r_{M_R}^\phi} & M_R^\phi
\end{array}
$$

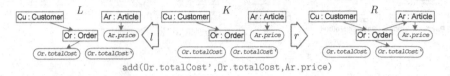

add(Or.totalCost',Or.totalCost,Ar.price)

Fig. 2. Graph transformation rule addArticle

Although it seems counterintuitive at a first glance that we require L^ϕ, K^ϕ and R^ϕ to share the same conjunction and label nodes, it does not mean that attribute values cannot be changed by a rule application, since attribute values can be modified by redirecting label edges.

To preserve model validity by a graph transformation rule application we introduce conditions that ensure that rules do not transform valid models into invalid ones.

Model Validity Preserving Graph Transformation Rules. A graph transformation rule $\mathbf{r}^\phi = \langle L \xleftarrow{l} K \xrightarrow{r} R, \phi \rangle$ typed over metamodel MM^ϕ is *model validity preserving* if: (i) *For each created object all attribute values are initialized.* Formally, for each created graph node $v \in V_R \backslash r_V(V_K)$ there exists exactly one label edge $e^{\mathbb{L}} \in E_R^{\mathbb{L}}$ for each corresponding attribute $e_{MM}^{\mathbb{L}} \in MM^\phi$: $s_{MM}^{\mathbb{L}}(e_{MM}^{\mathbb{L}}) = type_G(v)$ s.t. $type_{\mathbb{L}}(e^{\mathbb{L}}) = e_{MM}^{\mathbb{L}}$ assigning a value to the attribute, i.e., $s_R^{\mathbb{L}}(e^{\mathbb{L}}) = v$. (ii) *For preserved objects, rules can only change attribute values by redirecting label edges.* Formally, for each label edge $e_1^{\mathbb{L}} \in E_L^{\mathbb{L}}$ in the LHS pattern whose source graph node is preserved by the rule application (i.e. $\exists v \in V_K$ s.t. $s_L^{\mathbb{L}}(e_1^{\mathbb{L}}) = l_V(v)$), there exists exactly one label edge $e_2^{\mathbb{L}} \in E_R^{\mathbb{L}}$ of the same type (i.e., $type_{\mathbb{L}}(e_1^{\mathbb{L}}) = type_{\mathbb{L}}(e_2^{\mathbb{L}})$) in the RHS pattern such that $s_R^{\mathbb{L}}(e_2^{\mathbb{L}}) = r_V(v)$. Similarly, for each label edge in the RHS pattern with preserved source graph node, there exists exactly one label edge with similar source and same type in the LHS pattern. Note that for object deletion model validity is preserved by the dangling edge condition for the double pushout approach [1].

Example. Figure 2 shows the rule addArticle that adds an article Ar to the order Or of a customer Cu, and calculates the new total cost Or.totalCost' of order Or by adding the price Ar.price of the added article Ar to the actual total cost Or.totalCost of the order Or. The total cost value is updated by redirecting the label edge from the actual value Or.totalCost to the new value Or.totalCost'. Morphisms are implicitly specified in all the figures of the running example by matching node identifier. The result of applying the rule to user u, order o, and article pc in the model of Figure 1c is depicted in Figure 1d.

Consistency Guaranteeing Rules. A rule $\mathbf{r}^\phi$ with a set of precondition NACs NAC_L is *consistency guaranteeing* w.r.t a negative constraint NC^ϕ, iff for any arbitrary model M_L^ϕ and all possible applications of rule $\mathbf{r}^\phi$ that result in model M_R^ϕ it holds that M_R^ϕ is consistent w.r.t. the negative constraint NC^ϕ.

3 Constructing Precondition NACs with Attributes

In this section, we extend the results of constructing precondition NACs from negative constraints presented in [1] to symbolic graph transformation. The construction of precondition NACs are carried out by (i) constructing a postcondition NAC from the negative constraint and the RHS pattern of a GT-rule (Sec. 3.1) and (ii) back-propagating the postcondition NAC into an equivalent precondition NAC (Sec. 3.2). In Section 3.3 we show that the construction ensures consistency guarantee.

3.1 Construction of Postcondition NACs from Negative Constraints

For each non-empty subgraph of a negative constraint that is also a subgraph of the RHS pattern of a GT-rule, a postcondition NAC is constructed by gluing the graph parts of the negative constraint and the RHS together along the common subgraph. The logic part is obtained as the conjunction of the formulas of the RHS pattern and the negative constraint, where the label nodes that are glued along the common subgraph are replaced in both formulas with a common label.

Formally, the postcondition NACs $n_R^\phi : \langle R, \phi_R \rangle \to \langle N_R, \phi_{N_R} \rangle \in NAC_R$ for the RHS pattern $\langle R, \phi_R \rangle$ of a rule and a negative constraint $\langle NC, \phi_{NC} \rangle$ is derived as the gluings $\langle R, \phi_R \rangle \overset{n_R^\phi}{\to} \langle N_R, \phi_{N_R} \rangle \overset{q^\phi}{\leftarrow} \langle NC, \phi_{NC} \rangle$ such that the pair of match morphisms (n_R^ϕ, q^ϕ) is jointly epimorphic and $\mathcal{D} \models \phi_{N_R} \Leftrightarrow (\overline{n}_R(\phi_R) \wedge \overline{q}(\phi_{NC}))$.

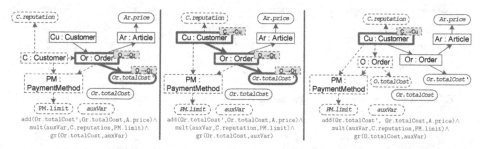

Fig. 3. Postcondition NACs derived for rule addArticle and neg. constr. limitOrder

Example. Figure 3 depicts all postcondition NACs derived from the rule addArticle (Fig. 2) and the negative constraint limitOrder (Fig. 1b). Solid nodes and edges belong to the RHS of rule addArticle. Dashed elements are from the negative constraint limitOrder, and the common subgraph is drawn bold.

The mapping of the RHS pattern of rule addArticle and the constraint limitOrder are implicitly denoted by the mapping of the node identifiers. For the common subgraph, we used the labels from the RHS of rule addArticle and denoted the mapping from the constraint by the grey boxes. E.g., ⌐O→Or⌐ denotes that node O of the constraint is mapped to node Or in the postcondition NAC.

3.2   Constructing Precondition NAC from Postcondition NAC

Each postcondition NAC constructed in the previous step is back-propagated to the LHS as a precondition NAC by reverting the modifications of the graph part specified by the symbolic GT-rule while preserving the logic formula.

Formally, for a GT-rule $\mathbf{r}^\phi = \langle L \xleftarrow{l} K \xrightarrow{r} R, \phi \rangle$ with a postcondition NAC $n_R^\phi : \langle R, \phi_R \rangle \to \langle N_R, \phi_{N_R} \rangle$, the precondition NAC $n_L^\phi : \langle L, \phi_L \rangle \to \langle N_L, \phi_{N_L} \rangle$ is derived as follows: (i) Construct $n_K : K \to N_K$ by the pushout complement of the pair (r, n_R) in **E*-graphs**. (ii) If (r, n_R) has a pushout complement then $n_L : L \to N_L$ is constructed by the pushout of l and n_K in **E*-graphs**. (iii) The precondition NAC is then defined by $n_L^\phi : \langle L, \phi_L \rangle \to \langle N_L, \phi_{N_L} \rangle$ where ϕ_{N_L} is the same formula as ϕ_{N_R}.

$$L^\phi \xleftarrow{l^\phi} K^\phi \xrightarrow{r^\phi} R^\phi$$
$$\left. n_L^\phi \right\downarrow \quad \left. n_K^\phi \right\downarrow \quad \left\downarrow n_R^\phi \right.$$
$$N_L^\phi \xleftarrow{l_{N_L}^\phi} N_K^\phi \xrightarrow{r_{N_R}^\phi} N_R^\phi$$

Note that the label nodes and the logic formula remains invariant after symbolic transformation [11] (i.e., $V_{N_L}^{\mathbb{L}} = V_{N_K}^{\mathbb{L}} = V_{N_R}^{\mathbb{L}}$ and $\mathcal{D} \models \phi_{N_L} \Leftrightarrow \phi_{N_K} \Leftrightarrow \phi_{N_R}$).
Example. Figure 4 shows the construction of the precondition NAC from the postcondition NAC depicted in the middle of Fig. 3. The precondition NAC prevents the rule addArticle to add an article to an order if the new total cost Or.totalCost' exceeds the product of the used payment method limit PM.limit and the reputation C.reputation of the customer. Note that label node identifier can be chosen arbitrarily, hence label node C.reputation refers to the reputation attribute of customer Cu.

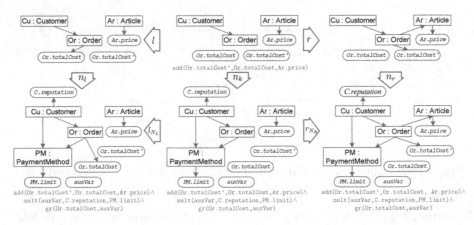

Fig. 4. Constructing a precondition NAC from a postcondition NAC

3.3 Proving the Correctness of the Construction Technique

In order to reuse the results from [1] to show that the presented construction is indeed sufficient and necessary to ensure consistency guarantee we have to prove the following properties for symbolic graphs:

1. **SymbGraphs**$_\mathcal{D}$ has a generalized disjoint union (binary coproducts).
2. **SymbGraphs**$_\mathcal{D}$ has a generalized factorization in surjective and injective parts for each symbolic graph morphism (weak $\mathcal{E}_\phi$-$\mathcal{M}'_\phi$ factorization).
3. Match morphisms $\mathcal{M}'_\phi$ are closed under composition and decomposition.
4. $\mathcal{M}'_\phi$ is closed under pushouts (PO) and pullbacks (PB) along $\mathcal{M}_\phi$-morphisms

Note that although we used typed graphs (i.e. graphs conform to a metamodel) in our running example and formalization we only provide proofs for untyped symbolic graphs as the proofs can be easily extended, since symbolic graphs are an adhesive HLR category [7] and consequently typed symbolic graphs form an adhesive HLR category (slice construction [1]).

Property 1 (SymbGraphs$_\mathcal{D}$ has binary coproducts.) *The diagram on the next page is a binary coproduct in **SymbGraphs$_\mathcal{D}$** if and only if it is a binary coproduct in **E*-graphs** and $D \models \phi_{1+2} \Leftrightarrow (\bar{i}_1(\phi_1) \land \bar{i}_2(\phi_2))$.*

Proof. In **E*-graphs** the coproduct is constructed componentwise as the disjoint union. Consequently, given symbolic graph morphisms f_1^ϕ and f_2^ϕ there exists E*-graph morphisms i_1, i_2, and c such that the diagram below commutes. The morphisms i_1^ϕ and i_2^ϕ are morphisms in **Symb-Graphs$_\mathcal{D}$** since $\mathcal{D} \models (\bar{i}_1(\phi_1) \land \bar{i}_2(\phi_2)) \Rightarrow \bar{i}_1(\phi_1)$ and $\mathcal{D} \models (\bar{i}_1(\phi_1) \land \bar{i}_2(\phi_2)) \Rightarrow \bar{i}_2(\phi_2)$. Also $c^\phi :$ $\langle G_{1+2}, \bar{i}_1(\phi_1) \land \bar{i}_2(\phi_2)\rangle \to \langle G_0, \phi_0 \rangle$ is a morphism in **SymbGraphs$_\mathcal{D}$**, as, by definition, $\mathcal{D} \models \phi_0 \Rightarrow$ $\bar{f}_1(\phi_1)$ and $\mathcal{D} \models \phi_0 \Rightarrow \bar{f}_2(\phi_2)$, $f_1 = c \circ i_1$, and $f_2 = c \circ i_2$, so $\mathcal{D} \models \phi_0 \Rightarrow$ $\bar{c}(\bar{i}_1(\phi_1)) \land \bar{c}(\bar{i}_2(\phi_2))$ that implies $\mathcal{D} \models \phi_0 \Rightarrow \bar{c}(\bar{i}_1(\phi_1) \land \bar{i}_2(\phi_2))$.

$$
\langle G_1, \phi_1 \rangle \xrightarrow{i_1^\phi} \langle G_{1+2}, \phi_{1+2} \rangle \xleftarrow{i_2^\phi} \langle G_2, \phi_2 \rangle
$$
$$
f_1^\phi \searrow \quad \downarrow c^\phi \quad \swarrow f_2^\phi
$$
$$
\langle G_0, \phi_0 \rangle
$$

Property 2 (SymbGraphs$_\mathcal{D}$ has weak $\mathcal{E}_\phi$-$\mathcal{M}'_\phi$ factorization.) *Given the symbolic morphisms $g^\phi : \langle G_0, \phi_0 \rangle \to \langle G_2, \phi_2 \rangle$, $e^\phi : \langle G_0, \phi_0 \rangle \to \langle G_1, \phi_1 \rangle$, and $m^\phi :$ $\langle G_1, \phi_1 \rangle \to \langle G_2, \phi_2 \rangle$ with $m \circ e = f$, where e is an epimorphism (i.e., surjective on all kinds of nodes and edges) and m of class $\mathcal{M}'$ of E*-graph morphisms, which are injective for graph nodes and all kind of edges. The symbolic morphisms e^ϕ and m^ϕ are the $\mathcal{E}_\phi$-$\mathcal{M}'_\phi$ factorization of g^ϕ if e and m are an epi-$\mathcal{M}'$ factorization of g in E*-graphs and $\mathcal{D} \models \phi_2 \Leftrightarrow \bar{e}(\phi_1)$.*

Proof. The category **E*-graphs** has weak epi-$\mathcal{M}'$ factorization [1]. Consequently, given symbolic graph morphism g^ϕ there exists an epimorphism e and morphism $m \in \mathcal{M}'$ in **E*-graphs** such that $g = m \circ e$. Obviously, morphism $e^\phi : \langle G_1, \phi_1 \rangle \to \langle G_2, \phi_2 \rangle$ is in **SymbGraphs$_\mathcal{D}$** since, by definition, $\mathcal{D} \models \phi_2 \Leftrightarrow$ $\bar{e}(\phi_1)$ implies $\mathcal{D} \models \phi_2 \Rightarrow \bar{e}(\phi_1)$. Morphism $m^\phi : \langle G_1, \phi_1 \rangle \to \langle G_2, \phi_2 \rangle$ is in **SymbGraphs$_\mathcal{D}$** since $\mathcal{D} \models \phi_2 \Rightarrow \bar{g}(\phi_1)$ and $g = m \circ e$, so $\mathcal{D} \models \phi_3 \Rightarrow \overline{m}(\bar{e}(\phi_1))$.

$$
\langle G_1, \phi_1 \rangle \xrightarrow{g^\phi} \langle G_3, \phi_3 \rangle
$$
$$
e^\phi \searrow \quad \nearrow m^\phi
$$
$$
\langle G_2, \phi_2 \rangle
$$

Property 3 ($\mathcal{M}'_\phi$ is closed under composition.) *If (i) $f^\phi : A^\phi \to B^\phi$ and $g^\phi : B^\phi \to C^\phi$ in $\mathcal{M}'_\phi$ then $g^\phi \circ f^\phi$ is in $\mathcal{M}'_\phi$, and if (ii) $g^\phi \circ f^\phi$ and g^ϕ are in $\mathcal{M}'_\phi$ then f^ϕ is in $\mathcal{M}'_\phi$.*

Proof. The property holds for **E*-graph** morphisms in $\mathcal{M}'$ that are injective for graph nodes and all kinds of edges [1]. Consequently, we have $f : A \to B \in \mathcal{M}'$, $g : B \to C \in \mathcal{M}'$ and $g \circ f \in \mathcal{M}'$. (i) Morphism $g^\phi \circ f^\phi \in \mathcal{M}'_\phi$, since $\mathcal{D} \models \phi_C \Rightarrow \overline{g}(\phi_B)$ and $\mathcal{D} \models \phi_B \Rightarrow \overline{f}(\phi_A)$ implies $\mathcal{D} \models \phi_C \Rightarrow \overline{g}(\overline{f}(\phi_A))$. (ii) Morphism $f^\phi \in \mathcal{M}'_\phi$, as $\mathcal{D} \models \phi_C \Rightarrow \overline{g}(\overline{f}(\phi_A))$ and $\mathcal{D} \models \phi_C \Rightarrow \overline{g}(\phi_B)$ implies $\mathcal{D} \models \phi_B \Rightarrow \overline{f}(\phi_A)$.

Property 4 ($\mathcal{M}'_\phi$ is closed under POs and PBs along $\mathcal{M}_\phi$-morphisms.) *$\mathcal{M}'_\phi$ is closed under pushouts and pullbacks along $\mathcal{M}_\phi$ morphisms if the pushout or pullback (1) with $h_1^\phi \in \mathcal{M}_\phi$, $h_2^\phi \in \mathcal{M}'_\phi$ or $g_2^\phi \in \mathcal{M}_\phi$, $g_1^\phi \in \mathcal{M}'_\phi$, respectively, then we also have $g_1^\phi \in \mathcal{M}'_\phi$ or $h_2^\phi \in \mathcal{M}'_\phi$ [1].*

Proof. In **E*-graphs** pushouts and pullbacks can be constructed componentwise [1]. Consequently the property holds for both: (i) choosing $\mathcal{M}$ as the class of morphisms injective for graph nodes and all kinds of edges and bijective for label nodes, and (ii) choosing $\mathcal{M}$ similar to $\mathcal{M}'$ (the class of morphisms injective for graph nodes and all kinds of

$$
\begin{array}{ccc}
\langle G_0, \phi_0 \rangle & \xrightarrow{h_1^\phi} & \langle G_1, \phi_1 \rangle \\
h_2^\phi \downarrow & (1) & \downarrow g_1^\phi \\
\langle G_2, \phi_2 \rangle & \xrightarrow[g_2^\phi]{} & \langle G_3, \phi_3 \rangle
\end{array}
$$

edges). Since in **SymbGraphs**$_\mathcal{D}$ pushouts and pullbacks exist along $\mathcal{M}_\phi$ and $\mathcal{M}'_\phi$ morphisms [11], $\mathcal{M}'_\phi$ is closed under pushouts and pullbacks along $\mathcal{M}_\phi$-morphisms (and $\mathcal{M}'_\phi$-morphisms).

After proving these properties for symbolic graphs, we can now apply results from [1] to show that the given construction ensures consistency guarantee.

Theorem 1 (Constructing NACs from negative constraints). *Given a symbolic graph transformation rule $\mathbf{r}^\phi = \langle L \xleftarrow{l} K \xrightarrow{r} R, \phi \rangle$ and the set of postcondition NACs NAC_R constructed from the rule $\mathbf{r}^\phi$ and the negative constraint NC^ϕ as defined in Section 3.1. The application of rule $\mathbf{r}^\phi$ satisfies the postcondition NAC iff model M_R^ϕ is consistent w.r.t. the negative constraint NC^ϕ.*

Proof. The proof follows from Theorem 7.13 in [1], and the properties 1–4.

Theorem 2 (Equivalence of the constructed precondition and postcondition NACs). *For each postcondition NAC n_R^ϕ over symbolic GT-rule $\mathbf{r}^\phi$, the precondition NAC n_L^ϕ constructed according to Section 3.2 is satisfied for each application of $\mathbf{r}^\phi$ iff the postcondition NAC n_R^ϕ is satisfied.*

We only provide a proof for the logic component, as the detailed proof of the construction for the category of **E*-graphs** can be found in [1].

Proof. Let the diagram below show the construction of the precondition NAC n_L^ϕ : $\langle L, \phi_L \rangle \to \langle N_L, \phi_{N_L} \rangle$ from the postcondition NAC $n_R^\phi : \langle R, \phi_R \rangle \to \langle N_R, \phi_{N_R} \rangle$ for rule $\mathbf{r}^\phi = \langle L \xleftarrow{l} K \xrightarrow{r} R, \phi \rangle$ according to Section 3.2. Assuming the construction is valid for **E*-graphs** (using the $\mathcal{M}$–$\mathcal{M}'$ PO–PB decomposition property [1]) we know that there exists an E*-graph morphism $x_L : N_L \to M_L \in \mathcal{M}'$ iff there exists morphism $x_R : N_R \to M_R \in \mathcal{M}'$ such that $x_R \circ n_R = m_R$ and $x_L \circ n_L = m_L$, and (1), (2), (3), (4) commute.

As the set of label nodes and the logic formula remains invariant after symbolic transformation [7] (i.e., $V_{M_L}^{\mathbb{L}} = V_{M_R}^{\mathbb{L}}$ up to isomorphism and $\mathcal{D} \models \phi_{M_L} \Leftrightarrow \phi_{M_R}$) we may consider $\overline{m}_L = \overline{m}_R$, and ϕ_{M_L} and ϕ_{M_R} to be the same formula abbreviated as ϕ''. Consequently we have to show that if there exists E*-graph morphisms x_R and x_L then $\mathcal{D} \models \phi'' \Rightarrow \overline{x}_L(\phi_{N_L})$ iff $\mathcal{D} \models \phi'' \Rightarrow \overline{x}_R(\phi_{N_R})$. This trivially holds, since the set of label nodes and

$$
\begin{array}{ccccc}
L^\phi & \xleftarrow{\;l^\phi\;} & K^\phi & \xrightarrow{\;r^\phi\;} & R^\phi \\
\downarrow & & \downarrow & & \downarrow \\
n_L^\phi \; (1) & n_K^\phi & (2) & n_R^\phi \\
& m_K^\phi & & \\
m_L^\phi \; N_L^\phi & \xleftarrow{} & N_K^\phi & \xrightarrow{} & N_R^\phi \; m_R^\phi \\
\downarrow & l_{N_L}^\phi & \downarrow & r_{N_R}^\phi & \downarrow \\
x_L^\phi \; (3) & x_K^\phi & (4) & x_R^\phi \\
\downarrow & & \downarrow & & \downarrow \\
M_L^\phi & \xleftarrow{\;l_{M_L}^\phi\;} & M_K^\phi & \xrightarrow{\;r_{M_R}^\phi\;} & M_R^\phi
\end{array}
$$

the logic formulas in the NACs N_L^ϕ and N_R^ϕ are also similar by construction (Sec. 3.2). Hence, we may consider $\overline{n}_L = \overline{n}_R$, and ϕ_{N_L} and ϕ_{N_R} to be the same formula, which implies that $\overline{x}_L = \overline{x}_R$.

4   Attributes in Search Plan Driven Pattern Matching

As demonstrated in Section 3, rule preconditions can be produced as symbolic graphs, whose graph part and logic formula describe structural and attribute restrictions, respectively. This section presents how a generated rule precondition can be actually checked by a tool in a practical setup as a pattern matching process. This paper extends the pattern matching approach for EMF models of [12] by attribute handling. The new process can be summarized as follows:

Section 4.1 A (declarative) pattern specification is derived from the symbolic graph representing the rule precondition. In this phase, the concept of declarative pattern specifications originates from [8], and the idea to describe attribute restrictions by predicates has been first proposed in [9], however, the *complete derivation process* is a novel contribution of this paper.

Section 4.2 Operations representing atomic steps in the pattern matching process are created from the pattern specification. In this phase, the concept to use operations in pattern matching for structural restrictions originates from [8,12], while the ideas of *attribute manipulating operations* and *their intertwinement with structure checking operations*, which results in a uniform process for both kinds of operations, are new contributions.

Section 4.3 The operations are filtered and sorted by a search plan generation algorithm [12] to prepare a valid (and efficient) search plan, which is then used, e.g., by a code generator to produce executable code for pattern matching as described in [13].

Due to space restrictions, the current paper only presents the new contributions of Sec. 4.1 and 4.2 in details. The techniques of Sec. 4.3, which have been described in other papers, are applicable for attributes without any change, consequently, this phase is only demonstrated on the running example.

4.1 Pattern Specification

Definitions in this subsection are from [8,12]. A *pattern specification* is a set of predicates over a set of variables as arguments. A *variable* is a placeholder for an object or an attribute value in a model. A *predicate* specifies a condition on a set of variables (which are also referred to as *arguments* in this context) that must be fulfilled by the model elements assigned to the arguments.

Four kinds of predicates are used in our approach. An *association predicate* refers to an association in the metamodel and prescribes the existence of a link, which conforms to the referenced association, and connects the source and the target object assigned to the first and second argument, respectively. An *attribute predicate*, whose concept stems from [9], refers to an attribute in the metamodel and ensures that the object assigned to the first argument has an attribute slot with the attribute value assigned to the second argument. An *attribute value predicate* places a restriction on attribute values as already discussed in Sec. 2.1. A *NAC predicate* refers to a NAC and ensures that the NAC is satisfied.

Deriving a Pattern Specification from a Pattern. A pattern specification is derived from a given pattern by the following *new* algorithm:

1. For each graph and label node in the pattern, a variable is introduced.
2. For each graph edge, an association predicate referring to the type of the graph edge is added to the pattern specification. The two arguments are the variables for the source and target graph nodes of the processed graph edge.
3. For each label edge, an attribute predicate of corresponding type is added to the pattern specification. The two arguments are the source graph node and the target label node of the processed label edge, respectively.
4. Each attribute value predicate conjuncted in the logic formula of the pattern is added to the pattern specification.
5. For each precondition NAC in the pattern, a NAC predicate is added to the pattern specification that has an argument for each node in the pattern.

Example. The pattern specification derived from the LHS pattern of rule addArticle (Fig. 4) consists of (i) the association predicate orders(Cu,Or) requiring an orders link between customer Cu and order Or, (ii) the attribute predicates totalCost(Or,Or.totalCost) and price(Ar,Ar.price) for the totalCost and price attributes of order Or and article Ar, respectively, (iii) the attribute value predicate add(Or.totalCost',Or.totalCost,Ar.price) (appearing in the logic formula of the LHS pattern), and (iv) the NAC predicate addArticle-NAC(Cu,Or,Ar,Ar.price,Or.totalCost,Or.totalCost').

4.2 Creating Operations

This subsection describes the process of creating operations from the predicates of the pattern specification. The definitions and the production of operations for association predicates are from [8,12], while the attribute and NAC handling operations are *novel contributions*. It should be highly emphasized that *the new process does not distinguish between the handling of attribute and structural restrictions any more. Consequently, all these operations are intertwined to an integrated pattern matching process.*

Definitions and Operations for Association Predicates. Let us assume that an (arbitrary) order is fixed for the variables in the pattern specification. An *adornment* represents binding information for *all variables* in the pattern specification by a corresponding character sequence consisting of letters B or F, which indicate that the variable in that position is *bound* or *free*, respectively. An *operation* represents an atomic step in the pattern matching process. It consists of a predicate, and an operation adornment. An *operation adornment* prescribes which arguments must be bound when the operation is executed.

For each association predicate, two operations are created with the corresponding adornments BB and BF. The operation adorned with BB verifies the existence of a link of corresponding type between the objects bound to the arguments. The operation with the BF adornment denotes a forward navigation.

Operations for Attribute, Attribute Value and NAC Predicates. For each attribute predicate, two operations are created with the corresponding adornments BB and BF. The operation adorned with BB checks that the (attribute) value of the corresponding attribute of the first argument is equal to the value of the second argument. The operation with adornment BF looks up the (attribute) value of the corresponding attribute of the first argument, and assigns this value to the second argument.

For each attribute value predicate, a set of used-defined operations is created. E.g., a user may define four operations for the attribute value predicate $add(x_1, x_2, x_3)$. The operation adorned with BBB checks whether the value of variable x_1 equals to the sum of the values of x_2 and x_3. The operation with FBB adornment assigns the sum of the values of x_2 and x_3 to variable x_1, while the operations adorned with BFB and BBF calculate the difference of the first and the other bound argument, and assign this difference to the free argument.

For each NAC predicate, an operation with only bound arguments is created that checks whether the corresponding NAC is satisfied.

Predicate	Op. Adornm.	Predicate	Op. Adornm.
Operations for association predicates		**Operations for attribute value predicates**	
orders(Cu,Or)	BB	add(Or.totalCost',Or.totalCost,Ar.price)	BBB
orders(Cu,Or)	BF	add(Or.totalCost',Or.totalCost,Ar.price)	FBB
Operations for attribute predicates		add(Or.totalCost',Or.totalCost,Ar.price)	BFB
totalcost(Or,Or.totalCost)	BB	add(Or.totalCost',Or.totalCost,Ar.price)	BBF
totalcost(Or,Or.totalCost)	BF		
price(Ar,Ar.price)	BB	**Operations for NAC predicates**	
price(Ar,Ar.price)	BF	addArticleNAC(Cu,Or,Ar,Ar.price,Or.totalCost,Or.totalCost')	BBBBBB

Fig. 5. Created operations for the LHS pattern of the addArticle rule

Example. Fig. 5 lists the operations derived from the LHS of rule addArticle.

4.3   Search Plan and Code Generation

The search plan and code generation techniques described in this subsection originate from [12] and [13], respectively. When pattern matching is invoked, variables can already be bound to restrict the search. The corresponding binding information of all variables is called *initial adornment* a_0. By using the initial adornment, a search plan generation algorithm [12] filters and sorts the operations to prepare a search plan, which is then processed by a code generator to produce executable program code.

A *search plan* is a sequence of operations, which handles each predicate of the pattern specification *exactly once*, and terminates in an adornment with only B characters, which means that all the variables are bound in the end.

Example. Let us suppose that customer Cu, order Or and article Ar are bound in the initial adornment, while the three attribute variables are free, and the search plan shown as comments on the right side of Fig. 6 has been generated. As both variables Cu and Or are initially bound, the operation orders$_{BB}$(Cu,Or) can be applied, which does not change the adornment. The second operation looks up the value of the totalCost attribute of the order stored in variable Or, and assigns this value to variable Or.totalCost, which gets bound by this act. Similarly, the third operation looks up the value of the price attribute of the article stored in variable Ar, and assigns this value to variable Ar.price. At this point, variables Or.totalCost and Ar.price are already bound, so their sum can be calculated and assigned to variable Or.totalCost' by the fourth operation. Finally, the NAC predicate is checked by the last operation. Note that each predicate is represented exactly once in the search plan and all variables are bound in the end, which means that the presented operation sequence is a search plan.

```
public Match addArticle_LHS(Customer Cu, Order Or, Article Ar){
   if(Cu.getOrders().contains(Or)){              // orders_BB(Cu,Or)
      double Or_totalCost=Or.getTotalCost();     // totalCost_BF(Or,Or.totalCost)
      double Ar_price=Ar.getPrice();             // price_BF(Ar,Ar.price)
      double Or_totalCost_p=Or_totalCost + Ar_price;  // add_FBB(Or.totalcost',Or.totalCost,Ar.price)
      if(!addArticleNAC(Cu,Or,Ar,                 // addArticleNAC_BBBBBB(Cu,Or,Ar,
         Ar_price,Or_totalCost,Or_totalCost_p)){  //        Ar.price,Or.totalCost,Or.totalCost')
        return new Match(Cu,Or,Ar,Ar_price,Or_totalCost,Or_totalCost_p);
      }
   }
   return null;
}
```

Fig. 6. Pattern matching code and the corresponding search plan

5   Related Work

The idea of constructing precondition application conditions for GT-rules from graph constraints was originally proposed in [4]. The expressiveness of constraints was extended in [14] that allows arbitrary nesting of constraints. In [15] the approach was generalized to the generic notion of high-level replacement systems.

Including attributes in the theory of graph transformation has been proposed in [16], where attributed graphs are specified by assigning to the label nodes terms of a freely generated term algebra over a set of variables. Although this approach can generate application conditions from attributed graph constraints, it comes with some technical difficulties (arising from the conceptual complexity of combining graphs with algebras) and it has limitations regarding expressiveness compared to symbolic graphs introduced in [11]. Compared to the original notion of symbolic graphs, which allows first order formulas expressing arbitrary constraint satisfaction problems (CSP), we can only handle CSPs for which we can generate valid search plans, which are basically those that have a *unique* solution. However, we can solve these CSPs in linear time in the number of predicates as every predicate is evaluated only once in a valid search plan. Despite this limitation, our approach remains still more expressive than attributed graphs, as these are restricted to (conditional) equations [11]. In [6] OCL preconditions for graph transformation rules are derived from graph constraints with OCL expressions. Consequently, complex expressions including cardinality constraints on collections are allowed. However, different concepts like graphs for expressing structural restrictions and OCL expression for attribute conditions are used that might complicate an efficient evaluation of the preconditions if different engines for the evaluation of graph conditions and OCL expressions are used. In our proposal, restrictions on graphs and attributes can be evaluated arbitrarily intertwined using a single engine. Moreover, as shown in [12] cost values can be assigned to (all) operations guiding the search plan generation process in optimizing the order of operations. A correctness proof is also not given in [6]. [17] suggested an approach based on Hoare-calculus for transforming postconditions to preconditions, which involved the handling of simple attribute conditions. However, implementation issues were not discussed in [17].

6   Conclusion

In this paper, we proposed an attribute handling approach for generating preconditions from graph constraints, whose correctness has been proven using the formalism of symbolic graphs. The presented technique generates preconditions that are transformed to pattern specifications, which are then processed by advanced optimization algorithms [12] to automatically derive search plans, in which the evaluation of attribute and structural restrictions can be intertwined.

One open issue is to analyze the generated NACs and to keep only the weakest preconditions, which could accelerate rule applications at runtime. Another interesting topic could be to determine whether symbolic graphs provide the right properties to construct precondition NACs from more complex constraints (e.g., nested constraints), however, we intentionally left this analysis for future work in favor for an implementation.

References

1. Ehrig, H., Ehrig, K., Prange, U., Taentzer, G.: Fundamentals of Algebraic Graph Transformation. Springer (2006)
2. Heckel, R.: Compositional verification of reactive systems specified by graph transformation. In: Astesiano, E. (ed.) ETAPS 1998 and FASE 1998. LNCS, vol. 1382, pp. 138–153. Springer, Heidelberg (1998)

3. Koch, M., Mancini, L.V., Parisi-Presicce, F.: A graph-based formalism for RBAC. ACM Trans. Inf. Syst. Secur. 5(3), 332–365 (2002)
4. Heckel, R., Wagner, A.: Ensuring consistency of conditional graph rewriting – a constructive approach. In: Corradini, A., Montanari, U. (eds.) Proc. of Joint COMPUGRAPH/SEMAGRAPH Workshop. ENTCS, vol. 2, pp. 118–126. Elsevier, Volterra (1995)
5. Habel, A., Heckel, R., Taentzer, G.: Graph grammars with negative application conditions. Fundamenta Informaticae 26(3/4), 287–313 (1996)
6. Cabot, J., Clarisó, R., Guerra, E., de Lara, J.: Synthesis of OCL pre-conditions for graph transformation rules. In: Tratt, L., Gogolla, M. (eds.) ICMT 2010. LNCS, vol. 6142, pp. 45–60. Springer, Heidelberg (2010)
7. Orejas, F., Lambers, L.: Delaying constraint solving in symbolic graph transformation. In: Ehrig, H., Rensink, A., Rozenberg, G., Schürr, A. (eds.) ICGT 2010. LNCS, vol. 6372, pp. 43–58. Springer, Heidelberg (2010)
8. Horváth, Á., Varró, G., Varró, D.: Generic search plans for matching advanced graph patterns. In: Ehrig, K., Giese, H. (eds.) Proc. of the 6th International Workshop on Graph Transformation and Visual Modeling Techniques, Braga, Portugal. Electronic Communications of the EASST, vol. 6 (March 2007)
9. Anjorin, A., Varró, G., Schürr, A.: Complex attribute manipulation in TGGs with constraint-based programming techniques. In: Hermann, F., Voigtländer, J. (eds.) Proc. of the 1st Int. Workshop on Bidirectional Transformations. ECEASST, vol. 49 (2012)
10. Shoenfield, J.R.: Mathematical logic, vol. 21. Addison-Wesley, Reading (1967)
11. Orejas, F., Lambers, L.: Symbolic attributed graphs for attributed graph transformation. In: Ermel, C., Ehrig, H., Orejas, F., Taentzer, G. (eds.) Proc. of the ICGT. Electronic Communications of the EASST, vol. 30 (2010)
12. Varró, G., Deckwerth, F., Wieber, M., Schürr, A.: An algorithm for generating model-sensitive search plans for pattern matching on EMF models. Software and Systems Modeling (2013) (accepted paper)
13. Varró, G., Anjorin, A., Schürr, A.: Unification of compiled and interpreter-based pattern matching techniques. In: Vallecillo, A., Tolvanen, J.-P., Kindler, E., Störrle, H., Kolovos, D. (eds.) ECMFA 2012. LNCS, vol. 7349, pp. 368–383. Springer, Heidelberg (2012)
14. Habel, A., Pennemann, K.-H.: Nested constraints and application conditions for high-level structures. In: Kreowski, H.-J., Montanari, U., Orejas, F., Rozenberg, G., Taentzer, G. (eds.) Formal Methods (Ehrig Festschrift). LNCS, vol. 3393, pp. 293–308. Springer, Heidelberg (2005)
15. Ehrig, H., Ehrig, K., Habel, A., Pennemann, K.-H.: Constraints and application conditions: From graphs to high-level structures. In: Ehrig, H., Engels, G., Parisi-Presicce, F., Rozenberg, G. (eds.) ICGT 2004. LNCS, vol. 3256, pp. 287–303. Springer, Heidelberg (2004)
16. Ehrig, H., Prange, U., Taentzer, G.: Fundamental theory for typed attributed graph transformation. In: Ehrig, H., Engels, G., Parisi-Presicce, F., Rozenberg, G. (eds.) ICGT 2004. LNCS, vol. 3256, pp. 161–177. Springer, Heidelberg (2004)
17. Poskitt, C.M., Plump, D.: Hoare-style verification of graph programs. Fundamenta Informaticae 118(1), 135–175 (2012)

From Core OCL Invariants
to Nested Graph Constraints*

Thorsten Arendt[1], Annegret Habel[2], Hendrik Radke[2], and Gabriele Taentzer[1]

[1] Philipps-Universität Marburg, Germany
{arendt,taentzer}@informatik.uni-marburg.de
[2] Universität Oldenburg, Germany
{habel,radke}@informatik.uni-oldenburg.de

Abstract. Meta-modeling including the use of the Object Constraint Language (OCL) forms a well-established approach to design domain-specific modeling languages. This approach is purely declarative in the sense that instance construction is not needed and not considered. In contrast, graph grammars allow the stepwise construction of instances by the application of transformation rules. In this paper, we consider meta-models with Core OCL invariants and translate them to nested graph constraints for typed attributed graphs. Models and meta-models are translated to instance and type graphs. We show that a model satisfies a Core OCL invariant iff its corresponding instance graph satisfies the corresponding nested graph constraint. The aim of this work is to establish a first formal relation between meta-modeling and the theory of graph transformation including constraints to come up with an integrated approach for defining modeling languages in an optimal way in the future.

Keywords: Meta modeling, OCL, graph constraints, application conditions.

1 Introduction

The trend towards model-based and model-driven software development causes a need of new, mostly domain-specific modeling languages with well-designed tool support. Therefore we need methods and techniques to define modeling languages and their tooling precisely and also intuitively. A comprehensive language definition needs the declarative as well as the constructive paradigm to specify language properties, to construct and recognize language instances as well as to modify them. Nowadays, modeling languages are typically defined by meta-models following purely the declarative approach. In this approach, language properties are specified by the Object Constraint Language (OCL) [1].

* This work is partly supported by the German Research Foundation (DFG), Grant HA 2936/4-1 (Meta modeling and graph grammars: integration of two paradigms for the definition of visual modeling languages).

H. Giese and B. König (Eds.): ICGT 2014, LNCS 8571, pp. 97–112, 2014.

In contrast, graph grammars have shown to be suitable and natural to specify visual languages in a constructive way, by using graph transformation [2]. Recently, nested graph constraints [3] have been developed to include also the declarative element into graph grammars. To ensure that a graph grammar fulfills a set of graph constraints, they can be translated to application conditions of graph rules such that all graphs fulfilling the constraints in the beginning keep on fulfilling them after applying graph rules being extended by translated application conditions.

While typed attributed graphs form an adequate formalization of instance models that are typed over a meta-model [4], the relation of OCL constraints to nested graph constraints has not been considered yet. We are interested in investigating this relation, since the translation of graph constraints to application conditions for rules opens up a way to combine declarative and constructive elements in a formal approach. By translating OCL to nested graph constraints, such an integration of declarative and constructive elements becomes possible also in the meta-modeling approach. It shall open up a way to translate OCL constraints to application conditions of model transformation rules making applications as e.g. auto-completion of model editing operations to consistent models possible.

As a basis, models and meta-models (without OCL constraints) are translated to instance and type graphs. In this paper, we investigate the relation of meta-models including OCL constraints and nested graph constraints for typed attributed graphs. It turns out that Core OCL invariants [5], i.e. Boolean expressions over navigations based on the type system, can be well translated to nested graph constraints. The aim of this work is to establish a first formal relation between meta-modeling and the theory of graph transformation to come up with an integrated approach for defining modeling languages in an optimal way in the future.

This paper is structured as follows: The next section presents OCL in a nutshell focusing on Core OCL invariants. Section 3 shows typed attributed graphs and graph morphisms as well as nested graph conditions. Section 4 presents our main contribution of this paper, the translation of Core OCL invariants to nested graph constraints. Section 5 discusses how Core OCL invariants can be translated to equivalent application conditions of graph rules. Section 6 compares to related work and concludes the paper.

2   Core OCL Invariants

In this section, we recall Core OCL constraints presenting a small example first and formally defining their syntax and semantics thereafter, according to the work by Richters [6] that went into the OCL specification by the OMG [1]. For illustration purposes, we use the following meta-model for simple Petri nets to recall OCL.

Example 1. A Petri net (*PetriNet*) is composed of places (*Place*) or transitions (*Transition*) which are linked together by arcs (*ArcTP* for linking exactly one transition to one place; *ArcPT* for linking exactly one place to one transition). Places and transitions can have an arbitrary number of incoming (*preArc*) and outgoing (*postArc*) arcs. Finally, Petri net markings are defined by the *token* attribute of places. However, this meta-model allows to build invalid models. For example, one can model a transition having no incoming arc, i.e., the transition can never be fired. Therefore, we complement the meta-model with invariants formulated in OCL.

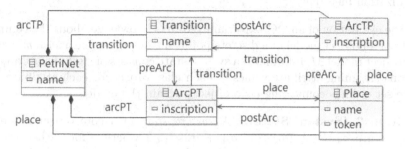

1. A transition has incoming arcs.
   `context Transition inv: self.preArc -> notEmpty()`
2. The number of tokens on a place is not negative.
   `context Place inv: self.token >= 0`
3. Each two places of a Petri net have different names.
   `context Petrinet inv: self.place -> forAll(p1:Place |`
   `self.place -> forAll(p2:Place | p1 <> p2 implies p1.name <>`
   `p2.name))` or alternatively
   `context Petrinet inv: self.place -> forAll(p1:Place,p2:Place |`
   `p1 <> p2 implies p1.name <> p2.name)`

Now, we consider Core OCL invariants in more detail. The **Core OCL** comprises the OCL type system and the language concepts that realize model navigation. The only kind of collections we consider are sets which conform well with using OCL for meta-modeling. Furthermore, we concentrate on selected Boolean-typed set operations only (*isEmpty, notEmpty, exists,* and *forAll*). This also means that user-defined operations are not allowed.

For Core OCL, we straiten the kind of object models being allowed: attributes have primitive types only, there are no operations defined, associations are binary, roles are the default ones indicating source and target, and multiplicities are not set, i.e. range between 0 and *. (It is obvious, however, that multiplicities can be expressed by Core OCL invariants.)

Definition 1 (Core Object Model). Let $DSIG = (S, OP)$ be a data signature with $S = \{Integer, Real, Boolean, String\}$ and a family of corresponding operation symbols OP. A *core object model* over $DSIG$ is a structure $M = (CLASS, ATT, ASSOC, associates, r_{src}, r_{tgt}, \prec)$ where $CLASS$ is a finite set of classes, $ATT = \{ATT_c\}_{c \in CLASS}$ is a family of attributes $att : c \to S$ of class c, $ASSOC$ is a set of associations, $associates$ is a function that maps each association to a pair of participating classes with $associates : ASSOC \to (CLASS \times CLASS)$, r_{src} and r_{tgt} are functions that map each association to a source respectively target role name with $r_{src}, r_{tgt} : ASSOC \to String$ and $r_{src}(assoc) = c_1$ and $r_{tgt}(assoc) = c_2$ for each $assoc \in ASSOC$ with $associates(assoc) = (c_1, c_2)$, and $\prec$ is a partial order on $CLASS$ reflecting its generalization hierarchy.

Since the evaluation of an OCL invariant requires knowledge about the complete context of an object model at a discrete point in time, we define a *system state* of a core object model M. Informally, a system state consists of a set of class objects, functions assigning attribute values to each class object for each attribute, and a finite set of links connecting class objects within the model.

Definition 2 (System State). A *system state* of a core object model M is a structure $\sigma(M) = (\sigma_{CLASS}, \sigma_{ATT}, \sigma_{ASSOC})$ where   for each class $c \in CLASS$, $\sigma_{CLASS}(c)$ is a finite subset of the (infinite) set of object identifiers $oid(c) = \{\underline{c}_1, \underline{c}_2, \dots\}$, for each attribute $att : c \to t \in ATT_c^{\prec}$, $\sigma_{ATT}(att) : \sigma_{CLASS}(c) \to I(t)$ is an operation from class objects to some interpretation of the primitive data type t where $ATT_c^{\prec} := \bigcup_{c \prec c'} ATT_{c'}$ is the set of all owned and inherited attribute symbols of a class c, for each $assoc \in ASSOC$ with $associates(assoc) = (c_1, c_2)$, $\sigma_{ASSOC}(assoc) \subset \sigma_{CLASS}^{\prec}(c_1) \times \sigma_{CLASS}^{\prec}(c_2)$ where $\sigma_{CLASS}^{\prec}(c) := \bigcup_{c' \prec c} \sigma_{CLASS}(c')$ is the set of all objects with type or super type c. The set $States(M)$ consists of all system states $\sigma(M)$ of M .

Definition 3 (Core OCL Expressions). Let T be a set of types consisting of all basic types S, all class types $CLASS$, and the collection type $\mathtt{Set(t)}$ for an arbitrary $t \in T$. Let Ω be a set of operations on T consisting of OP, ATT, appropriate association end operations, and set operations. Let $Var = \{Var_t\}_{t \in T}$ be a family of variable sets. The family of *Core OCL expressions* over T and Ω is given by $Expr = \{Expr_t\}_{t \in T}$ of sets of expressions. An expression in $Expr$ is a *VariableExpression* $v \in Expr_t$ for all variables $v \in Var_t$, *OperationExpressions* $e := \omega(e_1, \cdots, e_n) \in Expr_t$ for each operation symbol $\omega : t_1 \times \cdots \times t_n \to t \in \Omega$ and for all $e_i \in Expr_{t_i} (1 \leq i \leq n)$, *IfExpressions* $e := \mathtt{if}\ e_1\ \mathtt{then}\ e_2\ \mathtt{else}\ e_3 \in Expr_{Boolean}$ for all $e_1, e_2, e_3 \in Expr_{Boolean}$ and *IteratorExpressions* $e := s \to exists(v \mid b) \in Expr_{Boolean}$ and $e := s \to forAll(v \mid b) \in Expr_{Boolean}$, for all $s \in Expr_{Set(c)}, v \in Var_c$, and $b \in Expr_{Boolean}$.
Let $Env = \{\tau \mid \tau = (\sigma, \beta)\}$ be a set of environments with system states σ and variable assignments $\beta : Var_t \to I(t)$ which map variable names to values. The semantics of a Core OCL expression $e \in Expr_t$ is a function $I[\![e]\!] : Env \to I(t)$ for $t \in CLASS$ or $t \in S$. The corresponding semantics definition can be found in [6] and adapted to Core OCL in [7].

As mentioned above, we concentrate on invariants being formulated in Core OCL. Therefore, we consider invariants and OCL-constraints as synonyms in the remainder of this paper.

Definition 4 (Core OCL Invariant). A *Core OCL invariant* is a Boolean Core OCL expression with a free variable $v \in Var_C$ where C is a classifier type. The concrete syntax of an invariant is: `context v:C inv : <expr>`. The set $Invariant_M$ denotes the set of all Core OCL invariants over M.

Remark 1. The following properties hold for Core OCL invariants: (1) Navigation expressions to collections are not contained in other navigation expressions, e.g., `somePetriNet.arcTP.place -> notEmpty()` is replaced by `somePetriNet.arcTP -> exists(a:ArcTP | a.place -> notEmpty())`. (2) Iterator expressions are completed, i.e. the iterator variable is explicitly declared. Moreover, a variable declaration is always complete, i.e. consists of a variable name and a type name. (3) If `nav op nav` occurs for the same navigation expression nav and $op(nav, nav) = true$ then `nav op nav` can be replaced by `true`. (4) Note that constraints `v1 = v2.r` and `v1 <> v2.r` (for objects v1,v2 and reference r) are not possible since the result of `v1` is an object and `v2.r` yields a set of objects.

3   Nested Graph Conditions

In the following, we recall the formal definition of typed, attributed graphs as presented in [8]. They form the basis to define attributed graph conditions. Attributed graphs as defined here allow to attribute nodes only while the original version [8] supports also the attribution of edges.

Definition 5 (A-graphs). An *A-graph* $G = (G_V, G_D, G_E, G_A, src_G, tgt_G, src_A, tgt_A)$ consists of sets G_V and G_D, called graph and data nodes (or vertices), respectively, G_E and G_A, called graph and node attribute edges, respectively, and source and target functions: $src_G \colon G_E \to G_V, tgt_G \colon G_E \to G_V$ for graph edges and $src_A \colon G_A \to G_V, tgt_A \colon G_A \to G_D$ for node attribute edges. Given two A-graphs G^1 and G^2, an *A-graph morphism* $f \colon G^1 \to G^2$ is a tuple of functions $f_V \colon G_V^1 \to G_V^2, f_D \colon G_D^1 \to G_D^2, f_E \colon G_E^1 \to G_E^2$ and $f_A \colon G_A^1 \to G_A^2$ such that f commutes with all source and target functions, e.g. $f_V \circ src_G^1 = src_G^2 \circ f_E$.

We assume that the reader is familiar with the basics of algebraic specification. The definition of attributed graphs generalizes largely the one in [9] by allowing variables and a set of formulas that constrain the possible values of these variables. The definition is closely related to symbolic graphs [10].

Definition 6 (Attributed graphs). Let $DSIG = (S, OP)$ be a data signature, $X = \{X_s\}_{s \in S}$ a family of variables, and $T_{DSIG}(X)$ the term algebra w.r.t. $DSIG$ and X. An *attributed graph* over $DSIG$ and X is a tuple $AG = (G, D, \Phi)$ where G is an A-graph, D is a $DSIG$-algebra with $\sum_{s \in S} D_s = G_D$, and Φ is a finite set of $DSIG$-formulas[1] with free variables in X. A set $\{F_1, \ldots, F_n\}$ of formulas can be regarded as a single formula $F_1 \wedge \ldots \wedge F_n$. An attributed graph $AG = (G, D, \emptyset)$ with an empty set of formulas is *basic* and is shortly denoted by $AG = (G, D)$.

---

[1] $DSIG$-formulas are meant to be $DSIG$-terms of sort BOOL. One may consider e.g. a set of literals.

Given two attributed graphs AG^1 and AG^2, an *attributed graph morphism* $f: AG^1 \to AG^2$ is a pair $f = (f_G, f_D)$ of an A-graph morphism $f_G: G^1 \to G^2$ and a *DSIG*-homomorphism $f_D: D^1 \to D^2$ such that for all $s \in S$, $f_{G,G_D}|_{D_s^1} = f_{D,s}$, $f_{G,G_D} = \sum_{s \in S} f_{D,s}$, and $\Phi^2 \Rightarrow f(\Phi^1)$ where $f(\Phi^1)$ is the set of formulas obtained when replacing in Φ^1 every variable x in G^1 by $f(x)$.

Remark 2. We are interested in the case where D_s^1 is a *DSIG*-term algebra and D_s^2 is a *DSIG*-algebra (without variables). In this case the *DSIG*-homomorphism assigns values to variables and terms.

Attributed graphs in the sense of [9] correspond to basic attributed graphs. The results for basic attributed graphs can be generalized to arbitrary attributed graphs: attributed graphs and morphisms form the category **AGraphs**. The category has pushouts and $\mathcal{E}'$-$\mathcal{M}$ pair factorization in the sense of [9].

Definition 7 (Typed attributed graph over $ATGI$). An *attributed type graph* $ATGI = (TG, Z, \Phi')$ consists of an A-graph and a final *DSIG*-algebra Z. A *typed attributed graph* $(AG, type)$ over $ATGI$, short $ATGI$-*graph*, consists of an attributed graph $AG = (G, D, \Phi)$ and a *morphism type* $: AG \to ATGI^2$. Given two $ATGI$-graphs $AG^1 = (G^1, type^1)$ and $AG^2 = (G^2, type^2)$, an $ATGI$-*morphism* $f: AG^1 \to AG^2$ is an attributed graph morphism such that $type^2 \circ f = type^1$.

Typed attributed graphs and morphisms form a category that has pushouts and $\mathcal{E}'$-$\mathcal{M}$ pair factorization.

Fact 1 ([7]). $ATGI$-graphs and morphisms form the category **AGraphs**$_{ATGI}$ with pushouts and $\mathcal{E}'$-$\mathcal{M}$ pair factorization.

In [7], also typed attributed graphs typed over attributed type graphs with inheritance [11] are considered.
Nested graph conditions [3] are nested constructs which can be represented as trees of morphisms equipped with quantifiers and Boolean connectives. In the following, we introduce $ATGI$-conditions as conditions over $ATGI$-graphs, closely related to attributed graph constraints [10] and E-conditions [12].

Definition 8 (nested graph conditions). A *(nested) graph condition* on typed attributed graphs, short $ATGI$-*condition*, over a graph P is of the form *true*, $\exists(a, c)$, or $\exists(P \sqsupseteq C, c)$[3] where $a: P \to C$ is an injective morphism and c is an $ATGI$-condition over C. Boolean formulas over $ATGI$-conditions over P yield $ATGI$-conditions over P, that is $\neg c$ and $\bigwedge_{i \in I} c_i$ are $ATGI$-conditions over P. Conditions over $\emptyset$ are also called *constraints*.

Notation. Graph conditions may be written in a more compact form: $\exists a$ abbreviates $\exists(a, true)$, $\forall(a, c)$ abbreviates $\neg\exists(a, \neg c)$, $\bigvee_{i \in I} c_i$ abbreviates $\neg \bigwedge_{i \in I} \neg c_i$,

---

[2] We usually set $\Phi' = $ false for $ATGI$ so that $\Phi' \Rightarrow \Phi$ is true regardless of Φ.

[3] Conditions of the form $\exists(P \sqsupseteq C, c)$ are syntactic sugar, i.e. they can be expressed in terms of the other constructs. See long version [7].

and $c \Rightarrow c'$ abbreviates $\neg c \vee c'$. For an injective morphism $a\colon P \hookrightarrow C$ in a condition, we just depict the codomain C, if the domain P can be unambiguously inferred, i.e. if it is known over which graph a condition is.

Example 2 (OCL constraints as graph constraints). OCL constraint
`context Place inv: self.token >= 0` in Example 1 is represented as an attributed graph constraint in full and in abbreviated form. The last graph in the condition is decorated by the formula $x \geq 0$, with the notation as in [12]. The attributing DSIG-algebra is the quotient term algebra $T_{\mathrm{DSIG}_{\equiv}}(X)$ where $\equiv$ is the congruence relation on $T_{DSIG}(X)$ induced by $\geq (x,0)$.

$$\neg\exists\left(\emptyset \rightarrow \boxed{\text{self:Place}}, \neg\exists\left(\boxed{\text{self:Place}} \rightarrow \boxed{\begin{array}{c}\text{self:Place}\\\hline\text{token} = \text{x}\end{array}} \mid x \geq 0\right)\right)$$

$$\text{or, in short form: } \forall\left(\boxed{\text{self:Place}}, \exists\,\boxed{\begin{array}{c}\text{self:Place}\\\hline\text{token} \geq 0\end{array}}\right)$$

Definition 9 (Semantics of nested graph conditions). Let $p\colon P \rightarrow G$ be a morphism. *Satisfiability* of a condition over P is inductively defined as follows: Every morphism satisfies *true*. Morphism p satisfies $\exists(P \rightarrow C, c)$ if there exists an injective morphism $q\colon C \hookrightarrow G$ such that the left diagram below commutes and q satisfies c. Morphism p satisfies $\exists(P \sqsupseteq C, c)$ if there exist injective morphisms $b\colon C \hookrightarrow P$ and $q\colon C \hookrightarrow G$ such that $q = p \circ b$ and q satisfies c (see right diagram below). Morphism p satisfies $\neg c$ if p does not satisfy c, and p satisfies $\bigwedge_{i\in I} c_i$ if p satisfies each c_i ($i \in I$). We write $p \models c$ if $p\colon P \rightarrow G$ satisfies the condition c over P.

$$\exists\quad P \xrightarrow{\ \ a\ \ } C \ \triangleleft c \qquad\qquad \exists\quad P \xleftarrow{\ \ b\ \ } C \ \triangleleft c$$
$$p \searrow \underset{=}{\ } \swarrow q \qquad\qquad\qquad\quad p \searrow \underset{=}{\ } \swarrow q$$
$$G \qquad\qquad\qquad\qquad\qquad\qquad G$$

Satisfiability of a constraint (i.e. a condition over $\emptyset$) by a graph is defined as follows: A graph G satisfies a constraint c, short $G \models c$, if the morphism $p\colon \emptyset \hookrightarrow G$ satisfies c.

4   Translation of Meta-Models with Core OCL Invariants

To translate Core OCL invariants, we first show how to translate the type information of meta-models, i.e. core object models, to attributed type graphs with inheritance [11] are considered. Thereafter, system states are translated to typed attributed graphs. Having these ingredients available, our main contribution, the translation of Core OCL invariants, is presented, together with two example translations. Finally, completeness and correctness of the translation are shown.

4.1   Type and State Correspondences

To define the translation of Core OCL invariants to graph constraints, we translate a given object model to its corresponding type graph.

Definition 10 (Type Correspondence). Let $DSIG = (S, OP)$ be a data signature and Z the final DSIG-algebra. Given a core object model $M = (CLASS, ATT, ASSOC, associates, \prec)$ over $DSIG$, it corresponds to an attributed type graph with inheritance $ATGI = ((TG, Z), Inh)$ with type graph $TG = (TG_V, TG_D, TG_E, TG_A, src_G, tgt_G, src_A, tgt_A)$ and inheritance relation Inh if there is a *correspondence relation* $corr_{type} = (corr_{CLASS}, corr_{ATT}, corr_{ASSOC})$ with bijective mappings

- $corr_{CLASS} : CLASS \to TG_V$ with $\forall c1, c2 \in CLASS$:
  $c_1 \prec c_2 \iff (corr_{CLASS}(c_1), corr_{CLASS}(c_2)) \in Inh$,
- $corr_{ATT} : ATT \to TG_A$ with
  $src_A(corr_{ATT}(att)) = corr_{CLASS}(c)$ for $c \in CLASS$ and
  $tgt_A(corr_{ATT}(att)) = x$ if $att : c \to s \in ATT_c$ and $\{x\} = Z_s$ with $s \in S$,
- $corr_{ASSOC} : ASSOC \to TG_E$ with $src_G \circ corr_{ASSOC} = corr_{CLASS} \circ pr_1$
  and $tgt_G \circ corr_{ASSOC} = corr_{CLASS} \circ pr_2$ with $associates(a) = \langle c1, c2 \rangle$,
  $pr_i(a) = c_i$ for $i = 1, 2, c1, c2 \in Class$ and $a \in ASSOC$.

To show the correctness of our Core OCL invariant translation, we also need to establish a correspondence relation between system states and typed attributed graphs.

Definition 11 (State Correspondence). Let M be a core object model and $ATGI$ an attributed type graph, both defined over data signature $DSIG = (S, OP)$. We assume that $I(s) = D_s$ for all sorts $s \in S$. Furthermore, let $corr_{type}(M) = ATGI$ be a type correspondence.
Given a system state $\sigma(M) = (\sigma_{CLASS}, \sigma_{ATT}, \sigma_{ASSOC})$, it corresponds to an attributed graph $AG = (G, D)$ with $G = (G_V, G_D, G_E, G_A, src_G, tgt_G, src_A, tgt_A)$ typed over $ATGI$ by clan morphism $type$ if there is a *state correspondence relation* $corr_{state} = (c_{CLASS}, c_{ATT}, c_{ASSOC}) : States(M) \to Graph_{ATGI}$ defined by the following bijective mappings:

- $c_{CLASS} : \sigma_{CLASS} \to G_V$ with
  $type_{G_V}(c_{CLASS}(o)) = corr_{CLASS}(c)$ with $o \in \sigma_{CLASS}(c)$ and $c \in CLASS$,
- $c_{ATT} : \sigma_{ATT} \to G_A$ with $src_A(c_{ATT}(a)) = c_{CLASS}(o)$ and
  $tgt_A(c_{ATT}(a)) = d$ as well as $type_{G_A}(c_{ATT}(\sigma_{ATT}(att))) = corr_{ATT}(att)$ and
  $a \in \sigma_{ATT}(att)$ if $att : c \to s \in ATT_c^{\prec}$, $\sigma_{ATT}(att) : \sigma_{CLASS}(c) \to D_s$,
  $o \in \sigma_{CLASS}(c)$, $c \in CLASS$ and $\sigma_{ATT}(att)(o) = d$,
- $c_{ASSOC} : \sigma_{ASSOC} \to G_E$ with
  $src_G \circ c_{ASSOC} = c_{CLASS} \circ pr_1$ and $tgt_G \circ c_{ASSOC} = c_{CLASS} \circ pr_2$
  with $l = (o1, o2) \in \sigma_{ASSOC}(assoc)$ and $pr_i(l) = o_i$ for $i = 1, 2$.
  Furthermore, $type_{G_E} \circ c_{ASSOC}(\sigma_{ASSOC}) = corr_{ASSOC}(ASSOC)$.

4.2   Translation of Core OCL Invariants

To get an initial understanding on how Core OCL invariants shall be translated
to graph conditions, we take a pattern-based approach. The principle idea is that
navigation expressions are translated to graphs and graph morphisms while the
usual Boolean operations correspond to each other directly. The subset-operator
of graph conditions is useful to correspond to iterating variables in iterator expressions. In Figure 1, basic OCL patterns and their corresponding graph constraint patterns are depicted. In these patterns, "Class", "v", "v1","v2", "b" ,
"c", and "r" are variables for model elements. Non-terminal <op> can be replaced
by some comparator such as $=, <>, <$. Non-terminal <expr> may be replaced
by any Core OCL expression.

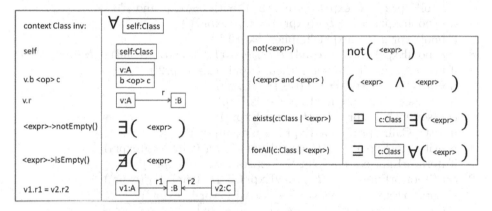

Fig. 1. Translation of basic OCL patterns to graph constraint patterns

In the following, we define the translation of Core OCL invariants as out-
lined in the beginning of this section. The translation is basically structured
along the definition of Core OCL expressions given in Def. 3. However, operation expressions, If expressions and iterator expressions are distinguished along
their result type yielding navigations (with $Set(t)$, $t \in CLASS$ as result type)
and (Boolean) expressions. In the following definition, rules 1 and 2 translate
the header of a CoreOCL invariant, rules 4 - 12 translate Boolean expressions,
rules 13 - 15 translate basic object comparisons, rules 16 - 17 translate attribute
value comparisons, and rules 18 - 19 translate basic navigation expressions and
variables.

Definition 12 (Constraint translation). Let $M = (CLASS, ATT, ASSOC,$
$associates, \prec)$ be a core object model with $ATGI = corr_{type}(M)$ being the corresponding attributed type graph and $t : Expr \rightarrow T$ a typing function which
returns the type of an OCL expression (for T see Section 2). Let furthermore
Invariant$_M$ be the set of Core OCL invariants over M as defined in Def. 4 and
GraphConstraint$_{ATGI}$ be the set of all graph constraints as defined in Definition 8. Then, the *translation functions*

- invariant translation: tr_I: Invariant$_M$ → GraphConstraint$_{ATGI}$
- expression translation tr_E: $Expr_{Boolean}$ → GraphConstraint$_{ATGI}$
- navigation translation tr_N: $Expr_c$ → Graph$_{ATGI}$ with $c \in CLASS$
- variable translation tr_V: Var_c → Graph$_{ATGI}$ with $c \in CLASS$

are defined as follows:

1. tr_I('context' C 'inv:' expr) $= \forall$ ($\boxed{\text{self:C}}$, tr_E(expr))
2. tr_I('context' var ':' C 'inv:' expr) $= \forall$ (tr_V(var), tr_E(expr))
3. tr_E (expr) $= tr_E$(setOpCallExpr) $\mid tr_E$(basicExpr)
   $\mid tr_E$(boolExpr) $\mid tr_E$(iteratorExpr)
4. tr_E(boolExpr) $= true$ if boolExpr ::= 'true'
5. tr_E(boolExpr) $= (\neg\ tr_E$(expr)) if boolExpr ::= '(' 'not' expr ')'
6. tr_E(boolExpr) $= (tr_E$(expr1) $op_g\ tr_E$(expr2)) with $op_g \in \{\wedge, \vee\}$
   if boolExpr ::= '(' expr1 op_b expr2 ')' with $op_b \in \{$ 'and', 'or'$\}$
7. tr_E(boolExpr) $= (\neg\ tr_E$(expr1) $\vee\ tr_E$(expr2))
   if boolExpr ::= '(' expr1 'implies' expr2 ')'
8. tr_E(boolExpr) $= ((tr_E$(cond) $\wedge\ tr_E$(expr1)) $\vee (\neg\ tr_E$(cond) $\wedge\ tr_E$(expr2)))
   if boolExpr ::= '(' 'if' cond 'then' expr1 'else' expr2')'
9. tr_E(setOpCallExpr) $= \neg\exists$ (tr_N(navExpr))
   if setOpCallExpr ::= navExpr '→' isEmpty()'
10. tr_E(setOpCallExpr) $= \exists$ (tr_N(navExpr))
    if setOpCallExpr ::= navExpr '→' notEmpty()'
11. tr_E(iteratorExpr) $= \exists$ (tr_N(navExpr) $\sqsupseteq tr_V$(var), tr_E(expr))
    if iteratorExpr ::= navExpr '→' exists (' var '|' expr ')'
12. tr_E(iteratorExpr) $= \forall$ (tr_N(navExpr) $\sqsupseteq tr_V$(var), tr_E(expr))
    if iteratorExpr ::= navExpr '→' forAll (' var '|' expr ')'
13. (a) tr_E(basicExpr) $= \exists$ ($\boxed{\text{v:t(v)}}$ $\boxed{\text{v2:t(v2)}}$ → $\boxed{\text{v,v2:t(v)}}$)[4]
       if basicExpr ::= v '=' v2
    (b) tr_E(basicExpr) $= \exists$ ($\boxed{\text{v:t(v)}}$ $\boxed{\text{v2:t(v2)}}$)
       if basicExpr ::= v '<>' v2
14. (a) tr_E(basicExpr) $= \exists$ (tr_V(v ':' t(v)) $\overset{as}{\underset{as2}{\rightrightarrows}}$ $\boxed{\text{:t(r)}}$)
       if basicExpr ::= v '.' r '=' v '.' r2, r is a role of as, r2 is a role of $as2$,
       and t(r) = t(r2) $\in CLASS$
    (b) tr_E(basicExpr) $= \exists$ (tr_V(v ':' t(v)) $\overset{as}{\rightarrow}$ $\boxed{\text{:t(r)}}$ $\overset{as2}{\leftarrow} tr_V$(v2 ':' t(v2)))
       if basicExpr ::= v '.' r '=' v2 '.' r2, r is a role of as, r2 is a role of $as2$,
       and t(r) = t(r2) $\in CLASS$
15. (a) tr_E(basicExpr) $= \exists$ ($\boxed{\text{:t(r)}}$ $\overset{as}{\leftarrow} tr_V$(v) $\overset{as2}{\rightarrow}$ $\boxed{\text{:t(r2)}}$)
       if basicExpr ::= v '.' r '<>' v '.' r2, r is a role of as, r2 is a role of $as2$,
       and t(r) = t(r2) $\in CLASS$

---

[4] Note that this a "non-injective" condition in the sense of [3], i.e. a condition with non-injective morphism. By [13], for each non-injective condition c and each morphism $p = m \circ e$ with e surjective and m injective, there is an injective condition Shift(e, c) in the sense of Definition 8 such that we have $p \models c \Leftrightarrow m \models$ Shift(c, c). Whenever a non-injective morphism occurs in a condition, we have to replace the whole condition by Shift(e, c).

(b) $tr_E(\text{basicExpr}) = \exists\ (tr_V(v) \xrightarrow{as} \boxed{:t(r)} \quad tr_V(v2) \xrightarrow{as2} \boxed{:t(r2)}\ )$

if basicExpr ::= v '.' r '<>' v2 '.' r2, r is a role of as, r2 is a role of $as2$, and t(r) = t(r2) $\in CLASS$

16. $tr_E(\text{basicExpr}) = \exists\ (\boxed{\dfrac{v:t(v)}{attr\ op\ x}}\ )^5$ if basicExpr ::= v '.' attr op x and x is a constant or a variable

17. (a) $tr_E(\text{basicExpr}) = \exists\ (\boxed{\begin{array}{c} v:t(v) \\ \hline attr\ =\ x \\ attr2\ op\ x \end{array}}\ )$

   if basicExpr ::= v '.' attr op v '.' attr2, attr $\neq$ attr2, and x is a new variable with t(x) = t(attr) = t(attr2).

   (b) $tr_E(\text{basicExpr}) = \exists\ (\boxed{\begin{array}{c} v:t(v) \\ \hline attr\ =\ x \end{array}}\ \boxed{\begin{array}{c} v2:t(v2) \\ \hline attr2\ op\ x \end{array}}\ )$

   if basicExpr ::= v '.' attr op v2 '.' attr2, v $\neq$ v2, and x is a new variable with t(x) = t(attr).

18. $tr_N(\text{navExpr}) = tr_V(v) \xrightarrow{as} \boxed{:t(r)}$

   if navExpr ::= v '.' r and r is a role of as

19. $tr_V(\text{var}) = \boxed{v:t(v)}$

   for var::= v ':' t with t = t(v) or var::= v

where $expr, expr1, expr2, boolExpr, setOpCallExpr, basicExpr, iteratorExpr \in Expr_{Boolean}, var \in Var_c, navExpr, navExpr1, navExpr2 \in Expr_c, c \in CLASS,$ $v, v2 \in Var_t, t \in T_M, as, as2 \in ASSOC, r = r_{tgt}(as), r2 = r_{tgt}(as2), op \in \{<, >, \leq, \geq, =, <>\}, attr \in ATT_c$ and w.l.o.g., $corr_{CLASS}(c) = c$ for $c - t(r),$ $c = t(r2), c = t(v)$ or $c = t(attr), corr_{ASSOC}(as) = as, corr_{ASSOC}(as2) = as2,$ and $corr_{ATT}(attr) = attr.$

In the following, we show two example translations using OCL invariants of Example 1. Small numbers behind equality signs denote the rules being used.

Example 3 (Translation of OCL constraint 1).

$tr_I(\ '\text{context Transition inv: self.preArc}\rightarrow\text{notEmpty()')} =^1$

$\forall\ (\boxed{\text{self:Transition}}\ tr_E('\text{self.preArc}\rightarrow\text{notEmpty()'}\ ) =^{10}$

$\forall\ (\boxed{\text{self:Transition}},\ \exists\ (tr_N('\text{self.preArc'}))) =^{18}$

$\forall\ (\boxed{\text{self:Transition}},\ \exists\ (tr_V('\text{self:Transition'}) \xrightarrow{preArc} \boxed{:\text{ArcPT}})) =^{19}$

$\forall\ (\boxed{\text{self:Transition}},\ \exists\ (\boxed{\text{self:Transition}} \xrightarrow{preArc} \boxed{:\text{ArcPT}}))$

Example 4 (Translation of OCL constraint 3.1).

$tr_I(\ '\text{context Petrinet inv: self.place}\rightarrow\text{forAll(p1:Place | self.place} \rightarrow$ forAll(p2:Place | p1 <> p2 implies p1.name <> p2.name))')$ =^1$

$\forall\ (\boxed{\text{self:Petrinet}},\ tr_E('\text{self.place}\rightarrow\text{forAll(p1:Place | self.place} \rightarrow$ forAll(p2:Place | p1 <> p2 implies p1.name <> p2.name))')) $=^{12}$

$\forall\ (\boxed{\text{self:Petrinet}},\ \forall\ (tr_E('\text{self.place'}) \sqsupseteq tr_V('\text{p1:Place'}),$ $tr_E('\text{self.place}\rightarrow\text{forAll(p2:Place | p1<>p2 implies p1.name<>p2.name)'}))) =^{18,12}$

---

[5] Compare the short notation of attribute conditions in Example 2.

$\forall\ (\boxed{\text{self:Petrinet}},\ \forall\ (tr_V(\text{'self'})\ \xrightarrow{place}\ \boxed{\text{:Place}}\ \sqsupseteq\ \boxed{\text{p1:Place}},$
$\forall\ (tr_E(\text{'self.place'})\ \sqsupseteq\ tr_V(\text{'p2:Place'}),$
$tr_E(\text{'p1 <> p2 implies p1.name <> p2.name'})))) =^{19,18,6}$

$\forall\ (\boxed{\text{self:Petrinet}},\ \forall\ (\boxed{\text{self:Petrinet}}\ \xrightarrow{place}\ \boxed{\text{:Place}}\ \sqsupseteq\ \boxed{\text{p1:Place}},$
$\forall\ (tr_V(\text{'self'})\ \xrightarrow{place}\ \boxed{\text{:Place}}\ \sqsupseteq\ \boxed{\text{p2:Place}},$
$\neg\ tr_E(\text{'p1 <> p2'})\ \lor\ tr_E(\text{'p1.name <> p2.name'})))) =^{19}$

$\forall\ (\boxed{\text{self:Petrinet}},\ \forall\ (\boxed{\text{self:Petrinet}}\ \xrightarrow{place}\ \boxed{\text{:Place}}\ \sqsupseteq\ \boxed{\text{p1:Place}},$
$\forall\ (\boxed{\text{self:Petrinet}}\ \xrightarrow{place}\ \boxed{\text{:Place}}\ \sqsupseteq\ \boxed{\text{p2:Place}},$
$(\neg\ tr_E(\text{'p1 <> p2'})\ \lor\ tr_E(\text{'p1.name <> p2.name'}))))) =^{10,13.b,17.b}$

$\forall\ (\boxed{\text{self:Petrinet}},\ \forall\ (\boxed{\text{self:Petrinet}}\ \xrightarrow{place}\ \boxed{\text{:Place}}\ \sqsupseteq\ \boxed{\text{p1:Place}},$
$\forall\ (\ \boxed{\text{self:Petrinet}}\ \xrightarrow{place}\ \boxed{\text{:Place}}\ \sqsupseteq\ \boxed{\text{p2:Place}},$
$\neg\ \exists\ (\boxed{\text{p1:Place}}\ \boxed{\text{p2:Place}})\ \lor\ \exists\ (\ \boxed{\begin{array}{c}\text{p1:Place}\\ name = x\end{array}}\ \boxed{\begin{array}{c}\text{p2:Place}\\ name <> x\end{array}}\ ))))$

4.3 Correctness and Completeness

To be sure that the translation of Core OCL invariants is well-defined, we show its correctness and completeness. Moreover, we want to ensure that each translation terminates. The proofs of all the following results are given in [7] in their complete form. Here, we just present the main proof ideas.

Proposition 1 (Termination). The invariant translation tr_I as defined in Definition 12 terminates.

tr_I terminates since all invariants are finite and each application of a translation rule decreases the number of syntactic tokens in an invariant.

Theorem 1 (Completeness of translation). Given a core object model M and its corresponding attributed type graph $ATGI = corr_{type}(M)$, all Core OCL invariants over M are translated to some graph constraint over $ATGI$.

We have to show that all Core OCL invariants can be translated to graph constraints. The proof is performed by induction on the structure of Core OCL invariants. First, we start to translate Core OCL invariants and continue to show the completeness of the translation for Core OCL expressions.
To show that the translation of Core OCL invariants is correct, we consider their semantics and the semantics of graph constraints. If an invariant holds for a system state, the corresponding graph constraint is fulfilled by the corresponding graph.

Theorem 2 (Correct Translation of Core OCL invariants). Given an object model M and its corresponding attributed type graph $ATGI = corr_{type}(M)$, the following statement holds for all Core OCL invariants $inv \in \text{Invariant}_M$: For all environments $env = (\sigma, \beta) \in Env$

$$I \llbracket inv \rrbracket (env) = true \iff G = corr_{state}(\sigma) \models tr_I(inv).$$

The proof is performed by induction on the translation rules given in Def. 12.

5   From Core OCL Invariants to Application Conditions

After having translated Core OCL invariants to graph constraints, we connect this new result with the existing theory on graph constraints [3,14]. A main result shows how nested graph constraints can be translated to right, and thereafter, to left application conditions of transformation rules. In the following, we illustrate at an example how a Core OCL invariant is translated to a left application condition.

By the results in [3,13], for each category with pushouts and $\mathcal{E}'$-$\mathcal{M}$ pair factorization, nested conditions in this category can be shifted over morphisms and rules. By Fact 1, $ATGI$-graphs and morphisms form the category $\mathbf{AGraphs}_{ATGI}$ with pushouts and $\mathcal{E}'$-$\mathcal{M}$ pair factorization. Consequently, $ATGI$-conditions can be shifted over $ATGI$-morphisms and rules.

Lemma 1 (shift of $ATGI$-conditions over morphisms and rules [7]).
1. There is a Shift-construction such that, for each $ATGI$-condition c over P and for each $ATGI$-morphism $b\colon P \to P'$, $n\colon P' \to H$, $n \circ b \models c \iff n \models \text{Shift}(b, c)$.
2. There is a construction Left such that, for each $ATGI$-rule $\varrho = \langle L \hookleftarrow K \hookrightarrow R \rangle$, each $ATGI$-condition ac over R, and each direct transformation $G \Rightarrow_{\varrho,g,h} H$, we have $g \models \text{Left}(\varrho, \text{ac}) \iff h \models \text{ac}$.

In the following, we illustrate at an example how a Core OCL invariant is translated to a left application condition.

We present a simple rule to create places in a Petri net. The graph constraint from Example 2 shall be translated to a left application condition. The conditions are given in abbreviated form (i.e. whenever it is unambiguous, only the codomain of a morphism is shown); node mappings are obvious from their relative position. Edge labels are omitted for brevity.

Right application condition:

This condition states that new and existing places have to come with non-negative numbers of tokens. The left application condition looks as follows (after trivial simplifying):

$$\forall \left(\boxed{\text{:Petrinet}} \longrightarrow \boxed{\text{self:Place}} \longleftarrow \boxed{\text{:Petrinet}}, \exists \left(\boxed{\begin{array}{c}\text{self:Place}\\ \text{token} \geq 0\end{array}} \longleftarrow \boxed{\text{:Petrinet}} \right) \right)$$

This states that every place in the Petri net has a non-negative token count.

6 Related Work and Conclusion

In the literature, there are several significant approaches to define a formal semantics for OCL. The motivations for a formal OCL semantics are manifold and include defining a clear semantics, generating model instances, and performing formal verification of UML/OCL models. All main approaches are logic-oriented, in contrast to ours being the first one that relates OCL to a graph-based approach. In the following, we sketch logic-oriented approaches using the Key prover, the Alloy project, and Constraint Logic Programming, respectively.

In [15], Beckert et al. present a translation of UML class diagrams with OCL constraints into first-order logic; the goal is logical reasoning about UML models. The translation has been implemented as a part of the KeY system, but can also be used stand-alone. Formal methods such as Alloy [16] can be used for instance generation: After translating a class diagram to Alloy, an instance can be generated or it can be shown that no instances exist. This generation relies on the use of SAT solvers and can also enumerate all possible instances. In [17], UML models are automatically transformed to corresponding Alloy representations. Alloy models can then be analyzed automatically, with the help of the Alloy Analyzer. A recent work translating OCL to relational logic is presented in [18] covering more features than UML2Alloy. The USE tool [6,19] can be used for generating snapshots that conform to the model or for checking the conformity of a specific instance. In [20], Cabot et al. present UMLtoCSP, a tool that is able to automatically check correctness properties of UML class model with OCL constraints based on Constraint Logic Programming.

All these approaches have in common that they translate class models with OCL constraints into logical facts and formulas forgetting about the graph properties of class models and their instances. Hence, the reasoning is performed on the level of model elements. Translating OCL invariants to graph constraints allows to keep graph structures as units of abstraction while checking for satisfiability. Pennemann has shown in [14] that a theorem prover for graph conditions works more efficient than theorem provers for logical formulas being applied to graph conditions. The key idea is here that graph axioms are always satisfied by default when using a theorem prover for graph conditions. Furthermore, a translation of OCL to graph constraints yields a new visualization of OCL which can help understanding. And finally, our translation offers a way to translate Core OCL invariants to application conditions of transformation rules. This is a new form to apply an OCL translation which might lead to number of new applications including test model generation as well as auto-completion of model editing operations. The backward translation from graph conditions to OCL might also be interesting to come up with model transformation rules restricted by OCL. In future work, we plan to extend this work towards the whole range of OCL invariants being translated to more powerful graph conditions.

Acknowledgement. We are grateful to Christoph Peuser and the anonymous referees for their helpful comments on a draft version of this paper.

References

1. OMG: Object Constraint Language, http://www.omg.org/spec/OCL/
2. Bardohl, R., Minas, M., Schürr, A., Taentzer, G.: Application of Graph Transformation to Visual Languages. In: Handbook of Graph Grammars and Computing by Graph Transformation, vol. 2, pp. 105–180. World Scientific (1999)
3. Habel, A., Pennemann, K.H.: Correctness of High-Level Transformation Systems Relative to Nested Conditions. Mathematical Structures in Computer Science 19, 245–296 (2009)
4. Biermann, E., Ermel, C., Taentzer, G.: Formal foundation of consistent EMF model transformations by algebraic graph transformation. Software and System Modeling 11(2), 227–250 (2012)
5. Chiorean, D., Bortes, M., Corutiu, D.: Proposals for a Widespread Use of OCL. In: Workshop on Tool Support for OCL and Related Formalisms, Technical Report LGL-REPORT-2005-001, EPFL, 68–82 (2005)
6. Richters, M.: A Precise Approach to Validating UML Models and OCL Constraints. PhD thesis, Universität Bremen, Logos Verlag, Berlin (2002)
7. Arendt, T., Habel, A., Radke, H., Taentzer, G.: From Core OCL Invariants to Nested Graph Constraints: Extended version (2014), http://www.uni-marburg.de/fb12/forschung/berichte/berichteinformtk/pdfbi/bi2014-01.pdf
8. Ehrig, H., Ehrig, K., Prange, U., Taentzer, G.: Fundamental Theory of Typed Attributed Graph Transformation based on Adhesive HLR Categories. Fundamenta Informaticae 74(1), 31–61 (2006)
9. Ehrig, H., Ehrig, K., Prange, U., Taentzer, G.: Fundamentals of Algebraic Graph Transformation. EATCS Monographs of Theoretical Computer Science. Springer (2006)
10. Orejas, F.: Symbolic Graphs for Attributed Graph Constraints. J. Symb. Comput. 46(3), 294–315 (2011)
11. Bardohl, R., Ehrig, H., de Lara, J., Taentzer, G.: Integrating Meta-modelling Aspects with Graph Transformation for Efficient Visual Language Definition and Model Manipulation. In: Wermelinger, M., Margaria-Steffen, T. (eds.) FASE 2004. LNCS, vol. 2984, pp. 214–228. Springer, Heidelberg (2004)
12. Poskitt, C.M., Plump, D.: Hoare-Style Verification of Graph Programs. Fundamenta Informaticae 118(1-2), 135–175 (2012)
13. Ehrig, H., Golas, U., Habel, A., Lambers, L., Orejas, F.: M-Adhesive Transformation Systems with Nested Application Conditions. Part 1: Parallelism, Concurrency and Amalgamation. Mathematical Structures in Computer Science 24 (2014)
14. Pennemann, K.H.: Development of Correct Graph Transformation Systems. PhD thesis, Universität Oldenburg (2009)

15. Beckert, B., Keller, U., Schmitt, P.H.: Translating the Object Constraint Language into First-order Predicate Logic. In: VERIFY, Workshop at Federated Logic Conferences, FLoC (2002)
16. Jackson, D.: Alloy Analyzer website (2012), http://alloy.mit.edu/
17. Anastasakis, K., Bordbar, B., Georg, G., Ray, I.: On challenges of model transformation from UML to Alloy. Software and System Modeling 9(1), 69–86 (2010)
18. Kuhlmann, M., Gogolla, M.: From UML and OCL to Relational Logic and Back. In: France, R.B., Kazmeier, J., Breu, R., Atkinson, C. (eds.) MODELS 2012. LNCS, vol. 7590, pp. 415–431. Springer, Heidelberg (2012)
19. Gogolla, M., Bohling, J., Richters, M.: Validating UML and OCL models in USE by automatic snapshot generation. SoSyM 4(4), 386–398 (2009)
20. Cabot, J., Clarisó, R., Riera, D.: UMLtoCSP: A Tool for the Formal Verification of UML/OCL Models using Constraint Programming. In: 22nd IEEE/ACM International Conference on Automated Software Engineering (ASE), pp. 547–548 (2007)

Specification and Verification of Graph-Based Model Transformation Properties*

Gehan M.K. Selim[1], Levi Lúcio[2], James R. Cordy[1],
Juergen Dingel[1], and Bentley J. Oakes[2]

[1] School of Computing, Queen's University, Kingston ON K7L2N8, Canada
{gehan,cordy,dingel}@cs.queensu.ca
[2] School of Computer Science, McGill University, Montreal QC H3A2A7, Canada
levi@cs.mcgill.ca, bentley.oakes@mail.mcgill.ca

Abstract. We extend a previously proposed symbolic model transformation property prover for the DSLTrans transformation language. The original prover generated the set of *path conditions* (i.e., symbolic transformation executions), and verified atomic contracts (constraints on input-output model relations) on these path conditions. The prover evaluated atomic contracts to yield either *true* or *false* for the transformation when run on any input model. In this paper we extend the prover such that it can verify atomic contracts and more complex properties composed of atomic contracts. Besides demonstrating our prover on a simple transformation, we use it to verify different kinds of properties of an industrial transformation. Experiments on this transformation using our prover show a speed-up in verification run-time by two orders of magnitude over another verification tool that we evaluated in previous research.

Keywords: MDD, model transformation, verification, property prover.

1 Introduction

In Model-Driven Development (MDD), *models* are the basic blocks of software development, and *model transformations* are used to map between models conforming to different metamodels. Given their key role in MDD, verification of transformations is becoming of increasing interest to researchers [2,16].

In this study, we formulate and focus on the following research question: "How can we efficiently verify properties of transformations expressed as input-output model relations?". We focus on properties expressed as input-output model relations since they have been highly investigated in the literature, using both textual (e.g., [9,3]) and graphical (e.g., [5,20]) property languages. After a thorough review of studies addressing the same research question, we found several limitations in the state of the art. For example, several studies translate either the transformation (e.g., [9,10]) or the property (e.g., [11]) of interest to an intermediate format to facilitate verification, without proving the soundness of the

---

* This work is supported in part by NSERC, as part of the NECSIS Automotive Partnership with General Motors, IBM Canada and Malina Software Corp.

H. Giese and B. König (Eds.): ICGT 2014, LNCS 8571, pp. 113–129, 2014.

translation. Secondly, other studies propose incomplete verification techniques that do not account for all possible transformation executions (e.g., [9]). Finally, a large number of studies proposed *input-dependent* [2] verification techniques (e.g., Henshin [4], AGG [20]) that prove properties for transformations only when run on a specific input. More general, input-independent techniques are needed were property verification is to be performed only once for the transformation, and verification results are to be guaranteed for all possible inputs.

In an attempt to answer the above research question and overcome limitations of previous studies, we investigate verifying properties of transformations implemented in the graph-based model transformation language DSLTrans [7]. DSLTrans is non-Turing complete, i.e., DSLTrans cannot specify transformations that require unbounded loops (e.g., simulation transformations). We extend a symbolic model transformation property prover for DSLTrans [14,12] that was previously limited to verifying atomic contracts (i.e., constraints on input-output model relations). The extension we present in this paper supports a more expressive property language that facilitates verifying atomic contracts and compositions of atomic contracts in the form of propositional logic formulae. Moreover, our prover now handles rules that overlap in their application.

The contribution of this study, at a high level, is extending a DSLTrans property prover that is *input-independent* [2], i.e., verification results generated by the prover hold for all possible inputs. Our specific contributions are:

- We describe how our prover currently handles overlapping rules (Section 4).
- We introduce our new property language, and show how it can be used to express commonly occurring properties, e.g., multiplicity invariants. (Section 5).
- We apply our extended prover to an industrial case study [18] (Section 6).
- We demonstrate how our extensions of the prover led to a two orders of magnitude improvement in execution time over the verification tool we used in another study [17]. We also discuss the strengths and limitations of our prover (Section 7).

This study adds to the state of the art (Section 8) and is useful to transformation verification research in general. We provide some evidence for our prover's scalability and usefulness since verification using our prover does not have to be performed for each input. Thus, we motivate researchers to adopt our prover. Moreover, users of languages other than DSLTrans can benefit from our study in two ways: (1) the study can be used as a guide to develop input-independent verification tools for any language; (2) higher order transformations (HOTs) can be developed to convert transformations in other languages to DSLTrans to enable using our prover. To develop such HOTs, research has to be conducted to understand what class of transformations can be translated to DSLTrans.

Section 2 summarizes DSLTrans and it's simplest properties; Section 3 overviews our prover's architecture; Section 4 describes path condition generation; Section 5 discusses our prover's verification technique; Section 6 demonstrates an industrial case study; Section 7 discusses our prover's strengths and limitations; Section 8 reviews related work; Section 9 concludes and presents future work.

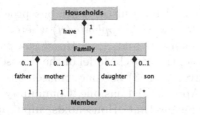

Fig. 1. Household Language

Fig. 2. Community Language

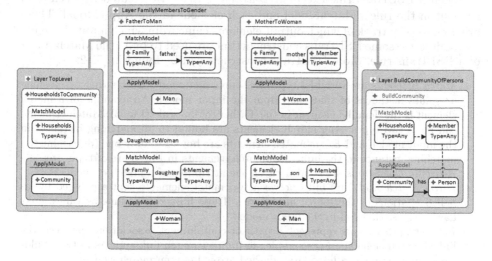

Fig. 3. The *Persons* Transformation expressed in DSLTrans

2   The DSLTrans Model Transformation Language

DSLTrans [7] is a graph-based transformation language that can be used to specify out-place (i.e., input-preserving) model transformations that are confluent and terminating by construction. DSLTrans rules are constructive – elements can be created but not deleted. The semantics of DSLTrans (currently defined using set theory) are in-line with, and can be defined using, pushout approaches. We demonstrate DSLTrans using a simple transformation as a running example.

Figs. 1 and 2 present two metamodels used to describe different representations of a set of people. The 'Household Language' represents people as members of families which in turn form a set of households. The 'Community Language' represents people as men or women who belong to a community.

Fig. 3 presents a DSLTrans transformation that aims to transform family members in the 'Household Language' (source metamodel) into men and women of a community in the 'Community Language' (target metamodel). In what follows, we refer to the transformation in Fig. 3 as the *Persons* transformation.

A DSLTrans transformation is composed of an ordered set of layers (e.g., 'TopLevel', 'FamilyMembersToGender', and 'BuildCommunityOfPersons' layers in Fig. 3) that are executed sequentially. A layer consists of a set of transformation rules that execute in a non-deterministic order but produce a deterministic result. Each rule is a pair (*MatchModel, ApplyModel*) where *MatchModel* is a pattern of source metamodel elements and *ApplyModel* is a pattern of target metamodel elements. For example, the MatchModel of the 'HouseholdsToCommunity' rule in the 'TopLevel' layer (Fig. 3) has one 'Households' class from the 'Household Language' and the ApplyModel has one 'Community' class from the 'Community Language'. This means that 'Households' input model elements will be transformed into 'Community' output model elements.

When a DSLTrans rule executes, *traceability links* are created between each element in the rule's MatchModel and each element in the ApplyModel. These are used to keep track of which output elements came from which input elements.

We describe some DSLTrans constructs that are used to build the MatchModel of a DSLTrans rule. More DSLTrans constructs can be found in [7,12].

- *Match Elements* are variables typed by source metamodel classes that can assume as values instances of that class from the input model. An example of a match element is the 'Family' element in the 'FatherToMan' rule (Fig. 3). Match elements can be of two types: *Any* match elements are bound to all matching instances in the input model, and *Exists* match elements are bound to only one (deterministic) matching instance in the input. All match elements in Fig. 3 are of type *Any*.
- *Attribute Conditions* are conditions on the attributes of a match element.
- *Direct Match Links* are links between two match elements that are typed by labelled relations of the source metamodel. These links can assume as values links having the same label in the input model.
- *Indirect Match Links* represent a path of containment associations between the linked match elements. For example, an indirect match link appears in the 'BuildCommunity' rule as a horizontal, dashed arrow between match elements.
- *Backward Links* link elements of the MatchModel and the ApplyModel of a rule, e.g., backward links are used in the 'BuildCommunity' rule and are denoted as vertical, dashed lines. Backward links are used to refer to traceability links between input and output model elements that are generated by the rules of previous layers.

Similar constructs can be used to build a rule's ApplyModel, as shown in Fig. 3.

- *Apply elements* are variables typed by target metamodel classes and linked by *apply links*. Apply elements that are not connected by backward links create output elements of the same type each time the MatchModel is found in the input. Apply elements that are connected by backward links are handled differently, e.g., 'BuildCommunity' rule connects 'Community' and 'Person' output elements that were formerly created from 'Households' and 'Member' input elements with a 'has' link.
- Apply elements can have *apply attributes* that can be set from references to one or more attributes of match elements.

AtomicContracts in DSLTrans: An *AtomicContract* is the simplest property expressible in our prover. Each *AtomicContract* is a pair (*pre, post*) that specifies a property of the form: "if the input model satisfies the precondition *pre*, then the output model should satisfy the postcondition *post*". A precondition is a constraint on the transformation's input in the form of a structural relation between input elements. A postcondition is a constraint on the transformation's output in the form of a structural relation between output elements. Preconditions and

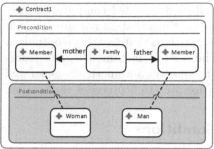
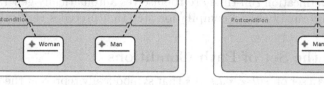

Fig. 4. *Contract1*; should hold   **Fig. 5.** *Contract2*; should not hold

postconditions are expressed using the same constructs as rules. Postconditions may also have traceability links to link postcondition elements to precondition elements. This signifies that the property will only match an output element that was previously created from an input element.

Figs. 4 and 5 demonstrate two *AtomicContracts* for the *Persons* transformation. Fig. 4 is interpreted as: "a mother and a father in a family will always be transformed to a woman and a man". Fig. 5 is interpreted as: "a family including a mother and a daughter will always be transformed to a man". Our prover should verify that *Contract1* (Fig. 4) will always hold for the *Persons* transformation, and *Contract2* (Fig. 5) will not always hold (with a counterexample).

3   The Symbolic Model Transformation Property Prover

Fig. 6 demonstrates our property prover's final architecture. Our prover takes four inputs: the DSLTrans transformation of interest, the transformation's source and target metamodels, and the property to verify. Verification is then carried out in two steps, as shown in Fig. 6. First, the prover generates the set of *path conditions* representing all possible executions of the input transformation (Section 4). Then, the prover verifies the input property on the generated set of path conditions and renders the property to be either *true* or *false* (with a counter example) for the transformation when run on any input model (Section 5).

We have chosen Python and T-Core [19] to implement our prover. T-Core is a Python library with primitives that support typed graph manipulation (e.g., graph matching/rewriting) and composition of these primitives into transformation blocks. The use of Python and T-Core allowed constructing our prover using MDD principles. In other words, all artifacts used at verification run-time are

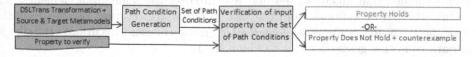

Fig. 6. The architecture of our symbolic model transformation property prover

models (instances of explicit metamodels), all model-related computations are implemented as transformations, and all computations that do not directly manipulate models are implemented as Python algorithms that have been optimized to minimize memory usage and run-time. The models, metamodels, and transformations used at verification run-time are themselves automatically generated by higher order transformations in a compilation step that precedes verification.

4   Generating the Set of Path Conditions

Our prover generates a set of *path conditions* that symbolically represent the possible transformation executions. For a transformation with n layers, our prover uses the transformation rules to build the path conditions in n iterations. In Fig. 7, we demonstrate how the path conditions for the *Persons* transformation are generated in iterations. We identify every rule in each layer of Fig. 3 with a pair of numbers, e.g., 4_2 corresponds to the fourth rule (ordered from top to bottom and then from left to right in Fig. 3) in the second layer (i.e., 'SonToMan' rule). We start off with the empty path condition, where we assume no transformation rule has been applied. To generate path conditions in iteration 1, the empty path condition is combined with all possible rule combinations of the first transformation layer. Similarly, to generate path conditions in iteration 2, each path condition from iteration 1 is combined with all *applicable* rule combinations of the second layer. A rule combination of the second layer that does not have backward links is always *applicable*, since it does not depend on rules from the first layer. Rule combinations of the second layer with backward links are combined with a path condition from iteration 1 only if the path condition generates the elements linked by backward links in the rule combination.

Each path condition thus accumulates a set of rules describing a possible path of rule applications through the transformation's layers. We refer to the accumulated MatchModels (or ApplyModels) of all the rules in a path condition as the path condition's *match pattern* (or *apply pattern*). Since our technique abstracts from how many times the rule executes for an input, a transformation rule only occurs once in each path condition. Thus, a path condition symbolically represents a set of concrete executions since each of the rules in a path condition can be concretely executed any number of times on an input model.

In Fig. 8, we show the path condition of the node with the dotted edge in Fig. 7. As shown from the numbers in the node, the path condition contains four combined rules (i.e., 'HouseholdsToCommunity', 'FatherToMan', 'MotherToWoman', 'BuildCommunity') and traceability links. When combining the rules, elements of the same type of the combined rules can be merged. This represents the fact that different rules may execute over the same input elements.

Only the path conditions from the last iteration are returned as the result since they capture all the possible *complete* transformation executions. Details on path condition generation can be found in [12].

Overlapping Rules: The industrial transformation presented later in Section 6 had *overlapping rules* which required treatment during path condition generation. Overlapping rules are defined as follows: when two rules in the same layer

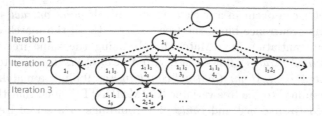

Fig. 7. Generation of the set of path conditions in iterations

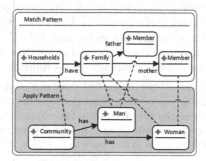

Fig. 8. A path condition of the Persons transformation

use match elements of the same metamodel classes of type *Any* or *Exists*, then
the MatchModel of one rule syntactically *subsumes* the MatchModel of the other
rule. For example, a rule having a MatchModel containing an *Any* match element of class 'A' is subsumed by a MatchModel of another rule that contains an
Exists match element of class 'A' and an *Any* match element of class 'B'.

Our path condition generation algorithm was extended to handle overlapping
rules. This extension led to a pronounced decrease in the number of generated
path conditions in our case study, since a set of rules in a *subsumption* relation
(described above) can often be merged into a smaller set of rules. Depending on
whether rules overlap totally or partially, rule merge may be done before or during path condition generation. This extension leads to an improved management
of the combinatorial explosion in path condition generation.

5   Verification of the Property of Interest

We extended the technique proposed in [14] for verifying *AtomicContracts* of
DSLTrans transformations to enable the verification of more complex properties.
Our extended technique employs the following syntax and semantics.

Syntax: Our syntax is based on propositional logic. An *AtomicContract* (*pre,post*)
is the smallest unit in our property language. A propositional formula can be
built using one or more *AtomicContracts* and the operators $\neg_{tc}$ (*not*), $\vee_{tc}$ (*or*),
$\wedge_{tc}$ (*and*), and $\implies_{tc}$ (*implication*), where *tc* stands for "transformation contract". Assuming that (*pre,post*) is an element of the set of *AtomicContracts*
AC, the syntax of formulae is:

$$\varphi := (pre, post) \mid \neg_{tc}\varphi \mid \varphi \vee_{tc} \varphi \mid \varphi \wedge_{tc} \varphi \mid \varphi \implies_{tc} \varphi \tag{1}$$

Free variables can occur in any element e of an *AtomicContract*'s pre/ post-condition. This occurrence binds the free variable to all the matches found for e within an instantiation of a MatchModel. Using the same free variable in different *AtomicContracts* allows these *AtomicContracts* to refer to the same matched element, e.g., *AtomicContract cont1* in Fig. 9 binds a matched element of type 'Community' to the free variable 'COMMUNITY' such that this element can be referred to in *cont2* and *cont3*. The bindings of a set of free variables $\{var_1, \ldots, var_n\}$ (in elements $\{e_1, \ldots, e_l\}$ of an *AtomicContract*) to matched elements $\{m_1, \ldots, m_n\}$ in a path condition is expressed as a binding function $l = \{(var_1, m_1), \ldots, (var_n, m_n)\}$, i.e., $l \in \mathcal{P}(FV \times BE)$, where FV and BE are the sets of free variables and bound elements, and $\mathcal{P}$ is the power set operator.

Semantics: We define a function $eval_{Atomic}(pc, c)$ that evaluates an *Atomic-Contract* $c = (pre, post)$ for a path condition pc as follows:

1. If pc contains an isomorphic copy of *pre* but does not contain an isomorphic copy of *post*, then $eval_{Atomic}(pc, c)$ returns *false* (i.e., c does not hold for pc and the transformation) and an empty set of binding functions $L = \emptyset$.
2. Otherwise, $eval_{Atomic}(pc, c)$ returns *true* (i.e., c holds for pc) and a set of binding functions L for the free variables of c, where $L \subseteq \mathcal{P}(FV \times BE)$.

Thus, $eval_{Atomic}$ is defined as $eval_{Atomic} : PC \times AC \rightarrow \{true,\ false\} \times \mathcal{P}(FV \times BE)$, where PC is the set of path conditions of a transformation τ. Note that a set L of binding functions is returned since an *AtomicContract* may evaluate to true using different bindings of the free variables. Thus, L is constructed from all binding functions l_i returned by all possible subgraph isomorphisms.

Assuming that *FORMULAE* is the set of elements generated by the grammar in Eqn.(1), we evaluate a formula φ for a path condition $pc \in PC$ using a function $eval:PC \times FORMULAE \rightarrow \{true,\ false\} \times \mathcal{P}(FV \times BE)$ as follows:

$$eval(pc, \varphi) = \begin{cases} (res_1, L_1) & \text{if } \varphi \in AC, eval_{Atomic}(pc, \varphi) = (res_1, L_1) \\ (\neg res_1, L_1) & \text{if } \varphi = \neg_{tc}\psi, eval(pc, \psi) = (res_1, L_1) \\ ((res_1 \vee res_2) \wedge C(L_1, L_2), & \text{if } \varphi = \psi \vee_{tc} \phi, eval(pc, \psi) = (res_1, L_1), \\ \quad L_1 \cup L_2) & eval(pc, \phi) = (res_2, L_2) \\ ((res_1 \wedge res_2) \wedge C(L_1, L_2), & \text{if } \varphi = \psi \wedge_{tc} \phi, eval(pc, \psi) = (res_1, L_1), \\ \quad L_1 \cup L_2) & eval(pc, \phi) = (res_2, L_2) \\ ((res_1 \implies res_2) \wedge C(L_1, L_2), & \text{if } \varphi = \psi \implies_{tc} \phi, eval(pc, \psi) = (res_1, L_1), \\ \quad L_1 \cup L_2) & eval(pc, \phi) = (res_2, L_2) \end{cases}$$
$$(2)$$

where the semantics of the propositional operators $(\neg, \vee, \wedge, \implies)$ is standard, and $res_i \in \{true,\ false\}$. The consistency function $C : \mathcal{P}(FV \times BE) \times \mathcal{P}(FV \times BE) \rightarrow \{true,\ false\}$ checks for two sets of binding functions (e.g., L and L') that all free variables bound by a binding function in the first set L will always be bound to the same elements by a binding function of the second set L' as follows:

$$C(L, L') = \forall l \in L, \exists l' \in L' : \left(\forall v \in FV_l : ((v, m) \in l \wedge (v, m') \in l') \implies m = m'\right) \text{ and}$$
$$\forall l' \in L', \exists l \in L : \left(\forall v \in FV_{l'} : ((v, m') \in l' \wedge (v, m) \in l) \implies m' = m\right)$$
$$(3)$$

where $m, m' \in BE$, and FV_l, $FV_{l'}$ are the sets of free variables used in l and l' respectively. Based on the former definitions, we evaluate a formula φ for a

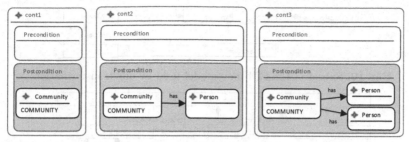

Fig. 9. Three *AtomicContracts* that can be used with different propositional operators to convey different properties for the *Persons* transformation

transformation τ (with path conditions PC) using a function $eval(\tau, \varphi)$:

$$eval(\tau, \varphi) = \begin{cases} true & \text{if } \forall pc \in PC : eval(pc, \varphi) = (true, L) \\ false & \text{otherwise} \end{cases} \quad (4)$$

where L is any set of binding functions. Thus, $eval(\tau, \varphi)$ renders a property φ to be *true* or *false* for a transformation τ by verifying φ for each path condition. Function $eval(\tau, \varphi)$ returns *true* only if for all path conditions of τ, φ holds and the bindings of all free variables consistently refer to the same elements.

Formulae of *AtomicContracts*: The new syntax and semantics allows us to formulate complex properties by composing propositional formulae of *Atomic-Contracts*. We demonstrate how the *AtomicContracts* in Fig. 9 (i.e., *cont1*, *cont2*, *cont3*) together with free variables can be used with different propositional operators to convey multiplicity invariants. A property that mandates that the *Persons* transformation will always generate an output where every community has one or more 'Persons' (i.e., a multiplicity invariant of '1..*') can be expressed as '*cont1* $\Longrightarrow_{tc}$ *cont2*'. In other words, if an element of type 'Community' is generated in the output, then this element must have at least one 'Person'. Whereas the property '*cont1* $\Longrightarrow_{tc}$ (*cont2* $\wedge_{tc} \neg_{tc}$*cont3*)' expresses a multiplicity invariant of '1..1' (i.e, if a 'Community' is generated in the output, then this 'Community' must have one 'Person' and not more).

6 Industrial Case Study

Previously in [18], we developed an industrial transformation that maps between subsets of a legacy metamodel for General Motors (GM) and the AUTOSAR metamodel. In that work, we focused on subsets of the metamodels that represent the deployment and interaction of software components. Later in [17], we proposed properties of interest for our GM-2-AUTOSAR transformation.

We use our prover to verify the properties proposed in [17] on the GM-2-AUTOSAR transformation [18] after reimplementing it in DSLTrans. In this section, we summarize the transformation [18] and its properties [17]. Then, we discuss formulating and verifying these properties using our prover.

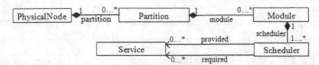

Fig. 10. Subset of the GM metamodel used by our transformation.

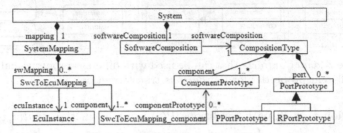

Fig. 11. Subset of the AUTOSAR metamodel used by our transformation

6.1   GM-2-AUTOSAR Model Transformation

The Source GM Metamodel: Fig. 10 illustrates the subset of the GM meta-model used in our transformation in [18]. A *PhysicalNode* may contain multiple *Partitions* (i.e., processing units). Multiple *Modules* can be deployed on a single *Partition*. A *Module* is an atomic, deployable, and reusable software element and can contain multiple *Schedulers*. A *Scheduler* is the basic unit for software scheduling. It contains behavior-encapsulating entities, and is responsible for providing/requiring *Services* to/from these behavior-encapsulating entities.

The Target AUTOSAR Metamodel: In AUTOSAR, an Electronic Component Unit (ECU) is a physical unit on which software is deployed. Fig. 11 shows the subset of the AUTOSAR metamodel [1] used by our transformation. The ECU configuration is modeled using a *System* that aggregates *SoftwareComposition* and *SystemMapping*. *SoftwareComposition* points to *CompositionType* which eliminates any nested software components in a *SoftwareComposition*. *SoftwareComposition* models the architecture of the software components (i.e., *ComponentPrototypes*) deployed on an ECU and their ports (i.e., *PPortPrototype*/ *RPortPrototype* for providing/ requiring data and services).

SystemMapping binds software components to ECUs using *SwcToEcuMappings*. *SwcToEcuMappings* assign *SwcToEcuMapping_components* to an *EcuInstance*. *SwcToEcuMapping_components*, in turn, refer to *ComponentPrototypes*.

Reimplementation of the GM-2-AUTOSAR Transformation in DSLTrans: We reimplemented the GM-2-AUTOSAR transformation [18] in DSLTrans so that we can verify it in our prover. Table 1 shows the rules in each transformation layer, and the input/output types that are mapped/generated by each rule. Rules of the first and third layers create output elements. Rules of the second layer generate associations between elements created by the the first layer (shown in the actual transformation using backward links). Thus, the input and output types shown for the rules of the second layer are types that have already been matched and created and for which the rules create associations.

Table 1. The rules in each layer of the GM-2-AUTOSAR transformation after reimplementing it in DSLTrans, and their input and output types

Layer	Rule Name	Input Types	Output Types
1	MapPhysNode2FiveElements	PhysicalNode	System, SystemMapping, SoftwareComposition, CompositionType, EcuInstance
	MapPartition	Partition	SwcToEcuMapping
	MapModule	Module	SwCompToEcuMapping_component, ComponentPrototype
2	MapConnPhysNode2Partition	PhysicalNode, Partition	SystemMapping, EcuInstance, SwcToEcuMapping
	MapConnPartition2Module	PhysicalNode, Partition, Module	CompositionType, ComponentPrototype, SwcToEcuMapping, SwCompToEcuMapping_component
3	CreatePPortPrototype	Scheduler	PPortPrototype
	CreateRPortPrototype	Scheduler	RPortPrototype

Table 2. Properties of interest for the GM-2-AUTOSAR transformation

Multiplicity Invariants: *(Properties defined on the target metamodel elements only)*
(M1) Each *CompositionType* is associated to at least one *ComponentPrototype*.
(M2) Each *SoftwareComposition* is associated to one *CompositionType*.
(M3) Each *SwcToEcuMapping* is associated to at least one *SwcToEcuMapping_component*.
(M4) Each *SwcToEcuMapping* is associated to one *EcuInstance*.
(M5) Each *System* is associated to one *SoftwareComposition*.
(M6) Each *System* is associated to one *SystemMapping*.
Security Invariant: *(Property defined on the target metamodel elements only)*
(S1) All the composite *SwcToEcuMappings* of a *System* must refer to *ComponentPrototypes* that are contained within the *CompositionType* lying under the same *System*.
Pattern Contracts: *(Properties that relate source and target metamodel elements)*
(P1) If a *PhysicalNode* is connected to a *Service* through the *provided* association (in the input), then the corresponding *CompositionType* will be connected to a *PPortPrototype* (in the output).
(P2) If a *PhysicalNode* is connected to a *Service* through the *required* association (in the input), then the corresponding *CompositionType* will be connected to a *RPortPrototype* (in the output).

To represent positive application conditions (PACs) in our transformation rules, we use a combination of *Any* and *Exists* match elements (Section 2). For example, rule 'MapPhysNode2FiveElements' in Table 1 maps every *PhysicalNode* to five elements, only if the *PhysicalNode* is eventually connected to *at least one Module*. Thus, the *MatchModel* of rule 'MapPhysNode2FiveElements' has a *PhysicalNode* (*Any*) match element connected to *Partition* and *Module* (*Exists*) match elements. Similarly, rule 'MapModule' maps every *Module* (represented as *Any* match element) only if it is contained in one *PhysicalNode* and one *Partition* (represented as *Exists* match elements). The *MatchModel* of rule 'MapPartition' also has a *Partition* (*Any*) match element connected to *PhysicalNode* and *Module* (*Exists*) match elements to represent a PAC. Thus, the rules in the first layer totally overlap if we abstract from the match element types (i.e., *Any* or *Exists*). The extension explained in Section 4 combines the rules of the first layer into one path condition which simplifies property verification. Partially overlapping rules (Section 4) also occur in layer 2 of our transformation.

6.2 GM-2-AUTOSAR Model Transformation Properties

In [17], we stated that properties could be *invariants* or *contracts*. Invariants are properties defined on the target metamodel elements only, while contracts relate source and target metamodel elements. Based on these definitions, we further defined four categories of properties in [17]: *Multiplicity Invariants*, *Uniqueness Contracts*, *Security Invariants*, and *Pattern Contracts*. For each category, we formulated several properties that are summarized in Table 2 and discussed in [17]. We omit Uniqueness Contracts in this study since they require reasoning about attribute values, which is not yet implemented in our property prover.

Multiplicity invariants ensure that the transformation's output preserves the multiplicities in the AUTOSAR metamodel. The security invariant mandates that a *System* does not refer to a *ComponentPrototype* that is not allocated in that *System*. Pattern contracts require that if a pattern of elements is found in the input, then a corresponding pattern of elements must be found in the output.

6.3   Verifying Properties of the GM-2-AUTOSAR Transformation

We demonstrate the formulation of pattern contracts (e.g., *P1* and *P2* in Table 2) in our prover by showing the formulation of *P1* in Fig. 12 as an example. *P1* mandates that if a *PhysicalNode* is connected to a *Service* through the *provided* association in the input (as in the precondition of Fig. 12), then the corresponding *CompositionType* will be connected to a *PPortPrototype* in the output (as in the postcondition). As explained in Section 2, using a traceability link in Fig. 12 mandates that *P1* will only match *CompositionType*s that were previously created from *PhysicalNode*s. We demonstrate the formulation of '1..1' multiplicity invariants (e.g., *M2*, *M4*, *M5*, *M6*) by showing *M6* as an example. *M6* ensures that if a *System* is created in the output, then this *System* must be connected to one *SystemMapping* (and not more). Using the *AtomicContracts* in Fig. 13, *M6* can be expressed as $AC2 \Longrightarrow_{tc} (AC3 \wedge_{tc} \neg_{tc} AC4)$. Variable 'SYSTEM' mandates that if $AC2$ holds for a specific *System*, then $AC3$ should hold and $AC4$ should not hold for the same *System*. Changing the former formula to $AC2 \Longrightarrow_{tc} AC3$ expresses a '1..*' multiplicity invariant (e.g., *M1*, *M3*). Using the *AtomicContracts* in Fig. 14, the security invariant *S1* can be expressed as $AC5 \Longrightarrow_{tc} AC6$. Variables 'SYSTEM' and 'COMPONENTPROTOTYPE' mandate that if $AC5$ holds for a specific *System* and *ComponentPrototype* then $AC6$ should also hold for the same *System* and *ComponentPrototype*.

Verification Results: We used our prover to verify the properties in Table 2. The transformation was found to violate *M1* and *M3*, i.e., our prover uncovered the same bugs that we found in the ATL transformation implementation using another tool in [17]. After examining the counter examples (not shown due to space limitations), we identified and fixed the two bugs. The properties were reverified on the updated transformation, and they all returned *true*. (i.e., our transformation will always satisfy the properties in Table 2).

To assess our prover's performance, we measured the time taken to generate path conditions and to verify the properties (Table 2) of the GM-2-AUTOSAR transformation after fixing the bugs. The prover took on average 0.6 seconds to generate the path conditions. Table 3 (first row) shows the time taken (in seconds) to verify the properties in Table 2 using the generated path conditions. We do not include the time taken for path condition generation in Table 3 since

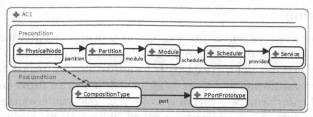

Fig. 12. One *AtomicContract* that is used to express property *P1*

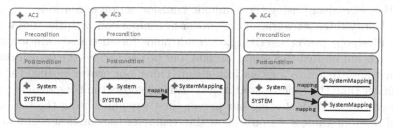

Fig. 13. Three *AtomicContracts* that are used to express property *M6*

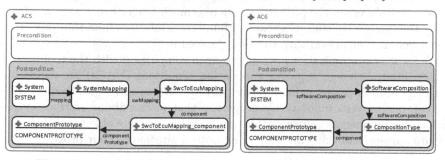

Fig. 14. Two *AtomicContracts* that are used to express property *S1*

Table 3. Time taken (in seconds) to verify the properties in Table 2 using our property prover (first row) and using a tool based on model finders [17] (second row)

Property	M1	M2	M3	M4	M5	M6	S1	P1	P2
Verification Time (our property prover)	.013	.017	.013	.017	.017	.019	.017	.02	.02
Verification Time ([17] at scope 6)	76	73.4	75	75	75.5	74.5	114	256	251

it is performed once for the transformation. The longest time taken to verify a property was 0.02 seconds (*P1*, *P2*). Thus, our prover can verify an industrial transformation's properties in a short time. More experiments are needed before we can claim that our prover scales to transformations of varying complexities.

Our property prover and the transformation used in [7,14] is available at [13]. The industrial transformation is not included for confidentiality reasons.

7   Discussion

We discuss the strengths and limitations of our prover by comparing it to a tool that we used to verify the GM-2-AUTOSAR transformation in [17]. The tool we used in [17] verifies ATL (textual) transformations by translating them to a relational representation and then using model finders to prove properties for the translated transformation within a scope (i.e., maximum number of objects per class). In contrast, the prover described in this study verifies DSLTrans (graphical) transformations in their native form (i.e., without translating them to another formalism) using the symbolic transformation executions.

We identify three strengths of our prover in comparison with the tool we used in [17]. First, our prover's verification result holds for all transformation executions and is not limited to a scope. Second, our prover verifies the transformation without translating it to another formalism. Third, our prover verified the properties faster than the tool we used in [17]. Table 3 shows the time taken to verify the properties in Table 2 using our prover (first row) and using the tool in [17] (second row). In Table 3, we only show the results for the smallest scope we used in [17] (i.e., 6). As shown in Table 3, our prover takes significantly shorter time to exhaustively verify the properties, whereas much longer times were needed to verify the same properties in a scope of 6 in [17]. Thus, we claim that our prover scales well in comparison with the tool we used in [17].

We identify two limitations of our prover in comparison with the tool we used in [17]. First, although negative application conditions (NACs) are expressible in DSLTrans, our prover cannot verify transformations with rules having NACs. Second, our prover cannot verify properties that reason about attribute values such as the uniqueness contracts (Section 6.2) that we were able to verify in [17]. We are currently working on addressing both limitations in our prover.

8   Related Work

We review input-independent verification techniques proposed for (1) textual and (2) graphical transformations, and (3) property languages similar to ours.

(1) Büttner et al. [9] and Cabot et al. [10] translated a transformation and its metamodels into a transformation model and used model finders (e.g., USE Validator) and constraint solvers (e.g., UMLtoCSP) to verify properties. Anastasakis et al. [3] and Baresi and Spoltini [6] translated a transformation into an Alloy model and used the Alloy Analyser to verify the Alloy model within a scope. Troya and Vallecillo [21] translated a transformation into Maude and used Maude's analysis capabilities to verify the transformation. Orejas and Wirsing [15] translated graphs to triple algebras to verify (e.g., using Maude) propositional formula of properties. The study claimed that verifying graph transformations is difficult, and hence the need for the translation to algebra.

(2) Becker et al. [8] verified if a transformation can generate *forbidden* patterns by checking if the backward application of each rule to each forbidden pattern can produce a valid input, and returns this input as a counterexample. Asztalos et al. [5] implemented a VMTS-based verification tool for in-place transformations. VMTS transformations are expressed as graphical rules scheduled by a

control flow graph. The tool assigns conditions to each edge in the control flow graph that are guaranteed to hold for the transformation (on any input) at this edge. Assigning conditions is performed by analyzing individual rules to generate their strongest post-conditions and iteratively propagating these conditions using inference rules. Eventually, the final edge in the control flow graph is assigned a condition p_{final} which will always hold for any input. A property p is then verified by evaluating $p_{final} \longrightarrow p$. Besides being semi-automated, another limitation of the tool is that a property's verification result may be *undecidable* due to (a) the lack of the necessary inference rules or (b) the need to collectively analyze the control flow graph instead of analyzing rules separately.

Tools such as Henshin [4] and AGG [20] have the drawback of being input-dependent, i.e., they verify transformations when run on a specific input. Similar to model checkers (e.g., Groove), Henshin [4] generates a state space that simulates all possible transformation executions for a specific input and verifies the generated state space. AGG [20] verifies a property on the input and reverifies it on the output of each rule application. AGG does not check all transformation executions; only the first found execution is verified. AGG, however, performs other types of analysis, e.g., critical pair analysis and graph parsing.

(3) Büttner et al. [9] expressed properties in OCL and verified them using model finders. PaMoMo [11] is a graphical language used to express contracts and complex properties that manipulate contracts. These properties can be compiled into OCL and injected into any OMG-based transformation implementation (e.g., ATL) for automated verification. The property languages used by Asztalos et al. [5] and AGG [20] are similar to ours; i.e., their graph-based property languages are used in their native graphical format and properties are contracts that can be used to build propositional formulae. The difference is that both studies [5,20] do not introduce a construct equivalent to our free variables which allow *AtomicContracts* in the same formula to refer to a specific element.

Difference Between Our Study and Related Work: Our study differs from related work in one or more of the following aspects: (i) Verification is performed on an intuitive, graphical language that does not require a mathematical background to be used, e.g., Maude [21,15]. (ii) We used our prover to verify a simple and an industrial transformation. (iii) We demonstrated several property kinds that our prover can conclusively verify (unlike [5]) as opposed to verifying specific property kinds, e.g., forbidden patterns [8]. (iv) Verification is based on generating the symbolic executions. (v) We have proved the *soundness* and *completeness* of our technique in [12]. Many studies translated a transformation into another formalism and verified properties on the translated transformation [9,10,3,6,21,15]. Such approaches should prove the soundness of the translated transformation before verifying properties. Moreover, such approaches should translate the verification result back to the original formalism for comprehension. Other studies proposed incomplete techniques that are restricted to a scope [9] or that do not guarantee that the transformation is fault-free, e.g., testing.

While textual property languages (e.g., OCL [9]) have been used for specifying properties, we believe that a graphical property language is useful as more researchers adopt graph transformations due to their intuitive, graphical format. Approaches where graphical properties are translated into a textual formalism (e.g., [11]) have two drawbacks: (a) the soundness of the translation should be proved before verifying the translated properties; (b) the translated properties in [11] cannot be used to automatically verify graphical transformations.

9   Conclusion and Future Work

In this study we extended a symbolic model transformation property prover [14,12] that initially only verified *AtomicContracts*. The extended prover now verifies *AtomicContracts* and propositional formulae of *AtomicContracts* for DSLTrans transformations. We have also extended the original path condition generation algorithm by treating overlapping rules. Further, we demonstrated our property prover on an industrial case study [18]. We showed that the prover is of practical use and features fast property proving times when compared with another prover. We also discussed the strengths and limitations of our prover.

For future work, more experiments on bigger transformations are needed to test the prover's scalability. Moreover, as mentioned in Sections 6.2 and 7, we plan to handle rules with NACs and attribute values when generating path conditions to facilitate verifying properties that reason about attribute values.

References

1. AUTOSAR Consortium. AUTOSAR System Template (2007),
   http://autosar.org/download/R3.1/AUTOSAR_SystemTemplate.pdf
2. Amrani, M., Lúcio, L., Selim, G., Combemale, B., Dingel, J., Vangheluwe, H., Le Traon, Y., Cordy, J.R.: A Tridimensional Approach for Studying the Formal Verification of Model Transformations. In: VOLT, pp. 921–928 (2012)
3. Anastasakis, K., Bordbar, B., Küster, J.: Analysis of Model Transformations via Alloy. MoDeVVa, 47–56 (2007)
4. Arendt, T., Biermann, E., Jurack, S., Krause, C., Taentzer, G.: Henshin: Advanced Concepts and Tools for In-Place EMF Model Transformations. In: Petriu, D.C., Rouquette, N., Haugen, Ø. (eds.) MODELS 2010, Part I. LNCS, vol. 6394, pp. 121–135. Springer, Heidelberg (2010)
5. Asztalos, M., Lengyel, L., Levendovszky, T.: Formal Specification and Analysis of Functional Properties of Graph Rewriting-Based Model Transformation. Software Testing, Verification and Reliability 23(5), 405–435 (2013)
6. Baresi, L., Spoletini, P.: On the Use of Alloy to Analyze Graph Transformation Systems. In: Corradini, A., Ehrig, H., Montanari, U., Ribeiro, L., Rozenberg, G. (eds.) ICGT 2006. LNCS, vol. 4178, pp. 306–320. Springer, Heidelberg (2006)
7. Barroca, B., Lúcio, L., Amaral, V., Félix, R., Sousa, V.: DSLTrans: A Turing Incomplete Transformation Language. In: Malloy, B., Staab, S., van den Brand, M. (eds.) SLE 2010. LNCS, vol. 6563, pp. 296–305. Springer, Heidelberg (2011)
8. Becker, B., Beyer, D., Giese, H., Klein, F., Schilling, D.: Symbolic Invariant Verification for Systems with Dynamic Structural Adaptation. In: ICSE (2006)

9. Büttner, F., Egea, M., Guerra, E., De Lara, J.: Checking Model Transformation Refinement. In: ICMT, pp. 158–173 (2013)
10. Cabot, J., Clarisó, R., Guerra, E., de Lara, J.: Verification and Validation of Declarative Model-to-Model Transformations Through Invariants. Systems and Software 83(2), 283–302 (2010)
11. Guerra, E., de Lara, J., Kolovos, D., Paige, R.: A Visual Specification Language for Model-to-Model Transformations. In: VL/HCC, pp. 119–126. IEEE (2010)
12. Lúcio, L., Oakes, B., Vangheluwe, H.: A Technique for Symbolically Verifying Properties of Graph-Based Model Transformations. Technical Report SOCS-TR-2014.1, McGill U (2014)
13. Lúcio, L., Selim, G.: DSLTrans Property Prover and Example Transformation, http://msdl.cs.mcgill.ca/people/levi/ police_station_verification_example.zip
14. Lúcio, L., Vangheluwe, H.: Model Transformations to Verify Model Transformations. In: VOLT (2013)
15. Orejas, F., Wirsing, M.: On the Specification and Verification of Model Transformations. In: Palsberg, J. (ed.) Mosses Festschrift. LNCS, vol. 5700, pp. 140–161. Springer, Heidelberg (2009)
16. Rahim, L.A., Whittle, J.: A Survey of Approaches for Verifying Model Transformations. SoSyM, 1–26 (2013)
17. Selim, G.M.K., Büttner, F., Cordy, J.R., Dingel, J., Wang, S.: Automated Verification of Model Transformations in the Automotive Industry. In: Moreira, A., Schätz, B., Gray, J., Vallecillo, A., Clarke, P. (eds.) MODELS 2013. LNCS, vol. 8107, pp. 690–706. Springer, Heidelberg (2013)
18. Selim, G.M.K., Wang, S., Cordy, J.R., Dingel, J.: Model Transformations for Migrating Legacy Models: An Industrial Case Study. In: Vallecillo, A., Tolvanen, J.-P., Kindler, E., Störrle, H., Kolovos, D. (eds.) ECMFA 2012. LNCS, vol. 7349, pp. 90–101. Springer, Heidelberg (2012)
19. Syriani, E., Vangheluwe, H.: De-/re-constructing Model Transformation Languages. EASST 29 (2010)
20. Taentzer, G.: AGG: A Graph Transformation Environment for Modeling and Validation of Software. In: Pfaltz, J.L., Nagl, M., Böhlen, B. (eds.) AGTIVE 2003. LNCS, vol. 3062, pp. 446–453. Springer, Heidelberg (2004)
21. Troya, J., Vallecillo, A.: A Rewriting Logic Semantics for ATL. JOT 10(5), 1–29 (2011)

A Static Analysis of Non-confluent Triple Graph Grammars for Efficient Model Transformation

Anthony Anjorin[1,*], Erhan Leblebici[1], Andy Schürr[1], and Gabriele Taentzer[2]

[1] Technische Universität Darmstadt,
Real-Time Systems Lab, Germany
{anjorin,leblebici,schuerr}@es.tu-darmstadt.de
[2] Philipps-Universität Marburg,
Fachbereich Mathematik und Informatik, Germany
taentzer@mathematik.uni-marburg.de

Abstract. Triple Graph Grammars (TGGs) are a well-known bidirectional model transformation language. All actively developed TGG tools pose restrictions to guarantee efficiency (polynomial runtime), without compromising formal properties. Most tools demand *confluence* of the TGG, meaning that a choice between applicable rules can be freely made without affecting the final result of a transformation. This is, however, a strong restriction for transformations with inherent degrees of freedom that should not be limited at design time. eMoflon is a TGG tool that supports *non-confluent* TGGs, allowing different results depending on runtime choices. To guarantee efficiency, nonetheless, a local choice of the next source element to be translated, based on source context dependencies of the rules, *must not* lead to a *dead end*, i.e., to a state where no rule is applicable for the currently chosen source element to be translated, and the transformation is not yet complete. Our contribution in this paper is to formalize a corresponding property, referred to as *local completeness*, using graph constraints. Based on the well-known transformation of constraints to application conditions, we present a static analysis that guarantees dead end-freeness for non-confluent TGGs.

Keywords: Bidirectionality, triple graph grammars, static analysis.

1 Introduction and Motivation

Model synchronization is a crucial task in numerous application domains. In a current research project, we have investigated and implemented a tool for synchronizing two textual languages used in the domain of *Concurrent Manufacturing Engineering* (CME). The tool[1] is able to propagate changes incrementally from documents in one language to documents in the other, thus enabling a concurrent engineering workflow.

---

* The project on which this paper is based was funded by the German Federal Ministry of Education and Research, funding code 01IS12054. The authors are responsible for all contents.

[1] A screencast demonstrating the tool is available at www.emoflon.org

H. Giese and B. König (Eds.): ICGT 2014, LNCS 8571, pp. 130–145, 2014.

Triple Graph Grammars (TGGs)[13] are a formally founded rule-based bidirectional model transformation language, and were used in the CME research project to realize the synchronization of models formulated in different modelling languages. We have identified TGGs to be the typical performance bottle-neck in such transformation chains [11] meaning that improving the *efficiency* (i.e., achieving polynomial runtime in model size) of TGG-based transformations is a current and crucial challenge. To the best of our knowledge, all TGG tools strive to guarantee efficiency by posing certain restrictions on the class of supported TGGs. The specification for the synchronization tool in the CME research project consists of about a 100 TGG rules, which means that manually checking all such restrictions is practically infeasible. For specifications of this size and larger, an automated and comprehensive static analysis of all required restrictions becomes crucial.

A common strategy to achieve efficiency is to demand *confluence*, meaning that choices between applicable TGG rules do not influence the final result of the transformation. This improves efficiency as wrong choices that might lead to *dead ends*, i.e., states where no rule is applicable but the transformation is not yet complete, are no longer possible [8].

In many application scenarios such as for the CME research project, however, the required transformations often have an inherent degree of freedom, which cannot always be restricted at design time to ensure confluence. In the CME research project, for example, the end-user (or a configuration module) must guide the synchronization appropriately, making choices based on case-by-case preferences. Adjusting the underlying TGG and rebuilding the synchronization tool for each possible set of choices is simply infeasible.

To support non-confluent TGGs and nonetheless ensure efficiency without compromising formal properties, the TGG tool eMoflon (www.emoflon.org) [12] determines a sequence in which source elements can be translated based only on source context dependencies of the TGG rules. This strategy has been shown to be efficient in [9], if a local choice of the next source element to be translated cannot lead to a dead end in the transformation. "Local" means that the entire source model is never searched globally for the next translatable element.

There is currently no static analysis for this required property of non-confluent TGGs referred to as *local completeness* [9], meaning that an exception is thrown whenever the condition is violated at runtime. As only suitable tests can reveal this, local completeness violations are currently one of the most common and frustrating mistakes made by eMoflon users, especially beginners.

Our main contribution in this paper is to formulate the condition for local completeness as a set of graph constraints, thus providing a constructive formalization and a static analysis for local-completeness of non-confluent TGGs. We apply well-known techniques, e.g., for transforming constraints to sufficient and necessary application conditions [4], which have already been shown in previous work [1,8] to be applicable in general to TGGs.

The paper is structured as follows: in Sect. 2 we provide a running example and recall basic definitions and results for TGGs. Our main contribution is presented in Sect. 3, providing a static analysis for non-confluent TGGs. Section 4 gives an overview of related approaches, while Sect. 5 concludes the paper with a brief summary and discussion of future work.

2   Running Example and Preliminaries

As a running example, we consider an excerpt of a real-world *Platform-Independent Model* (PIM) to *Platform-Specific Model* (PSM) transformation from the CME domain. An example is depicted in Fig. 1. The PIM is represented as a *Cutter Location Source* (CLS) file, which specifies the manufacturing process as a series of operations. CLS files can be executed in a simulator that visualizes the specified manufacturing process. The PIM is used to generate machine-specific *G-code*, depicted as the PSM to the right of Fig. 1. G-code programs can be executed on appropriate machines that realize the manufacturing process. In practice, G-code programs are sometimes optimized manually. For example, the sequence of operations used to move the machine to its initial position can be shortened for a particular machine and set-up. Manual updates to G-code programs that can be expressed on the PIM level must be propagated back to CLS as they would otherwise be overwritten and lost during code generation. This propagation must be *incremental* as CLS files contain information, which is discarded during code generation and cannot be regained from G-code.

In our example, we consider only the most basic operation used to move the tip of the current machine. It is specified in two different ways: (i) the machine performs a linear interpolation between its current and the target location maintaining a constant *feedrate* (speed), and (ii) the machine is free to realize the movement to be as *rapid* as possible. In the CLS syntax, this basic operation is specified via a GOTO/ X, Y, Z operation, where (X, Y, Z) are the coordinates of the target location. As a safety feature, an additional RAPID command is required to indicate that the next GOTO is to be executed in rapid mode. After such a "rapid" GOTO, the machine reverts to the default feedrate mode.

In G-code, a series of "G" switches are used to influence how the current *and all following operations* are executed by the machine. The switches G0 and G1 correspond to rapid and feedrate mode, respectively. In contrast to the CLS

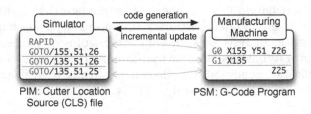

Fig. 1. An example of a PIM-to-PSM transformation in the CME domain

format, G-code is optimized for efficient interpretation, only requiring the actual changes from the previous line to be specified on each new line. On the second line, for example, G1 X135 is equivalent to G1 X135 Y51, G1 X135 Z26, or G1 X135 Y51 Z26. The G-code program depicted in Fig. 1 is, therefore, only one of 32 correct G-code programs! Is this flexibility required? Why not enforce the most efficient G-code program as depicted in Fig. 1? The reason is that efficiency is sometimes traded for maintainability of the templates used to generate G-code programs, i.e., values are repeated so that transformation templates can be reused in a different context. Before we specify TGG rules that appropriately capture this degree of freedom of the transformation, we have to establish a basic understanding of models, metamodels, and rule-based model transformation.

2.1 Consistency Specification with Triple Graph Grammars

In line with the algebraic formalization according to [3], *models* and *metamodels* are formalized as typed graphs and type graphs, respectively:

Definition 1 (Graph and Graph Morphism).
A graph $G = (V, E, s, t)$ consists of finite sets V of nodes and E of edges, and two functions $s, t : E \to V$ that assign each edge source and target nodes.
A graph morphism $h : G \to G'$, with $G' = (V', E', s', t')$, is a pair of functions $h := (h_V, h_E)$ where $h_V : V \to V', h_E : E \to E', h_V \circ s = s' \circ h_E \wedge h_V \circ t = t' \circ h_E$.

Definition 2 (Typed Graph and Typed Graph Morphism).
A type graph is a graph $TG = (V_{TG}, E_{TG}, s_{TG}, t_{TG})$.
A typed graph is a pair $(G, type)$ of a graph G together with a graph morphism type: $G \to TG$. Given $(G, type)$ and $(G', type')$, $g : G \to G'$ is a typed graph morphism iff $type = type' \circ g$.
$\mathcal{L}(TG) := \{G \mid \exists \, type : G \to TG\}$ *denotes the set of all graphs of type TG.*

Remark 1 (The Category of Typed Graphs).
Typed graphs and typed graph morphisms form a category **Graphs** *with the set $\mathcal{M}$ of injective typed graph morphisms (cf. [3]).*

Remark 2 (Attributed Typed Graphs with Inheritance).
As we formulate our definitions and theorems on the level of typed graphs and typed graph morphisms, they can be extended to attributed typed graphs with node type inheritance in a straightforward manner.

Next, *rule*-based model transformation is formalized, where rules have pre- and post-conditions with *constraints* that either hold globally for all models, or are used to guard rule application (*as application conditions*). As we only require creating (monotonic) rules for TGGs, the following definitions from [3] are thus simplified appropriately.

Definition 3 (Rule, Graph Grammar, and Derivation).
A rule is a typed graph morphism $r : L \to R \in \mathcal{M}$, where TG is a type graph and $L, R \in \mathcal{L}(TG)$. A graph grammar is a pair $GG = (TG, \mathcal{R})$ of a type graph TG and a finite set $\mathcal{R}$ of rules.

A direct derivation $G \overset{r@m}{\Longrightarrow} G'$ (or $G \overset{r}{\Longrightarrow} G'$) is given by a pushout
in **Graphs** (cf. diagram to the right).

$$\begin{array}{ccc} L & \overset{r}{\longrightarrow} & R \\ m \downarrow & PO & \downarrow m' \\ G & \overset{g}{\longrightarrow} & G' \end{array}$$

A derivation $G \overset{*}{\Longrightarrow} G'$ of length $n \geq 0$ in $GG = (TG, \mathcal{R})$ is a
sequence of n direct derivations $G \overset{r_1}{\Longrightarrow} G_1 \overset{r_2}{\Longrightarrow} \cdots \overset{r_n}{\Longrightarrow} G'$, with
$r_1, r_2, \cdots, r_n \in \mathcal{R}$. In case of length $n = 0$, we have $G' = G$.
$\mathcal{L}(GG, G_\emptyset) := \{G \in \mathcal{L}(TG) \mid G_\emptyset \overset{*}{\Longrightarrow} G\}$ denotes the language generated by a
graph grammar GG, where $G_\emptyset \in \mathcal{L}(TG)$ denotes the start typed graph.
$\mathcal{L}(GG) := \mathcal{L}(GG, \emptyset)$, where $\emptyset$ is the empty typed graph.

Definition 4 (Conditional Constraints).

A conditional constraint c is a typed graph morphism $c : P \to C$.
For conditional constraints c_i with $i \in I$ for some index set I, $\vee_{i \in I} c_i$ is also a
conditional constraint.
A typed graph G satisfies a conditional constraint c, denoted by $G \models c$, if either
$c : P \to C$ and $\forall p : P \to G \in \mathcal{M}$, $\exists q : C \to G \in \mathcal{M}$ such that $q \circ c = p$, or
$c = \vee_{i \in I} c_i$ and $\exists i \in I : G \models c_i$.
A graph grammar $GG = (TG, \mathcal{R})$ satisfies a conditional constraint c, denoted by
$GG \models c$, if for all derivations $G \overset{*}{\Longrightarrow} G'$ with $G \in L(GG, G_\emptyset)$, $G' \models c$.

Definition 5 (Conditional Application Conditions).

A conditional application condition over a typed graph L is a pair $ac = (a, \vee_{i \in I} c_i)$,
where $a : L \to P$ and $c_i : P \to C_i$ with $i \in I$, for some index set I.
A typed graph morphism $m : L \to G \in \mathcal{M}$ satisfies a conditional application
condition ac, denoted by $m \models ac$, if $\forall p : P \to G \in \mathcal{M}$ with $p \circ a = m$, $\exists i \in I$
and $q_i : C_i \to G \in \mathcal{M}$ such that $q_i \circ c_i = p$.
A conditional application condition for $r : L \to R$ is a conditional application
condition over L.
A rule r with a set of conditional application conditions $\mathcal{AC}$ is denoted by $(r, \mathcal{AC})$.
A graph grammar $GG = (TG, \mathcal{R})$ is a graph grammar without conditional application conditions, if all rules in $\mathcal{R}$ do not have conditional application conditions.
A conditional application condition ac is trivial, i.e., always satisfied, if $\exists i \in I$
such that $c_i : P \to P$ is the identity.
A Negative Application Condition (NAC) $ac = (a, \vee_{i \in I} c_i)$ is a conditional ap-
plication condition where $a : L \to P$ and I is the empty index set.
A NAC is, therefore, simply denoted by a typed graph morphism $a : L \to P$.

Given a TGG and a set of global constraints that must hold for all models, appropriate application conditions can be automatically generated for every TGG
rule, ensuring that the constraints are never violated.

Fact 1 (Construction of Application Conditions from Constraints).

Given a graph grammar $GG = (TG, \mathcal{R})$ without conditional application conditions, and a set $\mathcal{C}$ of conditional constraints.
There is a construction A producing a set of rules with conditional application
conditions $R' = A(GG, \mathcal{C}) = \{(r, \mathcal{AC}) \mid r \in \mathcal{R}\}$ s.t. $\forall c \in \mathcal{C} : GG' \models c$, with
$GG' = (TG, \mathcal{R}')$. The constructed conditional application conditions are suffi-
cient and necessary.

Proof. This is a special case of Thm. 7.23 in [3], proven on the basis of adhesive HLR categories, here for conditional constraints and monotonic rules. The diagram to the right shows the main steps of the construction, namely: (1) constructing all possible gluings $R+P$ of the right-hand side R of the rule and the premise P of each constraint, (2) producing for each gluing a post-condition via a pushout, where D_i represents all possible further gluings of elements in D, and finally (3) constructing a precondition from the post-condition by reversing the

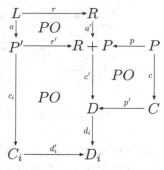

application of the rule (determining pushout complements P' and C_i). If this is not possible (i.e., a pushout complement does not exist) then the post-condition does not result in an equivalent pre-condition.

The central idea with TGGs [13,9] is to specify the consistency of source and target models by providing rules that define a language of consistent source and target models, connected by a *correspondence* model. Rules in a TGG thus describe the simultaneous evolution of *triples* of source, correspondence and target models, from which various operational transformations such as forward/backward transformations can be automatically derived. In the following, we denote *typed triple graphs* with single letters, e.g., G, which consist of typed graphs with an index $X \in \{S, C, T\}$, e.g., G_S, G_C, G_T.

Definition 6 (Typed Triple Graph and Typed Triple Morphism).
A triple graph $G = G_S \overset{\sigma_G}{\leftarrow} G_C \overset{\tau_G}{\rightarrow} G_T$ consists of graphs G_S, G_C and G_T, and graph morphisms $\sigma_G : G_C \rightarrow G_S$, $\tau_G : G_C \rightarrow G_T$.
A triple morphism $g = (g_S, g_C, g_T) : G \rightarrow G'$, $G' = G'_S \overset{\sigma_{G'}}{\leftarrow} G'_C \overset{\tau_{G'}}{\rightarrow} G'_T$
is a triple of graph morphisms $g_X : G_X \rightarrow G'_X, X \in \{S, C, T\}$,
s.t. $g_S \circ \sigma_G = \sigma_{G'} \circ g_C$ and $g_T \circ \tau_G = \tau_{G'} \circ g_C$.
Given a triple graph $TG = TG_S \overset{\sigma_{TG}}{\leftarrow} TG_C \overset{\tau_{TG}}{\rightarrow} TG_T$ called type triple graph, *a* typed triple graph *is a pair $(G, type)$ of a triple graph G and triple morphism $type : G \rightarrow TG$.*
Analogously to Def. 2, $\mathcal{L}(TG)$ denotes the set of all triple graphs of type TG.
Given $(G, type), (G', type') \in \mathcal{L}(TG)$, a typed triple morphism $g : G \rightarrow G'$ is a triple morphism such that $type = type' \circ g$.
The source-correspondence graph *$sc(G)$ of typed triple graph $G_S \overset{\sigma_G}{\leftarrow} G_C \overset{\tau_G}{\rightarrow} G_T$ is defined as $G_S \overset{\sigma_G}{\leftarrow} G_C \overset{\tau_G}{\rightarrow} Im(\tau_G)$, where $Im(\tau_G) \subseteq G_T$ is the codomain of τ_G.*
Given $g : G \rightarrow G'$ the source-correspondence morphism *$sc(g) : sc(G) \rightarrow sc(G')$ is defined by $sc(g) = (g_S, g_C, Im(g_C))$ with $Im(g_C) = \tau_{G'} \circ g_C \circ \tau_G^{-1}$ (τ_G^{-1} exists as it is bijective).*

Example 1. Figure 2 depicts a possible type triple graph and a typed triple graph, representing the consistent pair of PIM and PSM models of our running example (Fig. 1). A UML-like syntax is used and attributes (for coordinates in the running example) are excluded for presentation purposes. Finally, the types of edges in the typed triple graph are omitted as they can be uniquely determined

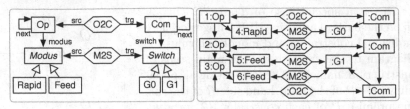

Fig. 2. A type triple graph (left), and a typed triple graph (right)

from the type triple graph. According to the type triple graph, a CLS model is represented as an ordered sequence of operations (Op), each equipped with a Modus, which can be either of type Rapid or Feed. The CLS specification in Fig. 1 is, therefore, represented by a sequence of three Ops: the first with a Rapid modus, and the subsequent two with Feed modi (the kind of CLS operation is given by the connected Mode of the Op). Similarly, a G-code model is represented by an ordered sequence of commands (Com) each equipped with a Switch either of type G0 or G1. The G-code model for the running example is a sequence of three Coms, where the first one has a G0 switch and the other two "share" a G1 switch. The sharing of the G1 switch represents the omission of unchanged information (the switch) on the last line of the G-code program (Fig. 1). Finally, the correspondence model consists of O2C and M2S links connecting related source and target elements.

As not all typed graph triples represent consistent pairs of CLS and G-code models, TGG rules are, therefore, required to further specify the actual language of meaningful CLS and corresponding G-code models. Due to the following Fact. 2, rules, derivations and graph grammars can be defined as in Def. 3, but for typed triple graphs and typed triple morphisms.

Fact 2 (The Category of Typed Triple Graphs).
*The class of all typed triple graphs and typed triple morphisms form a category called **TriGraphs** with the set M of injective typed triple morphisms.*

Proof. For the proof we refer the interested reader to Fact. 4.18 in [3].

Definition 7 (Triple Rules, Triple Graph Grammar).
Let TG be a type triple graph, and $L, R \in \mathcal{L}(TG)$.
A typed triple morphism $r : L \to R$ is a triple rule if r_S, r_C, and r_T are rules.
A triple graph grammar $TGG = (TG, \mathcal{R})$ consists of a type triple graph TG and a finite set $\mathcal{R}$ of triple rules.
The source-correspondence grammar *$sc(TGG)$ of a triple graph grammar $TGG = (TG, \mathcal{R})$ is a pair $(sc(TG), sc(\mathcal{R}))$, where $sc(\mathcal{R}) = \{sc(r) \mid r \in \mathcal{R}\}$ is the respective set of source-correspondence morphisms (rules).*
A conditional application condition $((a_S, a_C, a_T), \vee_{i \in I} c_i)$ is a conditional source application condition if a_C and a_T are identities (conditional target and correspondence application conditions are defined analogously).

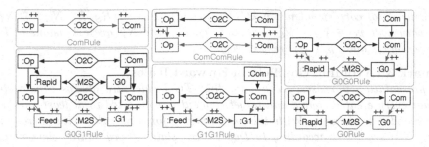

Fig. 3. Triple rules for running example

Example 2. The six TGG rules for the running example are depicted in Fig. 3. Every triple rule $r : L \to R$ is depicted in a compact syntax by denoting created elements, i.e., $R\backslash L$, with a ++ markup (and by displaying them in green). Com-Rule creates an Op and a Com connected by an O2C correspondence, depicted as a hexagon to improve readability. ComComRule also creates a corresponding pair of Op and Com elements, requiring a preceding triple as context to which the new elements are connected. The remaining rules handle the creation of corresponding Modi and Switches. The rules show that we have decided to allow possibly redundant repetitions of G0 switches with G0Rule, as well as reusing the G0 switch from the previous Com with G0G0Rule. This degree of freedom is, therefore, not restricted by the TGG and can be decided upon at runtime. For G1 switches, however, we have chosen to forbid redundant repetitions already at design time, i.e., a new G1 switch can only be created with G0G1Rule if it is impossible to reuse the G1 switch of the previous Com with G1G1Rule. The running example (Fig. 2) is consistent as it can be created by the following derivation: ComRule $\to$ G0Rule $\to$ ComComRule $\to$ G0G1Rule $\to$ ComComRule $\to$ G1G1Rule.

2.2   Operationalization of Triple Graph Grammars

Although TGG rules can be used directly to generate consistent source and target models, e.g., for testing purposes, TGGs are often *operationalized* to derive unidirectional forward and backward transformations [13,9]. To this end, a *forward rule* is derived from each TGG rule according to the following definition.

Definition 8 (Derivation of Forward Rules).
A TGG $= (TG, \mathcal{R})$ without conditional application conditions can be forward operationalized by deriving FWD(TGG) $:= (TG, \mathcal{R}_F)$, where $\mathcal{R}_F$ consists of forward rules. A forward rule $(r_F : FL \to FR, \mathcal{N}_F)$ for triple rule $r \in \mathcal{R}$ is defined by the diagram below (the triangle denotes a set of NACs):

$$
\begin{array}{ccc}
L = L_S \xleftarrow{\sigma_L} L_C \xrightarrow{\tau_L} L_T \\
r \downarrow \quad r_S\downarrow \quad r_C\downarrow \quad r_T\downarrow \\
R = R_S \xleftarrow{\sigma_R} R_C \xrightarrow{\tau_R} R_T
\end{array}
\implies
\begin{array}{ccc}
 & \overset{\mathcal{N}_F}{\triangledown} & \\
FL = R_S \xleftarrow{r_S \circ \sigma_L} L_C \xrightarrow{\tau_L} L_T \\
r_F\downarrow \quad id\downarrow \quad r_C\downarrow \quad r_T\downarrow \\
FR = R_S \xleftarrow{\sigma_R} R_C \xrightarrow{\tau_R} R_T
\end{array}
$$

$\mathcal{N}_F$ consists of correspondence NACs $(id, n_C, id) : FL \to (R_S \overset{\sigma_{FN}}{\leftarrow} N_C \overset{\tau_{FN}}{\to} N_T)$ for every node $v_C \in V_{R_C} \setminus V_{L_C}$ (cf. Def. 7). N_C and N_T are extensions of L_C and L_T by v_C and $\tau_{R_C}(v_C)$, respectively.

Remark 3 (Avoiding Conflicts in Forward Rules).

For presentation purposes, we assume that TGG rules are always constructed so that every created source and target element is connected to at least one new correspondence element. With this assumption, forward rules with correspondence NACs derived according to Def. 8 "translate" every source element exactly once and simplify the formalization as compared to introducing translation attributes [8], or bookkeeping [9]. To avoid unnecessary create/forbid conflicts, $\mathcal{N}_F$ can be further extended by application conditions to avoid obvious dead-ends when applying forward rules. We only give an intuition of how this works with our running example and refer to [9,8] for a description of filter NACs.

Example 3. Figure 4 depicts the forward rules derived from the triple rules in Fig. 3, where labels indicate the original triple rules (e.g., ComFwdRule derived from ComRule). Forward rules "translate" the source elements that are created in the respective triple rules by attaching correspondence elements to them, but only if the correspondence elements do not already exist, i.e., the elements have not already been translated. Similarly, context elements of source models in triple rules are required to be already translated in forward rules by demanding an attached correspondence element. ComFwdRule, for example, "translates" an Op by connecting it to a new O2C and a Com in the target model. This should only be possible if the Op has not been translated already (hence the correspondence NAC denoted here as a crossed out (forbidden) element). An additional source NAC in ComFwdRule forbids the existence of a previous Op, i.e., this forward rule can translate only the first Op in a sequence of connected Ops. If this is not prevented, the incoming edge from a previous Op would no longer be translatable as the original triple rules cannot create such an edge between two existing Ops. This is automatically detected and prevented by generating a source "filter" NAC (cf. Remark 3).

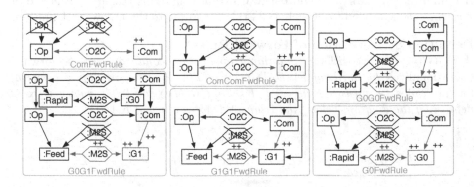

Fig. 4. Derived forward rules for the running example

The forward rule ComComFwdRule requires a translated previous Op by demanding an O2C as context, which can only exist if the connected Op has already been translated. All other forward rules are constructed analogously.

3 Efficient Model Transformation with TGGs

In addition to forward rules, a *control algorithm* is required to determine in what order source elements can be translated by forward rules. The challenge is to accomplish this translation efficiently without compromising formal properties. This has been shown in [9] for the algorithm depicted in Alg. 1. The algorithm consists of a main loop on Line 4 where a derivation with a source-correspondence rule is searched for. If none can be found, the loop is terminated and the resulting triple graph is expected to be consistent with respect to the TGG (Line 11). If this is not the case, an error is thrown on Line 12. Every derivation with a source-correspondence rule is extended to a derivation with a forward rule on Line 5. Note that this is done *locally* by extending m and not searching the whole triple graph for some suitable m' globally. If this extension is not possible, an error is thrown on Line 8. For non-confluent TGGs, the algorithm requires a choice (user or runtime configuration module) in the procedure chooseAndApplyFWDRule.

Algorithm 1. TGG-Based Forward Transformation

Require: $TGG = (TG, \mathcal{R})$ without application conditions, $\exists G \in \mathcal{L}(TGG)$.

1: **procedure** FWDTRANSFORM$(TGG, G_S \overset{\emptyset}{\leftarrow} \emptyset \overset{\emptyset}{\rightarrow} \emptyset) : G_S \leftarrow G_C \rightarrow G_T$
2: $\quad TGG_F = (TG, \mathcal{R}_F) \leftarrow \text{FWD}(TGG)$
3: $\quad G \leftarrow (G_S \overset{\emptyset}{\leftarrow} \emptyset \overset{\emptyset}{\rightarrow} \emptyset)$
4: $\quad$ **while** $\exists (G \overset{sc(r)@m}{\Longrightarrow} H)$ in $sc(TGG_F), r : FL \rightarrow FR \in \mathcal{R}_F$ **do**
5: $\quad\quad$ **if** $\exists (G \overset{r'@m'}{\Longrightarrow} H')$ in $TGG_F, r' : FL' \rightarrow FR' \in \mathcal{R}_F :$
6: $\quad\quad\quad \exists e : sc(FL) \rightarrow FL' \in \mathcal{M}, m = m' \circ e$ **then**
7: $\quad\quad\quad\quad G \leftarrow \text{CHOOSEANDAPPLYFWDRULE}(m)$
8: $\quad\quad$ **else** Error: No applicable forward rule at m
9: $\quad\quad$ **end if**
10: $\quad$ **end while**
11: $\quad$ **if** $G \in \mathcal{L}(TGG)$ **then return** G
12: $\quad$ **else** Error: No applicable source-correspondence rule $sc(r)$ for G
13: $\quad$ **end if**
14: **end procedure**

In the worst case, every source element is visited exactly once in the main loop, i.e., as often as there are elements in the source model (n_S times). Each time, a source-correspondence derivation must be determined and extended. This is bounded by $|\mathcal{R}_F| \cdot n^k$, where k is the number of elements of the largest forward rule, and n is the size of the resulting triple. With n^k as the upper bound for pattern matching, the resulting complexity is $O(n_S \cdot |\mathcal{R}_F| \cdot n^k) = O(n^k)$ (cf. [9]).

Our goal in this paper is to provide a static analysis for a class of TGGs for which this algorithm never fails (as on Lines 8 and 12). To avoid non-applicability of source-correspondence rules to not fully translated graphs (error on Line 12), we have to guarantee that every derivation $G \stackrel{*}{\Longrightarrow} H$ in $sc(TGG_F)$ can be prolonged if $G \notin \mathcal{L}(TGG)$, i.e., not all source elements have been translated. This is a well-known property that follows from *confluence* (of $sc(TGG)$ in this case) and can be checked statically via a *critical pair* analysis (cf., e.g., [3]), i.e., with well-known techniques already applied in the context of TGGs as in [8].

Definition 9 (Confluence, Source-Correspondence Confluence).
A pair $P_1 \stackrel{*}{\Longleftarrow} K \stackrel{*}{\Longrightarrow} P_2$ of derivations in a graph grammar is confluent *if there exists an X together with derivations $P_1 \stackrel{*}{\Longrightarrow} X$ and $P_2 \stackrel{*}{\Longrightarrow} X$. A graph grammar is* confluent *if all pairs of its derivations are confluent. A triple graph grammar TGG is* source-correspondence confluent *if $sc(TGG)$ is confluent.*

To prevent dead ends (error on Line 8), it must be possible to extend every source-correspondence derivation locally to a forward derivation. The following definition formulates a TGG property that prevents the errors thrown in Alg. 1.

Definition 10 (Efficiency of Triple Graph Grammars).
Let $TGG = (TG, \mathcal{R})$ without conditional application conditions and $FWD(TGG) = (TG, \mathcal{R}_F)$. TGG is forward efficient *for $G \in \mathcal{L}(TGG)$ if:*
(1) It is source-correspondence confluent, and with $TGG_F = (TG, \mathcal{R}_F)$

$$(2)\ \forall G' \in \mathcal{L}(TGG_F, G_S \stackrel{\emptyset}{\leftarrow} \emptyset \stackrel{\emptyset}{\rightarrow} \emptyset) : \exists (G' \stackrel{sc(r)@m}{\Longrightarrow} H)\ \text{with}\ r : FL \rightarrow FR \in \mathcal{R}_F$$

$$\Rightarrow \exists (G' \stackrel{r'@m'}{\Longrightarrow} H'), r' : FL' \rightarrow FR' \in \mathcal{R}_F\ \text{such that}\ m'\ \text{extends}\ m, \text{i.e.,}$$

$$\exists e : sc(FL) \rightarrow FL' \in \mathcal{M}\ \text{such that}\ m = m' \circ e.$$

Example 4. Considering the translation of the CLS model in Fig. 2, not all its direct derivations with source-correspondence rules can be extended locally to direct derivations with forward rules. For example, indicating the source elements translated by each application after the @ sign, the sequence sc(ComFwdRule)-@1:Op $\rightarrow_1$ sc(G0FwdRule)@4:Rapid $\rightarrow_2$ sc(ComComFwdRule)@2:Op $\rightarrow_3$ sc(ComComFwdRule)@3:Op $\rightarrow_4$ sc(G1G1FwdRule)@6:Feed $\rightarrow_5$ sc(G0G1FwdRule)@5:Feed is a derivation of source-correspondence rules that translates the CLS model. However, the fifth direct derivation sc(G1G1FwdRule)@6:Feed cannot be extended locally to a direct derivation with any forward rule. This is because the forward translation of 6:Feed before 5:Feed is not possible, although "parsing" them, i.e., the source-correspondence translation is. When translating 6:Feed, G1G1FwdRule requires a G1 switch, which can only be present if 5:Feed has already been translated. In general, this hidden dependency on 5:Feed cannot be accounted for when parsing the source model with source-correspondence rules. The TGG, therefore, does not fulfil the efficiency requirement of Def. 10 and can lead to a runtime error in Alg. 1. The formulation of Condition (2) in Def. 10, however, can only be checked at runtime. The following definition introduces the concept of a *local completeness constraint*, which can be statically checked, leading to a condition that will be shown to be sufficient for Condition (2) in Def. 10.

Definition 11 (Local Completeness Constraints, Local Completeness).
Let $TGG = (TG, \mathcal{R})$ be a triple graph grammar without conditional application conditions, $TGG_F = FWD(TGG) = (TG, \mathcal{R}_F)$.
For a forward rule $r_F : FL \to FR \in \mathcal{R}_F$, the set $lcc(r_F)$ of local completeness constraints *is defined as:*
$$lcc(r_F) := \{c \in \mathcal{M} \mid \exists r'_F : FL' \to FR' \in \mathcal{R}_F, c : sc(FL) \to FL'\}$$
The local completeness constraint *$lcc(TGG_F)$ is defined as:*
$$lcc(TGG_F) := \wedge_{r_F \in \mathcal{R}_F}(\vee_{c \in lcc(r_F)} c)$$
TGG_F is locally complete *if $TGG_F \models lcc(TGG_F)$.*

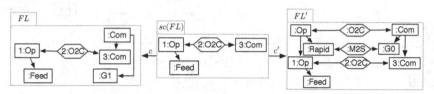

Fig. 5. Set of local completeness constraints $lcc(G1G1FwdRule)$

Example 5. Figure 5 depicts the set $\{c, c'\}$ of local completeness constraints for G1G1FwdRule, i.e., lcc(G1G1FwdRule). In the middle, the premise of the constraint is constructed as the source-correspondence graph $sc(FL)$ of the left-hand side FL of G1G1FwdRule. The two constraints result from the two possible conclusions constructed by determining all left-hand sides of forward rules (FL of G1G1FwdRule itself, and FL' of G0G1FwdRule) into which this premise can be injectively mapped. At least one of these constraints must be fulfilled, i.e., the local completeness constraint $lcc(TGG_F)$ is a disjunction of all such constraints. For the running example, the constraint demands that every occurrence of the source-correspondence context of G1G1FwdRule imply the context of at least one forward rule, i.e., in this case the context of G1G1FwdRule or of G0G1FwdRule. This means that an operation has to be already translated into a command with a preceding G0- or G1-command, *before* the feed of the operation is translated.

Local completeness constraints can be transformed to application conditions using the construction given in Fact. 1. In this manner, the forward rules of a TGG can be statically checked for local completeness by demanding that only trivial application conditions are generated. This idea is stated in the following.

Corollary 1 (Enforcing Local Completeness).
Let $TGG = (TG, \mathcal{R})$ be a triple graph grammar without conditional application conditions and $TGG_F = FWD(TGG) = (TG, \mathcal{R}_F)$. For every forward rule $r_F : FL \to FR \in \mathcal{R}_F$, there is a construction A producing application conditions $A(lcc(TGG_F), r_F)$. $(TG, \mathcal{R}'_F)$ with forward rules with the constructed application conditions $(r_F, A(lcc(TGG_F), r_F))$ is locally complete.

Proof. Follows directly from Fact. 1 as the given construction has been shown in [1] to be applicable for **TriGraphs**.

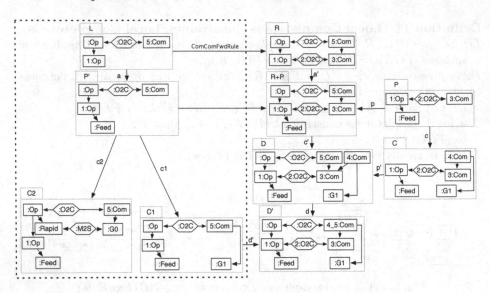

Fig. 6. Construction of local completeness application condition for ComComFwdRule

Example 6. To check if the TGG for the running example is locally complete, the application condition $(a, c_1 \vee c_2)$ depicted in Fig. 6 is constructed for the forward rule ComComFwdRule, from the local completeness constraints lcc(G1G1Fwd-Rule) (cf. Fig. 5). The steps of the construction are only shown for c_1. As the generated application condition is not trivial, the forward rules of the TGG are not locally complete. For the running example, the application condition states that if an Op has a Feed, then it can only be translated with ComComFwdRule, if the Modus of the previous Op has already been translated. Note that c_1 is a bit subtle, only demanding that a $G1$ be present and not caring how it is connected. The problematic sequence, obtained by translating 6:Feed with ComComFwdRule *before* 5:Feed, violates c_1 as the Com of the previous Op does *not* yet have a G1.

We can now present our main contribution, a static analysis to check forward efficiency as defined in Def. 10 using Algorithm 2.

Theorem 1 (Static Analysis of Non-Confluent TGGs).
$TGG = (TG, \mathcal{R})$ is forward efficient if $LCA(TGG) = true$.

Proof. $\forall G \in \mathcal{L}(TGG)$, we must show forward efficiency for TGG (Def. 10): $(LCA(TGG) = true) \implies TGG_F = \text{FWD}(TGG)$ is *source-correspondence conflu-ent* due to the check on Line 3 of Alg. 2, and is *locally complete* as the generated application conditions for local completeness are trivial (Line 5 of Alg. 2). Let $sc(r) : SL \to SR \in sc(\mathcal{R}_F), r \in \mathcal{R}_F$, with $TGG_F = (TG, \mathcal{R}_F)$.
$$\forall G' \in \mathcal{L}(TGG_F, G_S \xleftarrow{\emptyset} \emptyset \xrightarrow{\emptyset} \emptyset) : \exists (G' \xRightarrow{sc(r)@m} H)$$
$$\xRightarrow{Cor.1} \exists r' : FL' \to FR' \in \mathcal{R}_F, \exists c : SL \to FL' \in lcc(r) : G' \models c$$

$\overset{Def.4}{\Longrightarrow} \exists\, m' : FL' \to G', m = m' \circ c \overset{Def.3}{\Longrightarrow} \exists\, (G' \overset{r'@m'}{\Longrightarrow} H')$ in TGG_F

$\overset{Def.10}{\Longrightarrow} TGG_F$ is forward efficient.   □

Algorithm 2. Local Completeness Analysis (LCA)

Require: $TGG = (TG, \mathcal{R})$ without conditional application conditions.
1: **procedure** LCA(TGG) : Boolean
2:     $TGG_F \leftarrow$ FWD(TGG)
3:     **if** ISSOURCECORRCONFLUENT(TGG_F) **then**
4:         $\mathcal{AC}_{LC} \leftarrow$ CONSTRUCTLOCALCOMPAPPCONDITIONS(TGG_F)
5:         **return** ISTRIVIAL($\mathcal{AC}_{LC}$)
6:     **else return** $false$
7:     **end if**
8: **end procedure**

Example 7. The TGG for the running example can be made forward efficient by stating the hidden dependency on the previous modus explicitly in ComComRule. The corrected version of this rule is depicted to the left of Fig. 7. To the right, the derived application condition $(a^*, c1^* \vee c2^*)$ is now trivial, i.e., is always fulfilled as a Switch is either G0 or G1, meaning the previous problem is solved.

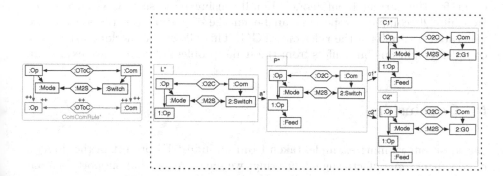

Fig. 7. Corrected TGG rule and trivial application condition for local completeness

4 Related Work

Schürr discusses the challenge of dealing with decision points in a TGG-based transformation process proposing two solutions in [13]: backtracking wrong decisions, or demanding confluence. For efficiency, the latter is the favoured strategy taken by all existing TGG approaches we are aware of. Hermann et al. [8] perform a critical pair analysis of forward rules and generate filter NACs to resolve critical pairs arising from obviously misleading (backtracking) paths. All other critical pairs (conflicts) must either be manually checked to be confluent, or removed by adjusting the TGG rules as required for confluence. More restrictively, Giese et al. [6] require confluence without filtering backtracking paths automatically. OCL constraints can be used to resolve critical pairs, but this is a manual process required for both forward and backward transformations. An algorithm to automate this process is yet to be provided by [6]. The TGG approach taken by Greenyer and Rieke [7] does not explicitly require confluence but, in case of decision points, the approach may fail in finding a valid transformation result, i.e., completeness is not guaranteed for non-confluent TGGs. Although confluence avoids backtracking (i.e., is used to show efficiency), solves completeness problems, and can be statically checked (cf. [8,6,7]), it can be too restrictive in practical scenarios (as in our running example) as it forces a TGG to be a bijection (a function in both directions).

The TGG algorithm in [9] is the only efficient and complete approach we are aware of that embraces *non-confluent* TGGs requiring local completeness and only source-correspondence confluence, as discussed in this paper. As the results in [9] do not provide any means to analyze this restriction statically, our contribution fills this gap by exploiting a constraint-based formalization of the required condition, making it amenable to well-known techniques.

Alternatives to a *static* analysis include analyses based on TGG rules *and* a concrete input model triple such as the dangling edge check in [9], and checks based on a precedence structure [10]. Although such analyses must be repeated for every new input model, they allow violations of properties that are not relevant for the current input model. Finally, using tools such as Groove [5] or Henshin [2], complex properties can be checked by exploring the state space generated by applying the rules of a TGG. This allows for checking arbitrarily complex conditions but suffers from the usual problem of state space explosion.

5   Conclusion

Based on our running example, taken from an industrial project in the domain of concurrent manufacturing engineering, we have argued that support for non-confluent TGGs is required for many practical scenarios. With the proposed static *local completeness analysis*, users can model degrees of freedom in consistency relations with TGGs, integrating runtime (user) interaction to decide between multiple applicable rules without compromising formal properties.

A limitation of our static analysis is that it is restricted to TGGs *without initial application conditions*. To handle constraints in the source and target metamodels, however, such application conditions are necessary and must also be taken into account. Further tasks include providing corresponding tool support to fully automate the proposed static analysis. This requires addressing the challenge of efficiently generating application conditions from constraints, filtering out redundant results, and presenting the results in a helpful manner.

References

1. Anjorin, A., Schürr, A., Taentzer, G.: Construction of Integrity Preserving Triple Graph Grammars. In: Ehrig, H., Engels, G., Kreowski, H.-J., Rozenberg, G. (eds.) ICGT 2012. LNCS, vol. 7562, pp. 356–370. Springer, Heidelberg (2012)
2. Arendt, T., Biermann, E., Jurack, S., Krause, C., Taentzer, G.: Henshin: Advanced Concepts and Tools for In-Place EMF Model Transformations. In: Petriu, D.C., Rouquette, N., Haugen, Ø. (eds.) MODELS 2010, Part I. LNCS, vol. 6394, pp. 121–135. Springer, Heidelberg (2010)
3. Ehrig, H., Ehrig, K., Prange, U., Taentzer, G.: Fundamentals of Algebraic Graph Transformation. Monographs in Theoretical Computer Science. An EATCS Series. Springer (2006)
4. Ehrig, H., Habel, A., Ehrig, K., Pennemann, K.H.: Theory of Constraints and Application Conditions: From Graphs to High-Level Structures. Fundamenta Informaticae 74(1), 135–166 (2006)
5. Ghamarian, A.H., de Mol, M., Rensink, A., Zambon, E., Zimakova, M.: Modelling and Analysis Using GROOVE. STTT 12 14(1), 15–40 (2012)
6. Giese, H., Hildebrandt, S., Lambers, L.: Toward Bridging the Gap Between Formal Semantics and Implementation of Triple Graph Grammars. Tech. Rep. 37, Hasso-Plattner Institute (2010)
7. Greenyer, J., Rieke, J.: Applying Advanced TGG Concepts for a Complex Transformation of Sequence Diagram Specifications to Timed Game Automata. In: Schürr, A., Varró, D., Varró, G. (eds.) AGTIVE 2011. LNCS, vol. 7233, pp. 222–237. Springer, Heidelberg (2012)
8. Hermann, F., Ehrig, H., Golas, U., Orejas, F.: Efficient Analysis and Execution of Correct and Complete Model Transformations Based on Triple Graph Grammars. In: Bézivin, J., Soley, M.R., Vallecillo, A. (eds.) MDI 2010, vol. 1866277, pp. 22–31. ACM Press (2010)
9. Klar, F., Lauder, M., Königs, A., Schürr, A.: Extended Triple Graph Grammars with Efficient and Compatible Graph Translators. In: Engels, G., Lewerentz, C., Schäfer, W., Schürr, A., Westfechtel, B. (eds.) Nagl Festschrift. LNCS, vol. 5765, pp. 141–174. Springer, Heidelberg (2010)
10. Lauder, M., Anjorin, A., Varró, G., Schürr, A.: Efficient Model Synchronization with Precedence Triple Graph Grammars. In: Ehrig, H., Engels, G., Kreowski, H.-J., Rozenberg, G. (eds.) ICGT 2012. LNCS, vol. 7562, pp. 401–415. Springer, Heidelberg (2012)
11. Leblebici, E., Anjorin, A., Schürr, A.: A Catalogue of Optimization Techniques for Triple Graph Grammars. In: Fill, H.G., Karagiannis, D., Reimer, U. (eds.) Modellierung 2014. LNI, vol. 225, pp. 225–240. GI (2014)
12. Leblebici, E., Anjorin, A., Schürr, A.: Developing eMoflon with eMoflon. In: Varró, D. (ed.) ICMT 2014. LNCS, vol. 8568, pp. 138–145. Springer, Heidelberg (2014)
13. Schürr, A.: Specification of Graph Translators with Triple Graph Grammars. In: Mayr, E.W., Schmidt, G., Tinhofer, G. (eds.) WG 1994. LNCS, vol. 903, pp. 151–163. Springer, Heidelberg (1995)

Transformation and Refinement
of Rigid Structures

Vincent Danos[1], Reiko Heckel[2], and Pawel Sobocinski[3]

[1] School of Informatics, University of Edinburgh, UK
[2] Department of Computer Science, University of Leicester, UK
[3] Electronics and Computer Science, University of Southampton, UK

Abstract. Stochastic rule-based models of networks and biological systems are hard to construct and analyse. Refinements help to produce systems at the right level of abstraction, enable analysis techniques and mappings to other formalisms. Rigidity is a property of graphs introduced in Kappa to support stochastic refinement, allowing to preserve the number of matches for rules in the refined system. In this paper: 1) we propose a notion of rigidity in an axiomatic setting based on adhesive categories; 2) we show how the rewriting of rigid structures can be defined systematically by requiring matches to be open maps reflecting structural features which ensure that rigidity is preserved; and 3) we obtain in our setting a notion of refinement which generalises that in Kappa, and allows a rule to be partitioned into a set of rules which are collectively equivalent to the original. We illustrate our approach with an example of a social network with dynamic topology.

1 Introduction

Graph transformations are a natural model for complex evolving networks including software architectures, social or technical networks, and chemical or biological systems. To address domain-specific requirements, modelling techniques have to tailor their notations, expressivity and analysis tools to a chosen class of problems. While benefiting from concepts and results of the general theory, domain-specific techniques can offer superior capabilities in the chosen domain. In order to avoid reinventing variants of the same concepts, an axiomatic approach to domain-specific graph transformation approaches is advisable.

Kappa [8], a stochastic rewriting approach for graphs representing molecular structures, is a case in point. For a particular class of (hyper)graphs and finely tuned constraints on rules and matches, its techniques for refinement, simulation and analysis [5,9] are significantly more powerful than those for standard (stochastic) graph transformations. An understanding of its relation with mainstream graph transformation is currently emerging (see also Sect. 6). The particular aim of the paper is to develop an axiomatic approach enabling the transfer of Kappa's refinement technique into transformation systems based on adhesive categories. This will allow a more general view of the domain-specific

H. Giese and B. König (Eds.): ICGT 2014, LNCS 8571, pp. 146–160, 2014.

constraints enabling Kappa's capabilities, which are at the heart of its success with biologists.[1]

One fundamental concept is stochastic rule refinement which, apart from a top-down method of developing models, enables techniques such as the thermodynamic approach [6] and the derivation of differential equations [14,7]. Refinements allow a rule to be replaced by a set of extended rules, jointly equivalent (in the sense of a stochastic bisimulation) to the original. We investigate the conditions under which Kappa-like refinement is possible in an adhesive setting. Alongside we present a model, based on typed attributed graphs, of a social network [11] as an interesting application for stochastic graph transformation.

The paper is organised as follows. We start with our general double-pushout setting: an adhesive ambient category of structures and its subcategories of patterns (e.g., left- and right-hand sides of rules) and states (objects to which the rewriting eventually applies). In Sec. 3 we turn to the fundamental notion of *rigidity*. This is a property of objects similar to the absence of V-structures in graphs [10], where no node is allowed to carry two or more edges unless they are distinguishable by their types or attributes, or those of their target nodes. We show how rigidity can be achieved canonically by placing negative constraints on structures. In Sec. 4 we show how to ensure that these negative constraints (and others) are invariant under rewriting, leading to the systematic extraction of match constraints based on a theory of matches as open maps [16]. With this material in place, we turn to rule refinements, generalising the notion of *growth policy* used in Kappa to specify them and illustrate this by a refinement of the social network model which is thermodynamically consistent in the sense of [6].

2   Structures, Patterns, and States

A type graph defines a structured vocabulary for instance graphs. However, depending on the interpretation of instances as states, patterns or arbitrary structures, they are subject to further constraints. *States* are the most constrained: negatively, by stating the absence of certain structures, or positively, requiring their presence. *Patterns* forming, e.g., the left- and right-hand sides of rules, are subject to negative constraints only, because they represent fragments of states, not deemed to be complete. *Structures* live in an adhesive ambient category for states and patterns. A category is *adhesive* [18] if it has pullbacks as well as pushouts for all pairs of morphisms where one is a mono, and where all such pushouts enjoy the van Kampen property. An example is the category of typed attributed graphs [12].

Assumption 1. *We assume an adhesive category* **C** *of* structures *equipped with*

1. *a full subcategory* **PC** *of* **C**, *called* pattern category, *closed under subobjects: for a monomorphism* $A \to B \in$ **C**, $B \in |$**PC**$|$ *implies* $A \in |$**PC**$|$.

---

[1]   The language was featured in Nature both in July and November 2009, and hailed as one of the future "mainstream components of modern quantitative biology" and the "harbinger of an entirely new way of representing and studying cellular networks" in Nature Methods in 2011.

2. *a full subcategory* $\mathbf{SC} \subseteq \mathbf{PC}$ *called* state category.

Due to 1, if a structure satisfies the constraints for patterns, all its substructures do. That means, such constraints are *negative*, demanding the absence of structure, not their presence. $\mathbf{SC}$ is defined by additional constraints on objects.

Remark 1. It follows that $\mathbf{PC}$ has pullbacks along pairs of morphisms where at least one is mono. They are constructed in $\mathbf{C}$ and, by closure of $\mathbf{PC}$ under subobjects, the pullback object is in $\mathbf{PC}$.

Rules are spans of monomorphisms. Transformations follow the double-pushout approach, with monomorphisms as matches [13]. We use a model of socially-driven evolution of opinions [11] to illustrate our concepts.

Example 1 (typed attributed graph transformation). Our ambient category $\mathbf{C}$ is that of attributed graphs over the type graph [12] in the top left of Fig. 1. The model features agents who vote for one of two parties and can be connected to other agents. We represent votes as node attributes $0, 1$. Connections are

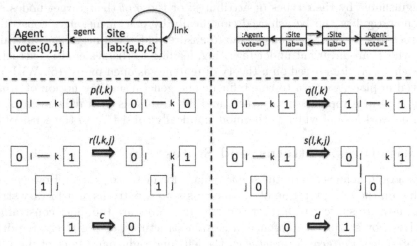

Fig. 1. type graph (top left), instance graph (top right) and rules in concrete syntax, where $l, k, j \in \{a, b, c\}$ (bottom)

identified by labels a, b, c on the sites they are attached to. Once restricted to rigid graphs, this will limit the number of an agent's connections to 3.

In the lower part of Fig. 1 rules are given in a condensed Kappa-like notation. Vote attributes are shown as labels inside Agents. Sites connected to Agents are shown by their labels only. Each Site label is attached to exactly one Agent, leaving agent edges implicit. Link edges are assumed to be symmetric, shown as undirected. To compare, the top right of Fig. 1 shows the left-hand side of rule p as full instance graph (omitting edge types, which can be inferred from sources and targets). Rules are given by rule schemata, e.g., $p(l, k)$ represents all rules obtained by choosing for l, k any labels from a, b, c. That means, l, k are not variables in the sense of attributed graphs, to be instantiated by matches, but metavariables to express rule schemata.

Rules model the coevolution of votes and connections: if two connected agents hold different votes, either one is converted to the opinion of the other (rules p, q), or the link between them is broken and one makes a new connection to an agent of the same opinion (rules r, s). We also allow spontaneous change of opinion (rules c, d).

We define patterns and states by constraints on structures. A *positive constraint* is a mono $c : P \to Q$, satisfied by an object G if for every mono $P \to G$ there is a mono $Q \to G$ which makes the triangle commute. A *negative constraint* is an object P, satisfied by G if there is no mono $P \to G$. In Assumption 1, the axiomatic treatment abstracts from the way patterns and states are specified, but constraints will be used in our running example. Negative ones play a role in ensuring rigidity.

Example 2 (constraints). Pattern and state constraints are given in the top and bottom of Fig. 2, resp. *Patterns* are subject to negative constraints, expressed by forbidden substructures. They include V-structures, parallel edges and loops, i.e., *V-S2A, V-S2S:* no Site is connected to two Agents nor Sites, *V-A2S:* no Agent is connected to two Sites with the same label, *V-vote:* no Agent has two vote attributes, *V-lab:* no Site has two labels and *PAR-S2A, PAR-S2S, LOOP-S2S:* there are no parallel edges or loops. The pattern category **PC** is the full subcategory of **C** satisfying the negative constraints.

States **SC** form the full subcategory of **PC** defined by constraints *SYM:* link edges are symmetric, *S2A:* every Site is connected to an Agent, *S-lab:* each Site has a label attribute, *A-vote:* each Agent has a vote, *A2S:* for all labels $l \in \{a, b, c\}$, each Agent has a Site labelled l. As before, l is a metavariable expanding *A2S* into three concrete constraints.

For the model presented we are interested in questions such as: What is the evolution over time of the number of agents holding certain votes, or of edges

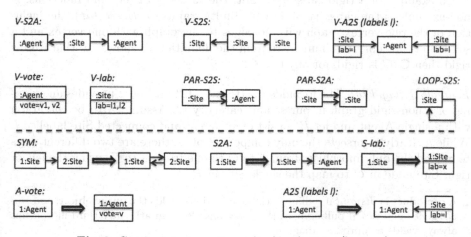

Fig. 2. Constraints on patterns (top) and states (bottom)

connecting agents of the same vs. those of different votes? What are their resulting long-term ratios? How do these correlate to initial conditions and rates assigned to rules? Kappa provides techniques to explore such questions by extracting differential equations (done manually in [11]) or deriving rates for rules from long-term probabilities of certain patterns. Our aim is to open up these techniques to a wider range of rewriting approaches. The notions in the following section provide the prerequisite.

3   Rigidity

The analysis techniques mentioned above use a notion of rule refinement that preserves the dynamics of the system based on a property of structures called *rigidity*. This helps to ensure that the number of matches for a set of extended rules is the same as for the original.

Assumption 2. *We assume that* **C** *is extensive [1]—that is,* **C** *has binary coproducts, which are disjoint and stable under pullback[2].*

An arrow $h : A \to B \in \mathbf{C}$ intersects all *components of* B iff for all B_1, B_2 and isos $\varphi : B \to B_1 + B_2$ where B_1 is not initial, the pullback object A_1 is not initial. An object C is *rigid* iff all morphisms $h : A \to B$ that

$$A_1 \longrightarrow B_1$$
$$\downarrow \qquad\qquad \downarrow$$
$$A \xrightarrow{h} B \xrightarrow{\varphi} B_1 + B_2$$

intersect all components of B behave like epis for morphisms between B and C, that is, for all $f, g : B \to C$ we have that $f \circ h = g \circ h$ implies $f = g$. A category **C** is *rigid* if all its objects are[3] (equivalently: if a morphism in **C** is epi iff it intersects all components of its target). The full subcategory of **C** of all rigid objects is denoted $Rg(\mathbf{C})$.

An example of a rigid category is **Set**, the category of sets and functions. A non-example is the category of directed multi graphs (V, E, src, tar): the inclusion of the one-vertex graph with no edges to the graph with one vertex and a self-loop is not epi. It is immediate from the definition to show, e.g. that if **C** is rigid then **C**/C is rigid, for any $C \in \mathbf{C}$.

Example 3 (rigidity). The forbidden pattern $V\text{-}A2S$ in Fig. 2 provides an example of a non-rigid graph in our sample category **C**. Assume A to be the graph with a single Agent and let B consist of an Agent and connected Site labelled l. While h clearly intersects the only component of B, there are two different ways to extend the only morphism from A to C to one from B to C, using the left or right Site node in C to map the single Site in B.

---

[2] A coproduct is *disjoint* if pulling back the injections yields the initial object, and *stable under pullback* if pulling back the injections along an arbitrary third morphisms always yields a coproduct diagram.

[3] This follows the tradition of regular, extensive and adhesive categories, in that finite colimits (here epis and initial object) are related with finite limits (here pullbacks).

We are interested in rigidity because it gives us a means for refinement via epis. If pattern A is refined by B via epi h, each occurrence of a A in C extends to an occurrence of B in C in at most one way. In our example category **PC** this is because, starting from an agent there is at most one connected site of each label, and starting from a site, there is at most one connected site, attached to a unique agent.

Lemma 1. *In rigid categories, epis are stable under pullbacks along coproduct injections.*

Proof. Suppose that f is epi and the diagram below left is a pullback.

$$
\begin{array}{ccc}
A_1 \xrightarrow{\ f_1\ } B_1 \\
\downarrow\qquad\ \downarrow \\
A \xrightarrow{\ f\ } B_1 + B_2
\end{array}
\qquad
\begin{array}{ccccc}
X_1 \xrightarrow{\ g_1\ } Y_1 & \longrightarrow & C_1 \\
\downarrow\qquad\ \downarrow & & \downarrow{i_1} \\
A_1 \xrightarrow{\ f_1\ } B_1 \xrightarrow{\ \varphi\ } C_1 + C_2 & & \\
\downarrow\qquad\ \downarrow & & \downarrow{i_1} \\
A \xrightarrow{\ f\ } B_1 + B_2 \xrightarrow{\varphi+B_2} (C_1+C_2)+B_2 \xrightarrow{\ \alpha\ } C_1 + (C_2+B_2)
\end{array}
$$

We need to show that f_1 is epi. Suppose that $\varphi : B_1 \to C_1{+}C_2$ is an isomorphism. Let Y_1 and X_1 be obtained by pulling back. Then, since pullbacks paste together and α is an isomorphism, X_1 is the pullback of $(\alpha(\varphi + B_2))f$ along the injection $C_1 \to C_1 + (C_2 + B_2)$, thus $X_1 \neq 0$, as required.

In the following we limit ourselves to working with rigid patterns.

Assumption 3. **PC** *is rigid, and the inclusion to* **C** *creates coproducts.*

That means, for objects A, B in **PC** their coproduct in **C** is also in **PC**. In case of negative constraints defining patterns, it means that their satisfaction is closed under coproducts. This is ensured if the constraints are connected. An object $A \in$ **C** is *connected* if for any coproduct $A_1 + A_2$ isomorphic to A, either $A_1 \cong A$ or $A_2 \cong A$.

Example 4 (rigid structures). The subcategory **PC** $\subseteq$ **C** defined by the pattern constraints in the top of Fig. 2 is indeed rigid, and since all constraints are connected, it has all coproducts, created by the inclusion to **C**.

These constraints arise in a canonical way, as minimal non-rigid objects in **C**: Disregarding *LOOP-S2S*, the pattern constraints in Fig. 2 represent all minimal non-rigid instance graphs (i.e., they are not rigid and have no proper subgraph that is). The only exception is *V-A2S* which has a non-rigid subgraph obtained by dropping site labels. However, in all our graphs (as patterns, rules, states) all sites carry a label to distinguish sites. *LOOP-S2S* is required because our model does not contain loops.

In general, constraints may arise from our knowledge of the domain, such as with the absence of cycles or non-labelled sites, but may also be derived systematically to ensure rigidity of the resulting category of patterns. One can think of this as a two-step process making explicit the distinction between constraints that are requirements-driven or derived canonically.

Proposition 1 (rigidity is closed under subobjects). *For monos* $e : C \to D$, *if* D *is rigid, so is* C. *That means,* $Rg(\mathbf{C})$ *is the full subcategory of* $\mathbf{C}$ *defined by the set of negative constraints consisting of all non-rigid objects in* $\mathbf{C}$ *that are without non-rigid proper subobjects.*

Proof. If C is not rigid, there exist A, B, h, f, g as above so that $f \circ h = g \circ h$, but $g \neq f$. This extends to a counterexample to D's rigidity because $e \circ f \circ h = e \circ g \circ h$ but, since e is mono, $g \neq f$ implies $e \circ g \neq e \circ f$. $\qquad\square$

For example, in the category of multi graphs (i.e., allowing multiple parallel edges), minimal non-rigid graphs are graphs of the form $1 \to 2 \leftarrow 3$ and $1 \leftarrow 2 \to 3$, graphs of two nodes connected by parallel edges, and graphs with a single node and two loops. No graph containing any of these as subgraphs is rigid, restricting us to graphs made up of chains and circles.

The following will be useful later, when considering epis as refinements.

Proposition 2 (epis and monos in rigid categories). *In a rigid category with all coproducts, for every arrow* $a : A \to B$ *there exists a maximal epi-mono factorisation* $A \xrightarrow{e} O \xrightarrow{f} B$, *i.e., for every factorisation* $A \xrightarrow{e'} O' \xrightarrow{f'} B$ *with* e' *epi, there is a morphism* $g : O' \to O$ *commuting the resulting triangles. It follows that* g *is epi, and uniquely determined by* e *and* e'. *It is mono if* f' *is.*

Proof. Let $\mathcal{C}(B)$ be the set of coproduct injections into B that have a non-initial pullback along f. Clearly, $\mathcal{C}(B)$ is closed under binary joins (taken in the lattice of coproduct injections) and is downwards closed by Lemma 1.

Since $\mathbf{C}$ has arbitrary coproducts, it is closed under arbitrary joins, in particular, it has a maximum element X. Thus $B \cong X + Y$ for some Y and, by assumption, $Y \notin \mathcal{C}(B)$.

Hence we have pullback diagrams, from which, using extensivity, $A \cong A' + 0$ and thus φ is an iso.

Let $e = a'\varphi^{-1}$, and $f = i$. It follows from the construction of the maximal element X that the factorisation is in fact the maximal one, i.e., for every other epi-mono factorisation $A \to X' \to B$ there is a unique $X' \to X$ commuting the two triangles. $\qquad\square$

$$
\begin{array}{ccc}
A' & \xrightarrow{a'} & X \\
\varphi \downarrow & & \downarrow i \\
A & \xrightarrow{a} & B \\
\uparrow & & \uparrow \\
0 & \longrightarrow & Y
\end{array}
$$

In our example category $\mathbf{PC}$ the maximal factorisation is given by cutting off all components of B not intersecting with A. This is the largest O while e is epi, but not the only one. For example, we can reduce O to O' until e' is surjective.

Morphisms in $\mathbf{PC}$ that are mono as well as epi are used as pattern refinements, extending the source by adding connected structure. The category $\mathbf{ME}(A)$ *of mono-epis under* A has as objects all morphisms in $\mathbf{PC}$ starting in A that are mono and epi. Morphisms in $\mathbf{ME}(A)$ between $b : A \to B$ and $c : A \to C$ are monos $B \to C$ in $\mathbf{PC}$ that commute the triangle.

Remark 2. It follows that arrows in $\mathbf{ME}(A)$ are epis and $\mathbf{ME}(A)$ is a preorder.

4   Matches as Open Maps

Given a rule $p : L \leftarrow K \rightarrow R$ in **PC** and an object $G \in$ **SC**, in order to apply p to G, we need a match $m : L \rightarrow G$. Apart from the standard gluing conditions, we restrict matches to a subclass of monos that satisfy suitable *reflection constraints*, i.e., matches have to reflect some of the structure of G. If this structure is not present in L, the match is invalid. Absence of structure in L acts like a negative application condition. In [16], such conditions were characterised categorically as *open maps* (used earlier to describe functional bisimulations [17]).

The idea is to use a subcategory **R** of **C** of "ordinary" morphisms to capture extensions of structure. An arrow $c : P \rightarrow Q$ in **R** can be seen as an implication saying that, for each occurrence of P in the source of an open morphism $m :$ $X \rightarrow Y$, if a corresponding occurrence of Q can be found in Y, then this must give rise to a compatible occurrence of Q in X. If P represents a prefix of a path (e.g., in an LTS) and Q a possible extension, this amounts to a reflecting property for the paths specified. In our case, $c : P \rightarrow Q$ represents the forbidden embedding of a pattern into a state graph.

Given a subcategory **R** of **C**, (historically called *path category*) a morphism $m : X \rightarrow Y$ is **R**-*open* if commutative squares mapping a $c : P \rightarrow Q$ in **R** to m have a *fill-in*, i.e., a morphism f such that the resulting triangles PQX and QXY commute. If **R** is understood, we refer to m as *open*.

Embeddings are monic **R**-open maps in **C** for a given path category **R**. The (wide) subcategory of **PC** with **R**-open monic morphisms only is called **PC$_R$**. We define the path category **R** by means of reflection constraints, i.e., morphisms generating **R** as the smallest subcategory of **C** containing the constraints.

Example 5. The reflection constraint in the top left of Fig. 3 states that there should not be links attached to an agent's sites in a graph if there are no such links in a pattern (i.e., a rule's left-hand side). As before, this is a family of constraints covering all possible labels. Consider the instance R-$AS2S(a,b)$ (shown right of the constraint); the match of the *new* rule is not open (shown right), as there is no fill-in of the square. In this case, we see that using that match would lead to a state (bottom right) violating the pattern constraint V-$S2S$ of Fig. 2, and non-rigid if $b = c$.

For a rule and negative constraints, it is possible to construct a set of reflection constraints such that, given a match into a graph satisfying the negative constraints, the match satisfies the reflection constraints (is open) iff the transformation does not lead to a graph violating the negative constraints. This provides us with a systematic way of defining the reflection constraints needed to guarantee preservation of rigidity.

We derive reflection constraints for a selection of minimal rules serving as generators for all rules allowed in a model. Reflection constraints that guarantee preservation of rigidity for this selection will be sufficient for all derived rules.

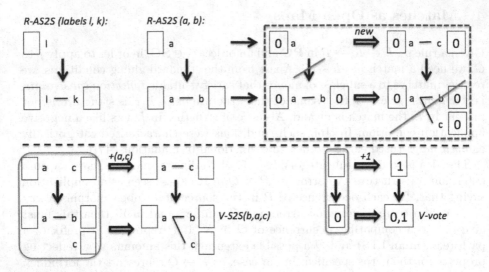

Fig. 3. Reflection constraints and preserving constraints (top) and derivation of reflection constraints from negative constraints and minimal rules (bottom)

A span $L \xleftarrow{l} K \xrightarrow{r} R$ is a *minimal creation rule* if R is atomic (i.e., cannot be obtained as a union of two proper subobjects), l iso and r mono but not iso; it is a *minimal deletion rule* if L is atomic, r iso and l mono but not iso. An object is *finite* if it has only finitely many subobjects.

Proposition 3. *In an adhesive category* **C** *with initial pushouts [12], if G and H are finite, every transformation $G \xRightarrow{p} H$ can be decomposed as a finite sequence of transformations $G = G_0 \xRightarrow{p_1} \ldots \xRightarrow{p_n} G_n = H$ via minimal rules p_i.*

Proof. Split the given transformation into a deletion and a creation phase, represented by monos $G \xleftarrow{g} D \xrightarrow{h} H$. By symmetry, it is enough to consider h. Since H is finite, there exists a (non-unique) finite chain of monos:

$$D = D_0 \xrightarrow{h_1} \ldots \xrightarrow{h_n} D_n = H$$

which compose to h such that none of the h_i is iso nor can it be decomposed into $h_i = h_{i2} \circ h_{i1}$ with h_{i2}, h_{i1} not iso. Initial pushouts over $h_{i+1} : D_i \longrightarrow D_{i+1}$ yield transformations $D_i \xRightarrow{p_{i+1}} D_{i+1}$ with $p_{i+1} : L_{i+1} \to R_{i+1}$. These are minimal rules because p_{i+1} is mono and R_{i+1} is atomic: assuming another mono $S \to R_{i+1}$ such that $S \cup L_{i+1} = R_{i+1}$, this results in a pushout pre-composable to the initial one. By initiality therefore, $S \cong R_{i+1}$. □

Hence, in order to guarantee that state properties are invariant, it is sufficient to ensure that they are preserved by a set of minimal rules that can implement all effects of the rules in a model. If we are concerned with negative constraints, it is even enough to consider minimal creation rules.

For a given set of negative constraints $\mathcal{N}$, let the *set of reflection constraints* $\mathbf{R}_{\mathcal{N}}$ be the set of morphisms $L \to O$ as in the diagram with:

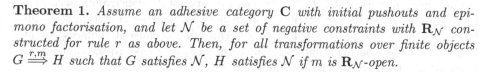

- $r : L \to R$ a minimal creation rule
- P in $\mathcal{N}$
- $R \to U \leftarrow P$ jointly epi (their pullback is a pushout)
- the pullback $L \cap P$ of $L \to R \to U \leftarrow P$ a proper subobject of $R \cap P$
- the pushout complement $LROU$ exists
- O satisfies all constraints in $\mathcal{N}$

Theorem 1. *Assume an adhesive category* $\mathbf{C}$ *with initial pushouts and epi-mono factorisation, and let* $\mathcal{N}$ *be a set of negative constraints with* $\mathbf{R}_{\mathcal{N}}$ *constructed for rule* r *as above. Then, for all transformations over finite objects* $G \overset{r,m}{\Longrightarrow} H$ *such that* G *satisfies* $\mathcal{N}$, H *satisfies* $\mathcal{N}$ *if* m *is* $\mathbf{R}_{\mathcal{N}}$-*open.*

Proof. Every counterexample, where the result of a transformation using r violates constraint P, can be reduced to U by an epi-mono factorisation, leading to the diagram above since open maps are closed under composition and prefix. $\square$

The reverse is not true because the construction of $\mathbf{R}_{\mathcal{N}}$ does not take into account the individual rules of the system, but works for arbitrary rules. For example, any identical rule would preserve negative constraints even on matches that are not $\mathbf{R}_{\mathcal{N}}$-open.

Example 6 (minimal rules and reflection constraints). At the bottom of Fig 3 we show the two minimal creation rules $+(a,c)$, $+1$ that all creation effects in our model are derived from. We only allow to add links between existing nodes (e.g., as part of rules rule changing links) and vote attributes (e.g., in rules changing the attribute). Therefore, deriving reflection constraints for these two we cover all possible cases where negative constraints could be violated.

We obtain two families of constraints, comprising respectively the left-hand side of the minimal rule and the graph it is applied to. The one derived from $+(a,c)$ is similar to $R\text{-}AS2S(a,b)$ above except for the additional agent with site labelled c. We adopt the somewhat stronger constraint which, as seen in the previous example, also addresses node creation, but observe that they are equivalent for all graphs that contain one more agent than $+(a,c)$'s left hand-side, with a c-labelled site attached. The second constraint prevents agents with undefined vote to be mapped to agents with defined vote attribute.

We define the notion of *rigid adhesive structure transformation (RAST)* in $\mathbf{PC}$ and $\mathbf{SC}$. We say that a rule $p = L \leftarrow K \to R$ is *derivable* from a set of rules M if there is a derivation: $L = G_0 \overset{p_1,m_1}{\Longrightarrow} \dots \overset{p_n,m_n}{\Longrightarrow} G_n = R$ with $p_i \in M$ such that p's span is the composition of the bottom spans of these steps via pullback.

Definition 1 (rigid adhesive structure transformation (RAST)). *Given* $\mathbf{PC}$ *and* $\mathbf{SC}$ *as before and a set of minimal rules* M *with creation rules* $M^+ \subseteq M$.

Let $\mathcal{N}$ be the subset of all objects in $|\mathbf{C}| \setminus |\mathbf{PC}|$ without proper subobjects in $\mathbf{PC}$, and $\mathbf{R}_{\mathcal{N}}$ the path category over reflection constraints derived for rules in M^+ and constraints $\mathcal{N}$.

The set of permissible rules in $\mathbf{PC}/\mathbf{SC}$ is given by the set of all rules derivable from M whose transformations via $\mathbf{R}_{\mathcal{N}}$-open maps preserve states, i.e., for all transformations $G \overset{p,m}{\Longrightarrow} H$ with p derivable from M and m $\mathbf{R}_{\mathcal{N}}$-open, $G \in \mathbf{SC}$ implies $H \in \mathbf{SC}$. A RAST model is a set of permissible rules, applied using double pushouts in $\mathbf{C}$ using open maps as matches.

Example 7 (RAST). In our model, all rules are derived from the minimal creation rules $+(a,b)$, $+1$, $+0$ and their inverse deletion rules. Note that, for example, deletion rule -1 by itself does not preserve state constraints because it removes the required vote attribute of an agent, violating constraints *A-vote*. However, the derived rule c in Fig. 1 *is* permissible because it replaces one attribute value by another. Using similar arguments it is easy to see that all our rules in Fig. 1 satisfy this requirement.

As demonstrated by the example and underlying theory, a Kappa-like RAST approach can be constructed systematically as follows.

- Define a type graph able to represent intended structures in rigid graphs.
- Define negative pattern constraints as minimal non-rigid objects (or stronger), as well as additional state constraints.
- Define permissible actions as subset of minimal rules.
- Derive reflection constraints from pattern constraints and minimal rules.
- Derive permissible rules based on minimal rules, state and reflection constraints.

Next we demonstrate how RAST models can be refined thanks to rigidity.

5   Growth Policies and Refinements

A pattern A can be seen as a predicate over objects in $\mathbf{SC}$, validated by a match $a : A \to G$. Refining the pattern, we want to replace A by a set of extensions $e : A \to E$, each representing a stronger predicate, such that their exclusive disjunction is equivalent to A. Intuitively, the set of solutions for A should split into disjoint subsets of solutions for the extended patterns. In other words, for every occurrence $a : A \to G$ there should be a single extension $e : A \to E$ with a unique decomposition of a as $b \circ e$, with $b : E \to G$.

If left-hand sides are taken to be patterns, this ensures that matches of the original rules are in one-to-one correspondence with matches of their refinements. In a stochastic transformation system, one can therefore replace a rule by its refinement without affecting the behaviour as expressed by its Markov chain.

Example 8 (rule refinement). A refinement for rule p of Fig. 1 is shown in Fig. 4. It is easy to check on a couple of examples that matches for p factor uniquely through exactly one of the extensions. Since matches are open maps, reflecting links, if an agent in a rule shows a site that is not connected, this site cannot be connected in the state graph. For example, the rule in the top right cannot be applied to a graph where the agent voting 1 has any further connections, because all three sites are present. Instead, the agent voting 0 in the same rule can have two more connections on the sites not mentioned in the rule. Each of the rules describes exactly one embedding of the original p into an immediate context.

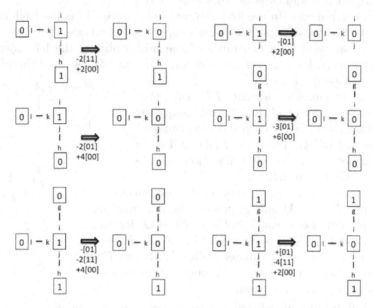

Fig. 4. A refinement for rule p. Labels $l, k, j, i, h, g \in \{a, b, c\}$; i, j, k pairwise distinct. Below each rule, the vector of change in the number of occurrences of the 3 different types of edges is indicated.

Growth policies specify refinements by stating how patterns can be extended. A *growth policy* Γ_A for an object A is a subset of $|\mathbf{ME}(A)|$ such that:
(1) Γ_A is consistent with states: For all $a : A \to G \in \mathbf{PC}$, $G \in |\mathbf{SC}|$, if $a = i \circ b$ is the maximal epi-mono factorisation of a, then $b \in \Gamma_A$.
(2) Γ_A is closed under pullbacks in $\mathbf{ME}(A)$: Given the outer diagram in $\mathbf{PC}$ and let $e : A \to E$ be formed using the pullback of f_1 and f_2 in $\mathbf{PC}$, $e_1, e_2 \in \Gamma_A$ implies $e \in \Gamma_A$.

Both assumptions are met by growth policies in Kappa [6]. (1) means that cutting off disconnected context of a state we obtain an epi satisfying the policy; (2) is a statement of the local nature of policies.

A *refinement of a pattern* A as defined by Γ_A is given by the set of representatives $a \in \Gamma_A$ of all isomorphism classes $[a : A \to B]$ such that for all $a' : A \to B' \in \Gamma_A$, $a' \to a$ implies $a \to a'$. In other words, up to iso a is a minimal object in Γ_A with respect to the preorder $\mathbf{ME}(A)$. We write $\Gamma(A)$ for the refinement, although it is only defined up to iso.

Theorem 2 (refinement). *Every mono* $a : A \to G \in \mathbf{PC}$ *with* $G \in |\mathbf{SC}|$
factors uniquely as $a = f \circ e$ *for a unique* $e \in \Gamma(A)$ *and monic* f.

Proof. We show the existence of a decomposition first. Given $a : A \to G$, let $O + G'$ be a coproduct object in $\mathbf{C}$ isomorphic to G, and such that O is the coproduct of all connected components of G intersecting with a, and G' is the coproduct of the remaining components. In the first diagram, o is obtained via pullback of a and in. Since $\mathbf{C}$ is extensive, pullbacks with coproduct injections yield the top row as a coproduct where, by neutrality of the initial object, the left injection is the identity on A. By construction o is epi, hence in $\mathbf{ME}(A)$, and therefore (by condition 1 above) o is in Γ_A.

$$
\begin{array}{ccccc}
A & \longrightarrow & A & \longleftarrow & 0 \\
\downarrow^{o} & & \downarrow^{a} & & \downarrow \\
O & \xrightarrow{in} & O + G' & \longleftarrow & G'
\end{array}
$$

Let e be the smallest element of Γ_A such that $e \to o$ (existing by condition 2, and unique by construction), and i be the corresponding (unique) morphism in $\mathbf{ME}(A)$. Then, $e \in \Gamma(A)$, and $in \circ i$ is mono since in is a coproduct injection and i is mono (by definition of $\mathbf{ME}(A)$).

Suppose now we are given another decomposition $f' \circ e' = a$ with e' in $\Gamma(A)$ and f' mono. The maximal epi-mono factorisation of a via O yields $j : E' \to O$. Its pullback with i in $\mathbf{PC}$ yields a decomposition $A \xrightarrow{e_0} E_0 \to G$ which is in Γ_A by closure under pullbacks. Since f' is mono, so is j and therefore $E_0 \to E$. Since, by assumption, e is minimal, e_0 and e are equal.

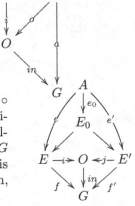

Instead of limiting morphisms a, f to be monic, we can choose any restriction to a class of morphisms which "behave like monos", i.e., which is closed under composition, identities, pullbacks and prefixes and includes coproduct injections. The case in point are of course open maps, as shown below.

Proposition 4. $\mathbf{PC_R}$ *has all pullbacks.*

Proof. Pullbacks preserve monos, pullbacks with monos preserve open maps [16]. Monos and open maps decompose so universal arrows are monic and open.

Assumption 4. *We assume that coproduct injections in* $\mathbf{PC}$ *are* **R**-*open.*

This holds if the target objects of all reflection constraints are connected. With this, the refinements can be applied to open monic matches as they are used in Kappa. Putting everything together we get our notion of refinement:

Corollary 1. *Every monic* **R**-*open* $a : A \to G \in \mathbf{PC}$ *with* $G \in |\mathbf{SC}|$ *factors uniquely as* $a = f \circ e$ *for a unique* $e \in \Gamma(A)$ *and monic* **R**-*open* f.

A refinement a of the left-hand side L extends to a rule p if p is applicable to a, i.e., the relevant pushout complement exists. While this is not guaranteed in general, reflection constraints can be used to restrict refinements of L. In our example rules do not delete nodes, so the pushout complement always exists.

Example 9 (growth policy for balanced refinement). The refinement in Fig. 4 is motived by the desire to track the number of links between agents with different votes vs. those of the same vote, counting occurrences of the following patterns.

[01]: $\boxed{0}$ ı ——— k $\boxed{1}$ *[00]:* $\boxed{0}$ ı ——— k $\boxed{0}$ *[11]:* $\boxed{1}$ ı ——— k $\boxed{1}$

Depending on the context, p may destroy or create different numbers of occurrences of these patterns. This is what motivates the refinement in Fig. 4: it replaces p with a set of jointly equivalent rules, each of which are *balanced* with respect to [01], [00], [11], i.e., they create or destroy a fixed number of occurrences of each pattern. The *balance* is given below the rule arrow, e.g., the rule in the top right destroys one occurrence of [01] and creates two of [00]. (Occurrences of [11] or [00] always come in pairs because of their symmetry.) The growth policy leading to this refinement is based on analysing partial overlaps of rules and patterns, requiring to extend a rule whenever a pattern overlaps a subrule in a way such that an application would create or destroy the pattern [6].

6   Related Work

Hayman and Heindel [15] developed a categorical generalisation of Kappa in the more general single-pushout setting, but did not consider growth policies and rule refinements. Rather than restricting matches to ensure preservation of constraints such as rigidity, constraints were incorporated into the construction of a transformation as a pushout of partial maps. Instead, we follow [16] in using a double-pushout approach which defines constraints on matches to preserve rigidity. Solutions for (causal) trace compressions introduced in [4] have been re-understood categorically in [3] using fibrations. A concrete fibrational approach to rule restriction was proposed in [2] in a DPO setting. It would be interesting to rethink and generalise this approach following the ideas in this paper.

7   Conclusion

Based on an analysis of the concept of rigid graphs at a categorical level we have defined a generic approach to rigid adhesive structure transformation and a Kappa-like notion of rule refinement based on growth policies. We have also developed a methodology for obtaining concrete instances of the generic approach in a systematic fashion, choosing an adhesive ambient category and defining appropriate constraints on graphs and rules. In addition, we have used a recent example from the literature on complex social networks to serve as an illustration of this methodology and show the concrete interest of our development.

We believe that this approach provides a platform to transfer analysis techniques and tools for stochastic systems from Kappa to adhesive structure transformations. For instance the (infinite) system of differential equations, which is central to the analysis of our example [11], can be now seen as a special case of the fragmentation technique developed in [7] and be produced by mechanical means (up to an arbitrary precision).

References

1. Carboni, A., Lack, S., Walters, R.: Introduction to extensive and distributive categories. Journal of Pure and Applied Algebra 84, 145–158 (1993)
2. Danos, V., Harmer, R., Winskel, G.: Constraining rule-based dynamics with types. Mathematical Structures in Computer Science 23(2), 272–289 (2013)
3. Danos, V., Feret, J., Fontana, W., Harmer, R., Hayman, J., Krivine, J., Thompson-Walsh, C.D., Winskel, G.: Graphs, rewriting and pathway reconstruction for rule-based models. In: FSTTCS. LIPIcs, vol. 18, pp. 276–288 (2012)
4. Danos, V., Feret, J., Fontana, W., Harmer, R., Krivine, J.: Rule-based modelling of cellular signalling. In: Caires, L., Vasconcelos, V.T. (eds.) CONCUR 2007. LNCS, vol. 4703, pp. 17–41. Springer, Heidelberg (2007)
5. Danos, V., Feret, J., Fontana, W., Krivine, J.: Scalable simulation of cellular signaling networks. In: Shao, Z. (ed.) APLAS 2007. LNCS, vol. 4807, pp. 139–157. Springer, Heidelberg (2007)
6. Danos, V., Harmer, R., Honorato-Zimmer, R.: Thermodynamic Graph Rewriting. In: D'Argenio, P.R., Melgratti, H. (eds.) CONCUR 2013 – Concurrency Theory. LNCS, vol. 8052, pp. 380–394. Springer, Heidelberg (2013)
7. Danos, V., Honorato-Zimmer, R., Jaramillo-Riveri, S., Stucki, S.: Deriving rate equations for site graph rewriting systems. In: Workshop on Static Analysis and Systems Biology, SASB, Seattle (2013)
8. Danos, V., Laneve, C.: Formal molecular biology. Theor. Comput. Sci. 325(1), 69–110 (2004)
9. Deeds, E., Krivine, J., Feret, J., Danos, V., Fontana, W.: Combinatorial complexity and compositional drift in protein interaction networks. PloS One e32032 (2012)
10. Dörr, H.: Efficient Graph Rewriting and Its Implementation. LNCS, vol. 922. Springer, Heidelberg (1995)
11. Durrett, R., Gleeson, J., Lloyd, A., Mucha, P., Shi, F., Sivakoff, D., Socoloar, J., Varghese, C.: Graph fission in an evolving voter model. Proceedings of the National Academy of Science 109, 3682–3687 (2012)
12. Ehrig, H., Ehrig, K., Prange, U., Taentzer, G.: Fundamentals of Algebraic Graph Transformation. EATCS Monographs Theor. Comput. Sci. Springer (2006)
13. Ehrig, H., Pfender, M., Schneider, H.: Graph grammars: An algebraic approach. In: IEEE Symposium on Switching and Automata Theory, pp. 167–180. IEEE (1973)
14. Feret, J., Danos, V., Harmer, R., Krivine, J., Fontana, W.: Internal coarse-graining of molecular systems. PNAS 106(16), 6453–6458 (2009)
15. Hayman, J., Heindel, T.: Pattern graphs and rule-based models: The semantics of Kappa. In: Pfenning, F. (ed.) FOSSACS 2013 (ETAPS 2013). LNCS, vol. 7794, pp. 1–16. Springer, Heidelberg (2013)
16. Heckel, R.: DPO Transformation with Open Maps. In: Ehrig, H., Engels, G., Kreowski, H.-J., Rozenberg, G. (eds.) ICGT 2012. LNCS, vol. 7562, pp. 203–217. Springer, Heidelberg (2012)
17. Joyal, A., Nielsen, M., Winskel, G.: Bisimulation from open maps. Inf. Comput. 127(2), 164–185 (1996)
18. Lack, S., Sobociński, P.: Adhesive Categories. In: Walukiewicz, I. (ed.) FOSSACS 2004. LNCS, vol. 2987, pp. 273–288. Springer, Heidelberg (2004)

Reversible Sesqui-Pushout Rewriting*

Vincent Danos[1], Tobias Heindel[1] , Ricardo Honorato-Zimmer[1],
and Sandro Stucki[2]

[1] School of Informatics, University of Edinburgh, UK
[2] Programming Methods Laboratory, EPFL, Lausanne, Switzerland

Abstract. The paper proposes a variant of sesqui-pushout rewriting
(SqPO) that allows one to develop the theory of nested application conditions (NACs) for arbitrary rule spans; this is a considerable generalisation
compared with existing results for NACs, which only hold for linear rules
(w.r.t. a suitable class of monos). Besides this main contribution, namely
an adapted shifting construction for NACs, the paper presents a uniform
commutativity result for a revised notion of independence that applies to
arbitrary rules; these theorems hold in any category with (enough) stable
pushouts and a class of monos rendering it weak adhesive HLR. To illustrate results and concepts, we use simple graphs, i.e. the category of binary
endorelations and relation preserving functions, as it is a paradigmatic example of a category with stable pushouts; moreover, using regular monos
to give semantics to NACs, we can shift NACs over arbitrary rule spans.

Introduction

Nested application conditions (NACs) for rules of graph transformation systems
(GTSs) are a popular and intuitive means to increase the versatility of graph
transformation. Tools such as AGG[1] and GROOVE[2] support a weakened form
of NACs, namely negative application conditions. So far, the theory of NACs is
fully developed only for double pushout (DPO) rewriting with so-called linear
rules, which means that transformation operations are restricted to deletion
and addition of nodes and edges; for the ubiquitous example of simple graphs,
linearity is even more restrictive, namely, it is not allowed to add or delete edges
between pairs of unchanged nodes.

We show that none of these restriction are necessary if we use a suitable combination of DPO and sesqui-pushout (SqPO) rewriting, which coincides with DPO
for the case of linear rules (in adhesive categories [16]). More precisely, we shall
extend the theory of NACs, notably the Translation Theorem [13, Theorem 6], to
arbitrary spans as rules, which means that we accommodate not only for the deletion of edges between preserved nodes in simple graphs but we can also handle the
operations of merging and cloning of nodes – at least, if rule applications are free

---

* This research was sponsored by the European Research Council (ERC) under grants
 587327 "DOPPLER" and 320823 "RULE".
[1] http://user.cs.tu-berlin.de/~gragra/agg/
[2] http://groove.sourceforge.net/groove-index.html

H. Giese and B. König (Eds.): ICGT 2014, LNCS 8571, pp. 161–176, 2014.

of *side-effects*. Absence of side-effects will be made formal by the definition of *reversible* SqPO (SqPO^r) rewriting, which is the new approach that we shall propose in this paper; its definition is quite natural: it merely amounts to restricting to those SqPO-diagrams that are also SqPO-diagrams "backwards" for the reversed rule. In the end, we obtain a variation of DPO rewriting, avoiding complications involving uniqueness of pushout complements (by use of final pullback complements [8]) and thus we do not need any restriction on rules or matches any more – having the best of both worlds.

Besides the extension of the theory of NACs to arbitrary rules, we provide a suitable notion of independence for SqPO^r rewriting and give the corresponding commutativity result, which specialises to the existing theory for the DPO approach (with linear rules). The only categorical requirements are pullbacks and (enough) stable pushouts since we do not rely on uniqueness of pushout complements any more, which allows to drop the restriction to (left-)linear rules. Roughly, reversible SqPO-rewriting combines the controlled rewriting mechanism of DPO rewriting with the expressive power of SqPO-rewriting while being in line with the usual notions and results about independence of adjacent rule applications.

We plan to put to use our stronger version of NACs in the context of formal modeling languages for systems biology, in particular the *Kappa* language [14]. The rewrite semantics of Kappa has been formalized as a GTS over a particular category of structured graphs [6,14]. However, it seems natural to reformulate these semantics using an adhesive category [16] and NACs: some of the extra structure present in the patterns of Kappa rules intuitively specifies (positive) application conditions; moreover, matches are required to preserve so-called *free sites*, which amounts to a simple family of NACs. Kappa also supports a quantitative analysis that approximates the evolution of the expected number of occurrences of a given set of observable graphs over time using a system of differential equations [5]; if we want to formalise observables as graphs with a NAC, we also need to keep track of the change in their occurrence counts for this quantitative analysis. This is the point where the shifting constructions for NACs as formulated in the literature [13,10] are the tool of choice. One might also consider extending this type of quantitative analysis to process calculi (via graphical encodings), in which case merging rules, as supported by the SqPO approach, become relevant, e.g. for encoding substitution rules. However, as we will see in Example 6, NAC translation may break down when using the SqPO approach; this is why we consider the SqPO^r approach.

The final contribution of this paper, is a first tentative solution to the failure of NAC translation for the SqPO approach (see Example 6): we first construct for each SqPO-diagram the "best approximation" by a SqPO^r-diagram using a suitable "minimally extended" rule instance, which intuitively just contains enough additional context to make the side-effects of SqPO-rewriting (notably deletion of dangling edges) explicit; we then translate NACs to all possible SqPO-rule-instances, for which we finally can use SqPO^r-rewriting. Even though this solution is not effective, we conjecture that it will be viable for graph transformation systems with bounded node degree.

We illustrate the new rewriting approach and our results through examples in the category of simple graphs; in fact, that our results apply to the category of simple graphs is interesting in itself. In summary, it should become clear that the proposal of the SqPO^r approach to rewriting is not merely triggered by the recent interest in reversible computation, but that it is of interest for the core theory of graph transformation and contributes to versatility in applications.

Structure of the Paper. We first recall the notion of (final) pullback complements [8] and sesqui-pushout (SqPO) rewriting [4] in Section 1, where we also state the relevant composition and decomposition results for the corresponding pullback squares, and recall related results on stable pushouts, which all together will be the technical backbone of our main theorems. We define reversible sesqui-pushout rewriting (SqPO^r) in Section 2 together with a notion of independence, for which we derive a uniform commutativity result (Theorem 1), assuming that (enough) stable pushouts exist. Then we recall the syntax and semantics of nested application conditions (NACs) in Section 3 and present our main result about NACs in Theorem 2, after describing the required categorical assumptions. In Section 4, we discuss how this result might be applied even to SqPO-rewriting. Related and future work is discussed in Section 5 where we also quickly discuss suitable categorical frameworks, before we conclude with a summary of our results in Section 6.

1   Preliminaries

A secondary theme of the present paper is the use of an algebraic approach to perform rewriting on simple graphs, which are ubiquitous in computer science and beyond, but are not as well-behaved w.r.t. algebraic graph rewriting as for example (multi-)hypergraphs; we use the following definition.

Definition 1 (Category of Simple Graphs). *A simple graph is a pair of sets* $G = (V_G, E_G)$ *where* $E_G \subseteq V_G \times V_G$ *is an endorelation over* V_G*; the elements of* V_G *are* nodes *or* vertices *and* E_G *contains all* edges *of the graph* G*. A graph morphism* $f \colon G \to H$ *is a function* $f \colon V_G \to V_H$ *such that*

$$E_H \supseteq (f \times f)(E_G) = \{ (f(e), f(e')) \mid (e, e') \in E_G \}.$$

The category of simple graphs, *denoted by* $\mathbb{G}$*, has simple graphs as objects and graph morphisms as morphisms; composition and identities are given by* $(f \circ g)(v) = f(g(v))$ *and* $\mathrm{id}_K(v) = v$ *for all morphisms* $f \colon G \to H$*,* $g \colon K \to G$ *and nodes* $v \in V_K$*.*

The category of simple graphs will serve as running example to illustrate the concepts and results of the paper; we have chosen to keep the list of preliminary categorical concepts as short as possible. A short discussion of suitable categories of graph-like structures is given later in Section 5.

1.1   Final Pullback Complements and Stable Pushouts

The crucial concept of sesqui-pushout rewriting that goes beyond standard text-books on category theory are *(final) pullback complements* [8]; we use the original definition in terms of the universal property illustrated on the right in Figure 1.

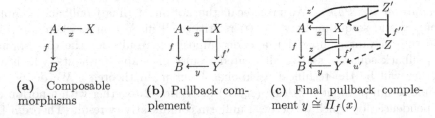

(a) Composable morphisms   **(b)** Pullback complement   **(c)** Final pullback complement $y \stackrel{.}{\cong} \Pi_f(x)$

Fig. 1. Pullback complements for composable morphisms and finality

Definition 2 (Final Pullback Complement (FPBC)). *Let* $B \leftarrow f- A \leftarrow x- X$ *be a pair of composable morphisms (cf. Figure 1(a)). A pullback complement for* $B \leftarrow f- A \leftarrow x- X$ *is a pair of composable morphisms* $B \leftarrow y- Y \leftarrow f'- X$ *such that* $A \leftarrow x- X -f' \rightarrow Y$ *is a pullback of* $A -f \rightarrow B \leftarrow y- Y$ *(cf. Figure 1(b)); it is final, or an* FPBC *for short, if for any morphism* $B \leftarrow z- Z$, *pullback* $A \leftarrow z'- Z' -f'' \rightarrow Z$ *of the co-span* $A -f \rightarrow B \leftarrow z- Z$, *and morphism* $u: Z' \rightarrow X$ *satisfying* $z' = x \circ u$, *there exists a unique morphism* $u': Z \rightarrow Y$ *such that* $y \circ u' = z$ *and* $f' \circ u = u' \circ f''$ *(cf. Figure 1(c), where the universally quantified morphisms are rendered as thick arrows and the dashed one denotes the unique morphism making the diagram commute). If* $B \leftarrow y- Y \leftarrow f'- X$ *is an* FPBC *for* $B \leftarrow f- A \leftarrow x- X$, *we write* $y \stackrel{.}{\cong} \Pi_f(x)$.

By the universal property, FPBCs are unique up to canonical isomorphism. If the composable pair $B \leftarrow y- Y \leftarrow f'- X$ is an FPBC of $B \leftarrow f- A \leftarrow x- X$, we mark this by a modified Freyd corner in the arising square as in Figure 1(c), i.e. we double the line that goes from the apex to the arrow f'; on several occasions, we shall refer to such squares as FPBC *squares*. The following example justifies the use of asymmetric notation.

Example 1 (Implicit Deletion as FPBC). Consider the FPBC square on the right; note that $y \stackrel{.}{\cong} \Pi_f(x)$ while $f \stackrel{.}{\not\cong} \Pi_y(f') \stackrel{.}{\cong} \mathrm{id}_B$.

A crucial property of final pullback complements is stability w.r.t. pullbacks[3]; we shall make extensive use of it in this paper and it also essential for the good behaviour of grammar morphisms (see [1]).

---

[3] This corresponds to the Beck-Chevalley condition in locally cartesian closed categories.

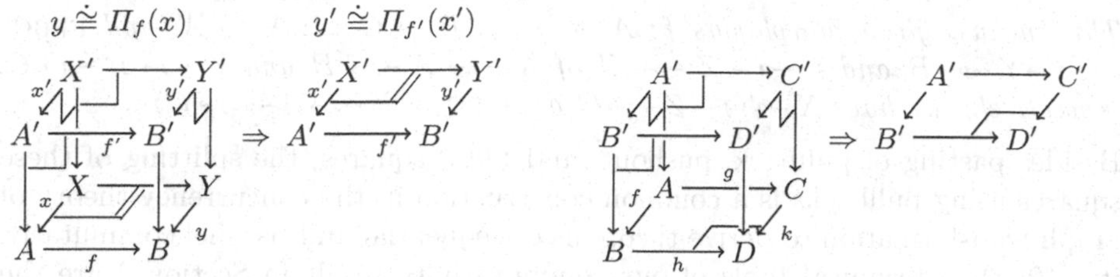

$$y \stackrel{.}{\cong} \Pi_f(x) \qquad\qquad y' \stackrel{.}{\cong} \Pi_{f'}(x')$$

(a) Stability of FPBCs   (b) Stable Pushout

Fig. 2. Stability under pullback

Lemma 1 (Stability of FPBC). *In every cube as in Figure 2(a) that has pullback squares on all faces, if the morphism $B \leftarrow y - Y$ (on the bottom face) is the second morphism of an* FPBC *for $B \leftarrow f - A \leftarrow x - X$, then the morphism $B' \leftarrow y' - Y'$ (on top) is the second morphism of an* FPBC *for $B' \leftarrow f' - A' \leftarrow x' - X'$.*

This lemma implies that, in categories with pullbacks, any FPBC square can be pulled back along a morphism with the "tip" of the square as codomain. Before we state the consequences that we shall use in the remainder of the paper, we define pullback stability for pushouts (as it follows the same pattern of diagrams).

Definition 3 (Stable Pushouts). *Let $B - h \rightarrow D \leftarrow k - C$ be a pushout of the span $B \leftarrow f - A - g \rightarrow C$ in a category $\mathbb{C}$; the pushout is* stable *if for every commutative cube as in Figure 2(b) on the left, the top square is a pushout square if all lateral faces are pullback squares.*

Even though there are categories of graph-like structures in which some pushouts are not stable (under pullback), we generally assume that all pushouts that we operate with are stable. Our running example $\mathbb{G}$ has all pushouts and these are stable.

1.2   Splitting and Composing Pushout and FPBC Squares

We now state the relevant lemmata that allow to compose and decompose pushout and FPBC squares where composition and decomposition are also known as *pasting* and *splitting*, respectively. The reader might want to skip forward to Section 2 and come back to the remainder of the present section to look up the details, especially at a first reading. The proofs do only use diagram chasing and the defining universal properties of pushouts, pullbacks, and FPBCs (and are rather unenlightening).

The first useful fact is composition of FPBC squares, similar to pasting of pushouts and pullbacks.

Lemma 2 (FPBC Composition). *We have (vertical) composition of* FPBCs *as follows:*

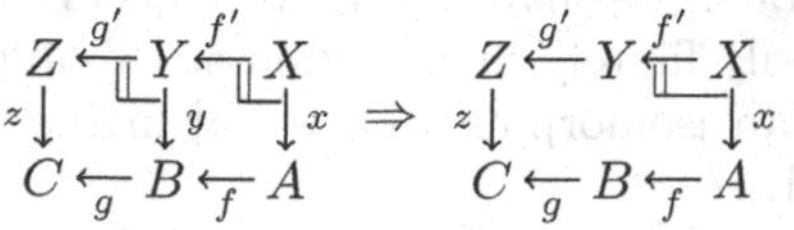

$$
\begin{array}{ccc}
Z \xleftarrow{g'} Y \xleftarrow{f'} X \\
{\scriptstyle z}\downarrow \quad {\scriptstyle y}\downarrow \quad {\scriptstyle x}\downarrow \\
C \xleftarrow{g} B \xleftarrow{f} A
\end{array}
\;\Rightarrow\;
\begin{array}{ccc}
Z \xleftarrow{g'} Y \xleftarrow{f'} X \\
{\scriptstyle z}\downarrow \quad\qquad {\scriptstyle x}\downarrow \\
C \xleftarrow{g} B \xleftarrow{f} A
\end{array}
$$

This means, given morphisms $f\colon A \to B$, $g\colon B \to C$, $x\colon X \to A$, *and* FPBCs $X -f'\to Y -y\to B$ *and* $Y -g'\to Z -z\to C$ *of* $X -x\to A -f\to B$ *and* $Y -y\to B -g\to C$, *respectively, we have* $X -g'\circ f'\to Z -z\to C$ *as* FPBC *of* $X -x\to A -g\circ f\to C$.

Besides pasting of pullback, pushout, and FPBC squares, the splitting of these squares using pullbacks is a common construction in the concurrency theory of graph transformation to derive theorems of sequential and parallel commutativity [12]. The technical tools of our commutativity result in Section 2 are the following two lemmata.

Lemma 3 (FPBC Splitting). *Let the leftmost diagram below be a pair of*

$$\begin{array}{ccc}
Z \xleftarrow{g'} Y \xleftarrow{f'} X \\
z\downarrow \quad y\downarrow \quad x\downarrow \\
C \xleftarrow{g} B \xleftarrow{f} A
\end{array} \,\&\, \begin{array}{ccc}
Z \xleftarrow{g'} Y \xleftarrow{f'} X \\
z\downarrow \quad \quad x\downarrow \\
C \xleftarrow{g} B \xleftarrow{f} A
\end{array} \,\&\, \begin{array}{cc}
B \xleftarrow{f} A \\
g\downarrow \quad \mathrm{id}\downarrow \\
C \xleftarrow{g\circ f} A
\end{array} \Rightarrow \begin{array}{ccc}
Z \xleftarrow{g'} Y \xleftarrow{f'} X \\
z\downarrow \quad y\downarrow \quad x\downarrow \\
C \xleftarrow{g} B \xleftarrow{f} A
\end{array}$$

pullback squares such that $C \leftarrow z- Z \leftarrow g'\circ f'- X$ *is an* FPBC *of* $C \leftarrow g\circ f- A \leftarrow x- X$ *and the span* $B \leftarrow f- A -\mathrm{id}\to A$ *is a pullback of* $B -g\to C \leftarrow g\circ f- A$; *then we have* FPBC *squares as in the rightmost diagram above, i.e.* $C \leftarrow z- Z \leftarrow g'- Y$ *is an* FPBC *of* $C \leftarrow g- B \leftarrow y- Y$ *and* $B \leftarrow y- Y \leftarrow f'- X$ *is an* FPBC *of* $B \leftarrow f- A \leftarrow x- X$.

If pushouts are stable, we have a similar result for pushouts that are also pullbacks.

Lemma 4 (Pushout splitting). *Let the leftmost diagram below be a pair of*

$$\begin{array}{ccc}
Z \xleftarrow{g'} Y \xleftarrow{f'} X \\
z\downarrow \quad y\downarrow \quad x\downarrow \\
C \xleftarrow{g} B \xleftarrow{f} A
\end{array} \,\&\, \begin{array}{ccc}
Z \xleftarrow{g'} Y \xleftarrow{f'} X \\
z\downarrow \quad \quad x\downarrow \\
C \xleftarrow{g} B \xleftarrow{f} A
\end{array} \,\&\, \begin{array}{cc}
B \xleftarrow{f} A \\
g\downarrow \quad \mathrm{id}\downarrow \\
C \xleftarrow{g\circ f} A
\end{array} \Rightarrow \begin{array}{ccc}
Z \xleftarrow{g'} Y \xleftarrow{f'} X \\
z\downarrow \quad y\downarrow \quad x\downarrow \\
C \xleftarrow{g} B \xleftarrow{f} A
\end{array}$$

pullback squares such that $Z -z\to C \leftarrow g\circ f- A$ *is a pushout of* $Z \leftarrow g'\circ f'- X -x\to A$ *that is pullback stable and* $B \leftarrow f- A -\mathrm{id}\to A$ *is a pullback of* $B -g\to C \leftarrow g\circ f- A$; *then the first two pullback squares are also pushout squares, i.e.* $Z -z\to C \leftarrow g- B$ *is a pushout of* $Z \leftarrow g'- Y -y\to B$ *and* $Y -y\to B \leftarrow f- A$ *is a pushout of* $Y \leftarrow f'- X -x\to A$.

Remark 1. If the morphism g in Lemma 3 (resp. Lemma 4) is a mono, the third assumption is trivially true (cf. [16, Lemma 4.6], for pushout splitting).

The generality of Lemmata 3 and 4 is tailored to fit exactly our new examples of independence in rewriting in Section 2.2, which feature both merging and cloning of nodes.

1.3   Finitely Powered Objects

In virtually all applications, objects of rewriting and rules are suitably finite. More precisely, in Section 4, we shall restrict to objects that are *finitely powered*, i.e. objects shall have only finitely many different subobjects, where a *subobject* of an object $A \in \mathbb{C}$ is an isomorphism class $[m]$ in the slice category $\mathbb{C}{\downarrow}A$, of some mono $m\colon M \rightarrowtail A$.

2 Reversible Sesqui-Pushout Rewriting

The central definition of the present paper is the reversible variant of sesqui-pushout rewriting (SqPO) [4]. It generalises the very controlled rewriting mechanism of DPO rewriting [3] to arbitrary rules as used in SqPO rewriting, i.e. any span can be used as a rule. As a result, we can perform DPO rewriting with duplication of entities as in SqPO rewriting; moreover, we also can lift the theory of application conditions and constraints [13], as we shall do in Section 3.

2.1 Definition and First Examples

The definition of reversible sesqui-pushout rewriting is trivial: we just require a double square diagram that is a sesqui-pushout diagram forwards and backwards. The benefits of this approach over the original one, besides being in line with the recent trend of reversible computation, will become clear when we discuss the theory of nested application conditions and its limitations in the SqPO approach.

Definition 4 (Reversible Sesqui-Pushout Rewriting). *A* rule *is any span of morphisms* $L \leftarrow\alpha- K -\beta\to R$, *i.e. any pair of morphisms sharing their domain; the* reversal *of a given rule* $\rho = L \leftarrow\alpha- K -\beta\to R$, *written* ρ^r, *is the rule* $R \leftarrow\beta- K -\alpha\to L$. *Let* $\rho = L \leftarrow\alpha- K -\beta\to R$ *be a rule, and let* $m \colon L \to A$ *be a morphism; an* SqPO-diagram *for* ρ *at* m *is a diagram as in* (1) *on the left*

$$
\begin{array}{ccc}
L \xleftarrow{\alpha} K \xrightarrow{\beta} R & \qquad & L \xleftarrow{\alpha} K \xrightarrow{\beta} R \\
{\scriptstyle m}\downarrow \quad {\scriptstyle o}\downarrow \quad \downarrow{\scriptstyle n} & & {\scriptstyle m}\downarrow \quad {\scriptstyle o}\downarrow \quad \downarrow{\scriptstyle n} \\
A \xleftarrow{\gamma} D \xrightarrow{\delta} B & & A \xleftarrow{\gamma} D \xrightarrow{\delta} B
\end{array}
\tag{1}
$$

such that $A \leftarrow\gamma- D \leftarrow o- K$ *is an* FPBC *of* $A \leftarrow m- L \leftarrow\alpha- K$ *and* $D -\delta\to B \leftarrow n- R$ *a pushout of* $D \leftarrow o- K -\beta\to R$; *in such a diagram, the morphism* m *is called* the match *for* ρ, *and* n *is the* co-match. *Diagram* (1) *is* reversible *or an* SqPO^r-diagram *if it is also an* SqPO-diagram *for* ρ^r *with match* n *and co-match* m, *i.e. we have a diagram as in* (1) *on the right; if we have such a* SqPO^r-diagram, *the match* m *is called* side-effect-free *for* ρ *or just* reversible. *We write* $A \models\langle\rho,m\rangle\Rightarrow B$ *or simply* $A \models\rho\Rightarrow B$ *if there exists a* SqPO^r-diagram *as in* (1) *on the right (where* $\rho = L \leftarrow\alpha- K -\beta\to R$ *is the rule and* $m \colon L \to A$ *is the match).*

If we have $A \models\langle\rho,m\rangle\Rightarrow B$, we also speak of a *rule application* (of ρ at m), or say that rule ρ rewrites A to B at match m.

Remark 2. In adhesive categories [16], the SqPO approach coincides with the double pushout approach [4] for linear rules, i.e. for rules consisting of pairs of monos.

As a consequence of this observation, new examples of SqPO rewriting either have non-linear rules or take place in a category that is not adhesive. This is another reason why we have chosen simple graphs as running example; it is the paradigmatic example of a category that is neither adhesive nor rm-adhesive [11].

Example 2 (Cloning and Merging in Graph Rewriting). Consider the rules described in (2) where all graph morphisms are uniquely determined except for the right morphism of the clone rule, which we take to be the identity.

$$\text{clone} = \boxed{\begin{array}{c} v \end{array}} \leftarrow \boxed{\begin{array}{c} u \\ w \end{array}} \rightarrow \boxed{\begin{array}{c} u \\ w \end{array}} \qquad \text{merge} = \text{clone}^r \qquad \text{loop} = \boxed{v} \leftarrow \boxed{v} \rightarrow \boxed{v} \qquad (2)$$

We have the following applications of these rules.

$$\boxed{v \rightarrow w} \ \models\text{clone}\Rightarrow \boxed{u \leftarrow v \rightarrow w} \ \models\text{loop}\Rightarrow \boxed{u \leftarrow v \rightarrow w} \ \models\text{merge}\Rightarrow \boxed{v \rightarrow x} \ \not\models\text{merge}\Rightarrow \boxed{z}$$

In each case, the match is uniquely determined by the effect on the graphs: first, we clone node w, obtaining its *clone* u with the same local connectivity, then we add a loop at node v, and finally we merge u with w. At the end of the above display, the merge-rule cannot be applied to the last graph as this would use an irreversible match: applying the reverse rule (cloning node z) would result in the completely connected graph on two nodes rather than the original graph.

The rule merge is not merely the reversal of clone, it may in fact serve as its inverse; more precisely, the composite rule clerge := $(\boxed{v} \leftarrow \boxed{vu} \rightarrow \boxed{v})$ has no effect if applied using SqPO^r rewriting, due to symmetry of SqPO^r-diagrams. Note that the category of simple graphs is somewhat peculiar as the clerge-rule can always be applied.[4]

2.2 Independence and Commutativity

The generality of arbitrary spans as rules necessitates an adaptation of the usual notion of independence of rewriting diagrams to obtain the expected commutativity result that allows to "switch" adjacent diagrams if they are *independent* (see Theorem 1 below).

Definition 5 (Independence). *Let* $\rho_i = L_i \leftarrow\alpha_i- K_i -\beta_i\rightarrow R_i$ $(i = 1, 2)$ *be spans*

$$
\begin{array}{ccc}
L_1 \xleftarrow{\alpha_1} K_1 \xrightarrow{\beta_1} R_1 & \quad L_2 \xleftarrow{\alpha_2} K_2 \xrightarrow{\beta_2} R_2 & \quad L_2 \xleftarrow{\text{id}} L_2 \xrightarrow{\text{id}} L_2 \\
m_1\downarrow \quad \downarrow o_1 \quad \downarrow n_1 & m_2\downarrow \quad \downarrow o_2 \quad \downarrow n_2 & \overline{m}_2:=\gamma_1 \circ b \downarrow \ (\dagger) \downarrow_b \ (\ddagger) \downarrow m_2 \\
A_1 \xleftarrow{\gamma_1} D_1 \xrightarrow{\delta_1} B_1 & A_2 \xleftarrow{\gamma_2} D_2 \xrightarrow{\delta_2} B_2 & A_1 \xleftarrow{\gamma_1} D_1 \xrightarrow{\delta_1} B_1
\end{array} \qquad (3)
$$

and let the diagram on the left in (3) *be a pair of* SqPO^r-*diagrams such that* $B_1 = A_2$. *A left witness for these diagrams is a morphism* $D_1 \leftarrow b- L_2$ *such that*

- $\overline{m}_2 := \gamma_1 \circ b$ *is a reversible match for* ρ_2, *and*
- $b \cong \delta_1^*(m_2)$ *and* $b \cong \gamma_1^*(\overline{m}_2)$, *i.e. we have the pullbacks on the right in* (3).

A right witness is a left witness for the "mirrored" situation, that is, a morphism $d\colon R_1 \rightarrow D_2$ *such that*

---

[4] In contrast, in the category of multi-graphs, clerge can only be applied to isolated nodes.

– $n_1 := \delta_2 \circ d$ *is a reversible match for* ρ_1^r, *and*
– $d \cong \gamma_2^*(n_1)$ *and* $d \cong \delta_1^*(n_1)$.

The pair of SqPO^r*-diagrams in* (3) *is independent if there exist a left and a right witness.*

The pullback requirements in (3) are vacuous if γ_1 and δ_1 are monos, i.e. commutativity of Squares (†) and (‡) is sufficient. The following example illustrates the subtle interplay of node merging and cloning with addition of loops.

Example 3. In the following pair of rule applications,

$$\boxed{v \rightarrow w} \models\text{clone}\Rightarrow \boxed{u \leftarrow v \rightarrow w} \models\text{loop}\Rightarrow \boxed{u \leftarrow v \rightarrow \circlearrowright w},$$

cloning node w is not independent of adding a loop at node w as the local structure changes. We can check formally that there is no left witness for the corresponding pair of SqPO^r-diagrams, because there is no suitable morphism making Square (†) in (3) a pullback square; nevertheless, the rules can be applied in the reverse order, but we obtain a different result.

$$\boxed{v \rightarrow w} \models\text{loop}\Rightarrow \boxed{v \rightarrow \circlearrowright w} \models\text{clone}\Rightarrow \boxed{u \leftarrow v \rightarrow w}$$

The role of Square (‡) in (3) features in the following pair of rule applications.

$$\boxed{u \leftarrow v \rightarrow w} \models\text{merge}\Rightarrow \boxed{v \rightarrow w} \models\text{loop}\Rightarrow \boxed{v \rightarrow w\circlearrowright}$$

Nodes u and w can only be merged as long as they have the same "neighbourhood", and adding a loop to w changes the local connectivity of w; formally, there is again no left witness, as we cannot find a suitable morphism making Square (‡) a pullback square. Adding a loop at either one of the nodes u and w makes merging u with w impossible when following the SqPO^r approach.

Remark 3. For the case of adhesive categories and linear rules (i.e. rules that consist of a pair of monos), commutativity of the squares on the right in (3) is enough (as mentioned above); moreover, in adhesive categories, the requirement that left witnesses are reversible matches is automatically true and we recover the usual definition of independence (of the DPO approach with linear rules).

The definition of independence is chosen sufficiently strong to obtain the following general theorem of commutativity of derivations; its proof is similar to the standard results of the literature (see, e.g. [12]), using the lemmata of Section 1.

Theorem 1 (Commutativity). *Let* $\mathbb{C}$ *be a category that has pullbacks and pushouts such that all pushouts are stable. Then for each pair of independent* SqPO^r*-diagrams with left and right witness as below on the left*

$$L_1 \xleftarrow{\alpha_1} K_1 \xrightarrow{\beta_1} R_1 \quad L_2 \xleftarrow{\alpha_2} K_2 \xrightarrow{\beta_2} R_2 \qquad L_2 \xleftarrow{\alpha_2} K_2 \xrightarrow{\beta_2} R_2 \quad L_1 \xleftarrow{\alpha_1} K_1 \xrightarrow{\beta_1} R_1$$
$$m_1 \downarrow \quad o_1 \downarrow \quad \overset{b}{\searrow} \quad \downarrow m_2 \quad \overset{d}{\searrow} \quad \downarrow o_2 \quad \downarrow n_2 \quad \rightsquigarrow \quad \overline{m_2} \downarrow \quad \downarrow o_2' \quad \downarrow n_2' \quad \downarrow m_1' \quad \downarrow o_1' \quad \downarrow n_1$$
$$A_1 \xleftarrow{\gamma_1} D_1 \xrightarrow{\delta_1} B_1 = A_2 \xleftarrow{\gamma_2} D_2 \xrightarrow{\delta_2} B_2 \qquad A_1 \xleftarrow{\gamma_2'} E_2 \xrightarrow{\delta_2'} C = C \xleftarrow{\gamma_1'} E_1 \xrightarrow{\delta_1'} B_2$$

there exists a corresponding pair of independent SqPO^r*-diagrams as above on the right with* $\overline{m}_2 = \gamma_1 \circ b$ *and* $\underline{n}_1 = \delta_2 \circ d$*, reversing the order of rule application.*

The commutativity result for pairs of independent diagrams is interesting in itself, as it applies to arbitrary rules, improves over previous results [4], and is completely symmetric. However, the main motivation to introduce the SqPO^r approach is to make the theory of applications conditions available for arbitrary rule spans.

3   On Nested Application Conditions

Application conditions for rules are an elegant and intuitive means to restrict the allowed matches of each rule individually. We give a short review of nested application conditions before we demonstrate that conditions can be moved freely between left and right-hand sides of rules as in previous work on DPO rewriting with linear rules [13,10]. Before we recall the full definition of NACs, we informally describe a simple example.

Example 4. We might want to apply the loop rule only at those matches that map to a node without any incoming edge, and we illustrate this application condition by adding a "forbidden" dashed edge.

Though most examples only require simple negative application conditions, one occasionally encounters situations where one wants the full generality of *nested application conditions* to restrict matches of certain rules, individually.

Definition 6 (Nested Application Condition). *A nested application condition (NAC) c on an object $P \in \mathbb{C}$, written $c \triangleright P$ or $P \triangleleft c$, is defined inductively as follows.*

Base Case *The trivial NAC is* $\mathtt{tt} \triangleright P$.
Inductive Steps *There are three constructors for NACs.*
    Existential Morphism $\exists(a, c') \triangleright P$ *is a NAC on P if $a\colon P \to Q$ in $\mathbb{C}$ is a morphism and $c' \triangleright Q$ is a NAC on Q.*
    Negation $\neg c' \triangleright P$ *is a NAC on P if $c' \triangleright P$ is so.*
    Conjunction $\bigwedge_{i \in \mathcal{I}} c_i \triangleright P$ *is a NAC if $\{c_i \triangleright P\}_{i \in \mathcal{I}}$ is a family of NACs on P indexed over a non-empty set $\mathcal{I} \neq \varnothing$.*

The negative application condition that we described in Example 4 can be formulated as $\neg \exists(\text{ⓥ} - \subseteq \to \text{ⓦ} \to \text{ⓥ}, \mathtt{tt})$ on the left-hand side of the loop-rule.

Concerning the semantics of NACs, several routes have been taken in the literature; the next definition strikes a compromise between generality and relevance for the present paper; we write $|P{\downarrow}\mathbb{C}|$ for the collection of all $\mathbb{C}$-morphisms with domain P, for any $P \in \mathbb{C}$.

Definition 7 (Semantics of NACs). *Let* X *be a collection of morphisms in* $\mathbb{C}$, *referred to as* splitting set; *for each* NAC $c \rhd P$, *its set of* instances, *denoted by* $[\![c \rhd P]\!] \subseteq |P{\downarrow}\mathbb{C}|$, *is defined by mutual recursion as follows.*

- *For every object* $P \in \mathbb{C}$, *we define* $[\![\mathtt{tt} \rhd P]\!] = |P{\downarrow}\mathbb{C}|$.
- *If* $c = \exists(a, c') \rhd P$ *with* $a \colon P \to Q$ *in* $\mathbb{C}$ *and* $c' \rhd Q$ *a* NAC, *we define*

$$[\![\exists(a, c') \rhd P]\!] = ([\![c' \rhd Q]\!] \cap X) \circ a := \left\{ n \circ a \mid (Q \xrightarrow{n} A) \in [\![c' \rhd Q]\!] \cap X \right\}.$$

- *If* $c = \neg c' \rhd P$ *with* $c' \rhd P$ *a* NAC, *we define*

$$[\![\neg c' \rhd P]\!] = |P{\downarrow}\mathbb{C}| \setminus [\![c' \rhd P]\!].$$

- *If* $c = \bigwedge_{i \in \mathcal{I}} c_i \rhd P$ *with* $\{c_i \rhd P\}_{i \in \mathcal{I}}$ *a family of* NACs, *we define*

$$\left[\!\!\left[\bigwedge_{i \in \mathcal{I}} c_i \rhd P\right]\!\!\right] = \bigcap_{i \in \mathcal{I}} [\![c_i \rhd P]\!].$$

We write $m \vDash^X c \rhd P$, *or* $m \vDash^X c$ *for short, if* $m \in [\![c \rhd P]\!]$.

Note that the splitting set features only at one place in this definition, namely in the clause that gives semantics to the $\exists$-constructor.[5] Every element of $[\![\exists(a, c') \rhd P]\!]$ is factored as an element of the splitting set X after a, thus "splitting off" from each candidate $f \in |P{\downarrow}\mathbb{C}|$ some morphism in X that satisfies c'. A second reason for the name *splitting set* is its function in Theorem 2, where we shall decompose rewriting diagrams using pushout splitting for which we need splitting sets to be *robust*.

Definition 8 (Robust Splitting Set). *A collection of monos* $\mathcal{M}$ *in a category* $\mathbb{C}$ *is a* robust splitting set *if*

- *the set* $\mathcal{M}$ *contains all identities and is closed under composition,*
- *the category* $\mathbb{C}$ *has pushouts and pullbacks along* $\mathcal{M}$,
- *the set* $\mathcal{M}$ *is stable under pushout and pullback, and*
- *pushouts along morphisms in* $\mathcal{M}$ *yield pullback squares.*

The reader that is familiar with adhesive categories and related concepts [9] will recognise these properties (see also Section 5).

For our running example $\mathbb{G}$, we chose the class of regular monos as splitting set, where a mono $m \colon G \to H$ is regular iff it reflects edges, i.e. if it satisfies the equation $(m \times m)(E_G) = E_H \cap (m(V_G) \times m(V_G))$.

Lemma 5. *Regular monos are a robust splitting set.*

The use of regular monos as splitting set has some subtle consequences.

---

[5] There is indeed a hidden existential quantifier in the definiens of $[\![\exists(a, c') \rhd P]\!]$, namely $\left\{ n \circ a \mid n \in [\![c' \rhd Q]\!] \cap X \right\} = \left\{ f \in |P{\downarrow}\mathbb{C}| \mid \exists n \in [\![c' \rhd Q]\!] \cap X . f = n \circ a \right\}$.

Example 5 (Cloning Nodes without Loops). Consider the clone-rule with the application condition that the node to be cloned does not have a loop.

The corresponding NAC for the clone-rule is $c_{\neg \circlearrowright} := \neg\exists(\textcircled{v}-\subseteq\rightarrow\textcircled{v}\!\!\text{)}, \mathtt{tt})$; however, (ab)using the fact that we have fixed regular monos as our splitting set, we could as well use the following NAC $c_b := \neg\exists(\textcircled{v}-\mathrm{id}\rightarrow\textcircled{v}, \mathtt{tt})$ (where the morphism is the identity on a single node). The two are equivalent, in the sense that $[\![c_{\neg\circlearrowright}]\!] = [\![c_b]\!]$. While this example is admittedly somewhat pathological, it illustrates the functioning of the regular monos as the splitting set.

Our main result about NACs is the following (cf. [13, Theorem 6] and [10, Lemma 3]).

Theorem 2 (NAC translation). *Let $\mathcal{M}$ be a robust splitting set in a category with pushouts; then for each rule $\rho = L \leftarrow\!\alpha\!- K -\beta\!\rightarrow R$, and every NAC $c \triangleright R$ (resp. $\bar{c}\triangleright L$), there exists a NAC $c_\rho \triangleright L$ (resp. $\bar{c}_{\rho^r} \triangleright R$) such that for every SqPO^r-diagram*

$$c_\rho \triangleright L \xleftarrow{\alpha} K \xrightarrow{\beta} R \triangleleft c$$
$$\Big\downarrow m \qquad \Big\downarrow o \qquad \Big\downarrow n$$
$$A \xleftarrow{\gamma} D \xrightarrow{\delta} B$$

we have $n \vDash^{\mathcal{M}} c$ iff $m \vDash^{\mathcal{M}} c_\rho$ (resp. $m \vDash^{\mathcal{M}} \bar{c}$ iff $n \vDash^{\mathcal{M}} \bar{c}_{\rho^r}$).

Much more interesting than repeating all minute details of the translation of NACs is the fact that there is no hope that exactly the same result could be obtained for SqPO-rewriting, as illustrated by the following example.

Example 6 (Failure of NAC Translation). Applying the loop-rule (using plain SqPO) to $\textcircled{v}$ or to $\textcircled{v}\!\!\text{)}$ we get the same result, and thus the same co-match; note that the match into $\textcircled{v}\!\!\text{)}$ is not reversible. Adding the NAC $\neg\exists(\textcircled{v}-\mathrm{id}\rightarrow\textcircled{v}, \mathtt{tt})$ to the loop-rule forbids application to the graph $\textcircled{v}\!\!\text{)}$ but it can still be be applied to $\textcircled{v}$. Thus, a corresponding equivalent application condition on the right-hand side of the loop-rule cannot exist, because either it admits the comatch into $\textcircled{v}\!\!\text{)}$ or not, but in both cases it does not distinguish between the two matches. Hence, we have shown that $\neg\exists(\textcircled{v}-\mathrm{id}\rightarrow\textcircled{v}, \mathtt{tt})$ does not have any exact counterpart on the right-hand side in the sense of Theorem 2.

It is not obvious how one could work around this counter-example in general. However, for the case of monic matches, the next section describes how Theorem 2 can be applied even to SqPO-rewriting, at least in favourable cases.

4   On Reversibility of Sesqui-Pushout Rewriting

After the generalisation of key results of DPO rewriting to SqPO^r rewriting, we shall describe a method that gives for each SqPO-diagram (with monic match and co-match) a minimal rule-instance that can be used to achieve the same rewriting effect using SqPO^r-rewriting. This method can be understood as a natural measure for how far away a SqPO-derivation is from being reversible. More precisely, we can divide each SqPO direct derivations into a SqPO-derivation (the *rule-instantiation*) over a SqPO^r-diagram such that vertical composition of the FPBC and pushout squares yield the original SqPO-diagram. Intuitively, the rule-instance contains just enough additional context to cover the side-effects of the original SqPO-diagram.

Theorem 3 (Instantiation Thoerem). *Let* $\mathbb{C}$ *be a category with pullbacks in which pushouts are stable; then for each* SqPO-*diagram with match* $m: L \to A$ *such that* A *is finitely powered, there exists a least subobject* $[m': L' \rightarrowtail A]$ *of* A *such that the match* m *factors as* $m = m' \circ i$ *(for a unique* i*) and the original* SqPO-*diagram is the vertical composition of an* SqPO-*diagram with match* i *over a* SqPO^r-*diagram with match* m'*, i.e. we have the following diagram.*

$$
\begin{array}{ccc}
L \xleftarrow{\alpha} K \xrightarrow{\beta} R & & L \xleftarrow{\alpha} K \xrightarrow{\beta} R \\
\Big\downarrow m \qquad \qquad \Big\downarrow n & & i\Big\downarrow \quad k\Big\downarrow \quad j\Big\downarrow \\
\qquad \quad o & \rightsquigarrow & L' \xleftarrow{\gamma'} K' \xrightarrow{\delta'} R' \\
& & m'\Big\downarrow \quad o'\Big\downarrow \quad n'\Big\downarrow \\
A \xleftarrow{\gamma} D \xrightarrow{\delta} B & & A \xleftarrow{\gamma} D \xrightarrow{\delta} B
\end{array}
$$

For the case of simple graphs, we illustrate the idea that the construction of the (proof of the) theorem just adds minimal context in the neighbourhood of nodes that are merged or deleted.[6]

Example 7. The below diagram (where i, m', and α are inclusions) is an example for a decomposition of an SqPO-diagram according to Theorem 3.

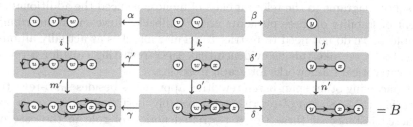

It is important to remember that we are working in the category of simple graphs because the outer double square diagram would not be an SqPO-diagram in the category of multi-graphs, as there would be an additional edge from y to z in B. The upper row is an SqPO-diagram and the lower row a SqPO^r-diagram; moreover, the subobject $[m']$ is minimal in the sense of Theorem 3, which we will explain using the metaphor of minimal context.

---

[6] For the case of multi-graphs, the situation is slightly different because merging is in general not the inverse operation to cloning.

We want to argue (informally) that $[m']$ is obtained by just adding enough context to the left-hand side L. First, we have to add the edges at node u, which have been left "dangling" in the left square/column of the outer SqPO-diagram; second, and slightly more intricate, we also need to add the node x and the edge (y, x) because merging nodes v and w (in the right column) has "side-effects" as the nodes v and w differ in their local structure w.r.t. node x in B. We do not need to add z, as its connectivity to nodes v and w is the same.

As announced before, we can now lift Theorem 2 to SqPO-rewriting, as follows. For any rule $\rho = L \leftarrow\alpha- K -\beta\rightarrow R$ with NAC $c \rhd L$ and rule application $A \models_{\langle \rho, m \rangle} \Rightarrow B$, we can use Theorem 3 to factor the match into $m = m' \circ i$, obtaining a rule instance $\rho' = L' \leftarrow\alpha'- K' -\beta'\rightarrow R'$. Now we can use [13, Corollary 3] to shift $c \rhd L$ from ρ to ρ' along i, i.e. there is a NAC $i(c) \rhd L'$ such that, for every morphism $f \colon L \to X$, we have $f \models^{\mathcal{M}} i(c)$ iff $f \circ i \models^{\mathcal{M}} c$; finally, we apply the construction of Theorem 2 to ρ' and thus obtain $c'_i \rhd R'$.

In favourable cases, there are only a finite number of rule-instances that matter for a given graph transformation system. In fact, we expect this to be the case for systems where all reachable graphs have bounded node degree and SqPO^r-diagrams are required to have monic matches and co-matches. If there are only finitely many (relevant) rule instances, we can "compile" a set of SqPO-rules into an equivalent set of SqPO^r-rules. This means that, at least in favourable cases, the theory of NACs might even be applicable for general SqPO-rewriting.

5   Related and Future Work

The properties of robust splitting sets (see Definition 8) are reminiscent of vertical weak adhesive HLR categories (see [9] for an overview of this and related concepts), which more or less exactly fulfil these requirements. To obtain the commutativity result of Theorem 1, even if we restrict matches and co-matches of SqPO^r-diagrams to the relevant class of monos, we need the additional requirement of stability of these pushouts under pullback. However, in our running example $\mathbb{G}$, we do not need to restrict to regular matches as actually all pushouts are stable; we take this as evidence that SqPO^r-rewriting calls for a revision of the categorical frameworks for graph rewriting.

Concerning other span-based rewriting approaches besides SqPO-rewriting, we mention [17,7,18]; SqPO^r-rewriting is probably best understood as a very restricted instance of these (for a restricted notion of rule in the case of [18]) since none of these approaches requires a double pushout diagram. A concrete example where the difference becomes apparent is the last forbidden application of the merge rule in Example 2: it is forbidden by SqPO^r-rewriting but would be allowed by all of the three other approaches [17,7,18].

As future work, a thorough comparison with recent proposals for the transformation of simple graphs is in place; e.g. we conjecture that for simple graphs, SqPO^r-diagrams comprise a pair of minimal pushout complements [2], using spans of (arbitrary) monos for rules. Finally, we plan a fundamental study of SqPO-rewriting and SqPO^r-rewriting in the span bi-category to unify the main algebraic approaches to rewriting by universal properties in the span-bicategory whenever the category of rewriting is suitably well-behaved, following up on previous work on adhesive categories [15].

6 Conclusion

We have proposed SqPO^r-rewriting as a symmetric variant of SqPO-rewriting, which ensures that each rule application has a corresponding inverse application using the reversed rule. We have established a commutativity result, which incorporates sequential and parallel commutativity into a single theorem. Furthermore, we have shown that nested application conditions can be expected to function in the same way as for double pushout rewriting with linear rules as our Theorem 2 generalises a corner stone of the theory of NACs if we have a suitable class of monos that endows the category of rewriting with the structure of a vertical weak adhesive HLR category. Finally we have given a construction that allows to give for each SqPO-diagram (with monic match and co-match) the "best approximation" by a SqPO^r-diagram using a minimally extended rule-instance, which makes Theorem 2 even applicable to SqPO-rewriting. All these results hold in the category of simple graphs with the class of regular monos as splitting set; we take this as a promising indicator that SqPO^r-rewriting should be studied in more detail, for the particular case of simple graphs but also in the context of categorical frameworks for graph rewriting.

Acknowledgements. Thanks to Ilias Garnier for help finding the term *splitting set* and feed-back on an early draft of this paper. We also thank the anonymous reviewers for their insightful comments.

References

1. Baldan, P., Corradini, A., Heindel, T., König, B., Sobociński, P.: Unfolding Grammars in Adhesive Categories. In: Kurz, A., Lenisa, M., Tarlecki, A. (eds.) CALCO 2009. LNCS, vol. 5728, pp. 350–366. Springer, Heidelberg (2009)
2. Braatz, B., Golas, U., Soboll, T.: How to delete categorically Two pushout complement constructions. Journal of Symbolic Computation 46(3), 246–271 (2011); Applied and Computational Category Theory
3. Corradini, A., Montanari, U., Rossi, F., Ehrig, H., Heckel, R., Löwe, M.: Algebraic Approaches to Graph Transformation. Part I: Basic Concepts and Double Pushout Approach. In: Handbook of Graph Grammars and Computing by Graph Transformation, pp. 163–245. World Scientific Publishing Co., Inc., River Edge (1997)
4. Corradini, A., Heindel, T., Hermann, F., König, B.: Sesqui-pushout Rewriting. In: Corradini, A., Ehrig, H., Montanari, U., Ribeiro, L., Rozenberg, G. (eds.) ICGT 2006. LNCS, vol. 4178, pp. 30–45. Springer, Heidelberg (2006)
5. Danos, V., Feret, J., Fontana, W., Harmer, R., Krivine, J.: Abstracting the differential semantics of rule-based models: Exact and automated model reduction. In: 2010 25th Annual IEEE Symposium on Logic in Computer Science (LICS), pp. 362–381 (July 2010)

6. Danos, V., Feret, J., Fontana, W., Harmer, R., Hayman, J., Krivine, J., Thompson-Walsh, C.D., Winskel, G.: Graphs, rewriting and pathway reconstruction for rule-based models. In: D'Souza, D., Kavitha, T., Radhakrishnan, J. (eds.) FSTTCS. LIPIcs, vol. 18, pp. 276–288. Schloss Dagstuhl - Leibniz-Zentrum fuer Informatik (2012)
7. Duval, D., Echahed, R., Prost, F.: Graph rewriting with polarized cloning. CoRR abs/0911.3786 (2009)
8. Dyckhoff, R., Tholen, W.: Exponentiable morphisms, partial products and pullback complements. Journal of Pure and Applied Algebra 49(1-2), 103–116 (1987)
9. Ehrig, H., Golas, U., Hermann, F.: Categorical Frameworks for Graph Transformation and HLR Systems Based on the DPO Approach. Bulletin of the EATCS 102, 111–121 (2010)
10. Ehrig, H., Habel, A., Lambers, L.: Parallelism and concurrency theorems for rules with nested application conditions. ECEASST 26 (2010)
11. Garner, R., Lack, S.: On the axioms for adhesive and quasiadhesive categories. Theory and Applications of Categories 27(3), 27–46 (2012)
12. Habel, A., Müller, J., Plump, D.: Double-pushout Graph Transformation Revisited. Mathematical Structures in Computer Science 11(5), 637–688 (2001)
13. Habel, A., Pennemann, K.H.: Correctness of high-level transformation systems relative to nested conditions. Mathematical Structures in Computer Science 19(2), 245–296 (2009)
14. Hayman, J., Heindel, T.: Pattern graphs and rule-based models: The semantics of kappa. In: Pfenning, F. (ed.) FOSSACS 2013. LNCS, vol. 7794, pp. 1–16. Springer, Heidelberg (2013)
15. Heindel, T., Sobociński, P.: Being Van Kampen is a universal property. Logical Methods in Computer Science 7(1) (2011)
16. Lack, S., Sobociński, P.: Adhesive and quasiadhesive categories. RAIRO – Theoretical Informatics and Applications 39(3), 511–545 (2005)
17. Löwe, M.: Refined graph rewriting in span-categories: A framework for algebraic graph transformation. In: Ehrig, H., Engels, G., Kreowski, H.-J., Rozenberg, G. (eds.) ICGT 2012. LNCS, vol. 7562, pp. 111–125. Springer, Heidelberg (2012)
18. Monserrat, M., Rosselló, F., Torrens, J., Valiente, G.: Single-pushout rewriting in categories of spans I: The general setting. Tech. rep., Informe d'investigació, Department of Software (LSI) Universitat Politècnica de Catalunya (1997)

On Pushouts of Partial Maps*

Jonathan Hayman[1] and Tobias Heindel[2]

[1] Computer Laboratory, University of Cambridge, UK
[2] School of Informatics, University of Edinburgh, UK

Abstract. The paper gives a sufficient condition for the existence of all pushouts in an arbitrary category of partial maps $\mathbb{C}_{*\mathcal{M}}$ that is necessary whenever the category of total maps $\mathbb{C} \subseteq \mathbb{C}_{*\mathcal{M}}$ has cocones of spans; the latter is the case in all slice categories of $\mathbb{C}$ and thus the condition is necessary locally. The main theorem is that, given an admissible class of monos $\mathcal{M}$ in a category $\mathbb{C}$ that has cocones of spans, the category of partial maps $\mathbb{C}_{*\mathcal{M}}$ has pushouts if and only if the category of total maps $\mathbb{C}$ has hereditary pushouts and right adjoints to inverse image functors (where both properties are w.r.t. $\mathcal{M}$). This result clarifies previous work by Kennaway on graph rewriting in categories of partial maps that implicitly assumed existence of cocones of spans in the category of total maps.

Introduction

The best-known approaches to algebraic graph transformation are single [1] and double [2] pushout rewriting (see also [3]). While the double pushout (DPO) approach has been studied extensively in a variety of categorical frameworks [4], all of which are variants of adhesive categories [5], the relation of single pushout (SPO) rewriting to adhesive categories has been much less extensively studied [6]. This is despite the fact that the work of Kennaway [7] has discussed the central concept of hereditary pushout, which is closely related to adhesive categories [8].

Kennaway's work [7] does not settle the question of what exactly is missing on top of hereditary pushouts to have all pushouts of partial maps. The answer to this question is the main contribution of this paper. Additionally, we identify the missing (implicit) assumption of cocones of spans in the statement of Theorem 3.2 of [7][1] which, taken literally, has a natural counterexample (see Example 3).

Motivated by the main theorem (stated in the abstract), we propose categories with hereditary pushouts and right adjoints to inverse image functors as the paradigmatic categorical framework for SPO rewriting, which is a class of categories that share key properties of the category of graphs (or of any quasi-topos) and provide enough structure to reason about the existence of pushouts of partial maps. This class includes most of the common categories of graph-like structures.

---

* This research was funded by the European Research Council (ERC) under grants 320823 "RULE" and Advanced Grant "ECSYM" and the Agence Nationale de la Recherche (ANR) under grant "AbstractCell".
[1] In the proof of Theorem 3.2(iii) in Ref. [7] on page 495, we can read "forming a commutative square in **C** with some arrows $N' \to D'$ and $O' \to D'$".

H. Giese and B. König (Eds.): ICGT 2014, LNCS 8571, pp. 177–191, 2014.

Finally, we describe how the encoding given in [9] of the Kappa Language [10] fits into this framework, making the connection to our motivating application for this work. We describe the categorical structure of the construction, which is reasonably easy (Lemma 4), mitigating the complexity on the concrete level of the category of so-called *pattern graphs*, which were introduced in [9] for the purpose of encoding. We reuse the category of pattern graphs as a key example, though, in principle, the usual category of hypergraphs (with term graphs as full subcategory [11,12]) would be very similar; however, the concrete conditions for pushouts of partial maps of term graphs (with regular domains of definition) are non-trivial, and thus, the objective advantage of pattern graphs (with *coherent* pattern graphs as full subcategory) is the full treatment of all details in [9] – leaving the treatment of term graphs as future work.

Structure of the Paper. We begin with a review of preliminary notions of category theory that are used to define categories of partial maps in Section 1, where we also define the category of pattern graphs [9], which we use as running example to illustrate the most important concepts. In Section 2, we develop and summarise results concerning pushouts of partial maps that will not only be essential to develop our main theorem but also serve to clarify our contribution. The main theorem itself is developed in Section 3; under a mild assumption, namely existence of cocones of spans, it gives a necessary and sufficient condition for the existence of pushouts of partial maps. Section 4 explains how the main theorem applies to the actual encoding of the Kappa language using pattern graphs, and we discuss further related work in Section 5 before we conclude.

1   Preliminaries

Assuming familiarity with basic concepts of category theory, we recall categories of partial maps based on admissible classes of monos [13]; we also define inverse and direct image functors. We shall reuse the authors' pattern graphs from Ref. [9] as running example category to illustrate the central concepts.

We use $\mathbb{C}, \mathbb{D}, \mathbb{X}$, etc. to range over categories, and $\mathbb{SET}$ is the category of sets and functions. We write $A \in \mathbb{C}$ if A is an object of the category $\mathbb{C}$ and $f \colon A \to B$ in $\mathbb{C}$ or $A -f\to B$ in $\mathbb{C}$ if f is a morphism in $\mathbb{C}$ with domain A and codomain B; finally, the identity on an object $A \in \mathbb{C}$ is denoted by id_A and $g \circ f$ is the composition of morphisms $f \colon A \to B$ and $g \colon B \to C$ in $\mathbb{C}$; we write $A' \rightarrowtail_m \to A$ if m is a mono. As usual, $\mathbb{C}(A, B)$ is the homset of morphisms with domain A and codomain B (assuming that $\mathbb{C}$ is locally small). We fix a category $\mathbb{C}$ to which all objects and morphisms belong, unless stated otherwise.

1.1   Pattern Graphs

A pattern graph is an edge labelled graph in which the targets of edges can be placeholders for nodes that satisfy a certain *specification*, represented by words over the set of edge labels; the idea is that a word $p = \lambda_1 \ldots \lambda_n$ of edge labels λ_i ($i = 1, \ldots, n$) stands for some node v that is at the start of a path $v = v_0, e_1, v_1 \ldots v_{n-1}, e_n, v_n$ where each edge e_i is labelled by λ_i ($i = 1, \ldots, n$). Example 1 shows an example and the formal definition is as follows.

Definition 1 (Pattern Graph (PG)). *Let Λ be a fixed set of* labels. *We denote the set of prefix-closed languages over Λ by $\wp_\leq(\Lambda^*) = \{\phi \subseteq \Lambda^* \mid pq \in \phi \text{ implies } p \in \phi\}$ where Λ^* is the monoid of words over Λ and $\varepsilon \in \Lambda^*$ is the empty word; elements of $\wp_\leq(\Lambda^*)$ are* specifications.

A pattern graph (PG) *is a pair $G = (V_G, E_G)$ where V_G is a set of* nodes *such that $V_G \cap \wp_\leq(\Lambda^*) = \varnothing$ and $E_G \subseteq V_G \times \Lambda \times (V_G \cup \wp_\leq(\Lambda^*))$ is a set of* edges. *A* basic graph *is a pattern graph (V_G, E_G) such that $E_G \subseteq V_G \times \Lambda \times V_G$.*

Example 1 (Pattern Graph). In the middle of (1),

  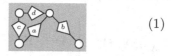
$$(1)$$

we have illustrated a pattern graph with two nodes (drawn as white circles) and two edges (rendered as labelled kites) with labels c and d; the c-edge has the specification $\{\varepsilon, a, ab\}$ as target, which is drawn as a question mark with two consecutive kites with labels a and b. We can think of this pattern graph as a collection of basic graphs, including the ones shown on the left and the right.

Definition 2 (Semantics of Specifications). *Let $G = (V, E)$ be a pattern graph. A node $v \in V$* satisfies $p \in \Lambda^*$, *written $v \models_G p$, if either p is the empty word ε or $p = \lambda p'$ (for some $\lambda \in \Lambda$ and $p' \in \Lambda^*$) and there exists $(v, \lambda, x) \in E$ such that either (i) $x \models_G p'$ and $x \in V$ or (ii) $p' \in x$ and $x \in \wp_\leq(\Lambda^*)$. A node $v \in V$* satisfies $\phi \in \wp_\leq(\Lambda^*)$, *written $v \models_G \phi$, if $v \models_G p$ for all $p \in \phi$.*

Pattern graphs congregate into a category where morphisms are functions between node sets that preserve the structure (w.r.t. suitable "instances" of specifications).

Definition 3 (Category of Pattern Graphs). *A* homomorphism *from a pattern graph G to a pattern graph H, denoted by $f : G \to H$, is a function $f : V_G \to V_H$ such that*

(i) $(f(u), \lambda, f(v)) \in E_H$ holds whenever $(u, \lambda, v) \in E_G$ and $v \in V_G$; and
(ii) for all edges $(u, \lambda, \psi) \in E_G$ with $\psi \in \wp_\leq(\Lambda^*)$, there exists $x \in V_H \cup \wp_\leq(\Lambda^*)$ such that $(f(u), \lambda, x) \in E_H$ and one of the following hold:
    (i) $x \in V_H$ and $x \models \psi$;
    (ii) $x \in \wp_\leq(\Lambda^*)$ and $\psi \subseteq x$.

A homomorphism $f : G \to H$ *is an* inclusion *if $f(v) = v$ holds for all $v \in V_G$, in which case we write $G \subseteq H$ and call G a* subgraph *of H.*

The category of pattern graphs, *denoted by $\mathbb{PG}$, has PGs as objects, homomorphisms as morphisms, the* identity *on a PG G is the function id_{V_G}, and* composition *of morphisms is function composition. Finally, $\mathbb{BG} \subseteq \mathbb{PG}$ is the full subcategory of basic graphs.*

1.2   Categories of Partial Maps

If $\mathbb{C}$ has pullbacks (along monos), we have an associated category of partial maps, which we denote by $\mathbb{C}_*$. It has the same objects as $\mathbb{C}$ and each homset $\mathbb{C}_*(A, B)$ contains partial maps, which are essentially pairs of a mono $A \leftarrowtail^m A'$ in $\mathbb{C}$ and a morphism $A' -f \to B$ in $\mathbb{C}$ (quotiented up to isomorphism at A').

Definition 4 (Spans and Partial Maps). *A* span *is a diagram of the form $A \leftarrow m- X -f \to B$ in $\mathbb{C}$; such a span is a* partial map span *if m is a mono. A* partial map *from A to B, denoted by $(m, X, f) \colon A \to B$, (m, f), or $(\!(f, m)\!)$, is an isomorphism class of a partial map span, i.e.*

$$(m, X, f) = \left\{ A \xleftarrow{\ n\ } Y \xrightarrow{\ g\ } B \ \middle| \ \begin{array}{l} \textit{There exists an isomorphism } i \colon Y -\cong\to X \\ \textit{such that } \ A \underset{n}{\overset{m}{\underset{\nwarrow}{\swarrow}}} \overset{X}{\underset{Y}{i\uparrow}} \overset{f}{\underset{g}{\searrow}} B \ \textit{ commutes.} \end{array} \right\}$$

for some representative partial map span $A \leftarrow m\!\!-\!\!< X -f\to B$. A partial map (m, f) is a total map *if m is an isomorphism.*

Partial maps in $\mathbb{SET}$ are essentially partial functions and a partial map from a PG G to a PG H corresponds to a pair of a subgraph $G' \subseteq G$ and a morphism from G' to H (where G' is the *domain of definition*); both correspondences amount to the standard choice of a representative span for each partial map.

Often one wants to restrict the class of monos that can be used in partial maps. For example, in [9], for the encoding of the Kappa language, it is crucial that the domains of definition in partial maps are *closed*.

Definition 5 (Closed Mono). *An inclusion $i \colon G -\subseteq\to H$ in $\mathbb{PG}$ is* closed *if $(v, \lambda, x) \in E_H$ and $v \in V_G$ imply $(v, \lambda, x) \in E_G$ (for all $v \in V_H$, $\lambda \in \Lambda$, and $x \in V_H \cup \wp_<(\Lambda^*)$); in this situation G is a* closed subgraph *of H. A mono $m \colon G' \rightarrowtail H$ is* closed *if it is isomorphic to a closed inclusion $i \colon G -\subseteq\to H$ (in $\mathbb{PG}/H$). The class of closed monos is denoted by $\mathfrak{Cl}$.*

Thus, each node v in a closed subgraph $G \subseteq H$ has the same successors as v in H, where a successor of v is any node w for which $(v, \lambda, w) \in E_H$ holds for some $\lambda \in \Lambda$.

To obtain categories of partial maps where the left legs of all partial map spans belong to a certain class $\mathcal{M}$ (as detailed in Definition 7), one has to ensure that $\mathcal{M}$ is *admissible* [13].

Definition 6 (Admissible Classes of Monos). *Let $\mathcal{M}$ be a class of monos in $\mathbb{C}$, the elements of which are called $\mathcal{M}$-morphisms, and we write $A' \hookrightarrow^m A$ if $m \in \mathcal{M}$. The class $\mathcal{M}$ is* stable *(under pullback) if for each pair of morphisms $B -f\to A \leftarrowtail^m C$ with $m \in \mathcal{M}$ and each pullback $B \leftarrow m'- D -f'\to C$ of $B -f\to A \leftarrowtail^m C$, the mono m' belongs to $\mathcal{M}$.*

The class $\mathcal{M}$ of monos is admissible, *if*

(i)   the category $\mathbb{C}$ has pullbacks along $\mathcal{M}$-morphisms;
(ii) the class $\mathcal{M}$ is stable under pullback;
(iii) the class $\mathcal{M}$ contains all identities;

(iv) the class $\mathcal{M}$ is closed under composition: if $(A \hookrightarrow^m B), (B \hookrightarrow^n C) \in \mathcal{M}$ then $(A \hookrightarrow^{n \circ m} C) \in \mathcal{M}$.

We now fix an admissible class $\mathcal{M}$ in $\mathbb{C}$. Examples of admissible classes (in any category) are regular monos and isomorphisms; open subspaces of topological spaces and downward closed subsets of partial orders induce more interesting examples, insofar as they are nontrivial proper subclasses of all monos, which we shall refer to as $\mathcal{M}$ono. Finally, closed monos are admissible.

Lemma 1 (Closed Monos are Admissible). *The class $\mathcal{Cl}$ is an admissible class of monos in $\mathbb{PG}$.*

The definition of admissible classes of monos exactly captures the conditions of a well-defined category of $\mathcal{M}$-partial maps [13].

Definition 7 (Partial Map Categories). *The category of $\mathcal{M}$-partial maps, denoted by $\mathbb{C}_{*\mathcal{M}}$, has the same objects as the category $\mathbb{C}$ and the morphisms between two objects $A, B \in \mathbb{C}_{*\mathcal{M}}$ are the elements of*

$$\mathbb{C}_{*\mathcal{M}}(A, B) = \{(m, X, f) : A \to B \mid A \leftarrow^m X \to^f B \ \& \ m \in \mathcal{M}\}$$

which contains all $\mathcal{M}$-partial maps from A to B.[2]

The identity on an object A is $(\mathrm{id}_A, A, \mathrm{id}_A)$; given two $\mathcal{M}$-partial maps $(m, X, f) : A \to B$ and $(k, Z, h) : B \to C$, their composition is $(k, Z, h) \circ (m, X, f) = (m \circ p, U, h \circ q)$ where $X \leftarrow^p U \to^q Z$ is some arbitrary[3] pullback of $X \to^f B \leftarrow^k Z$.

The covariant embedding of $\mathbb{C}$, denoted by $\Gamma : \mathbb{C} \to \mathbb{C}_{*\mathcal{M}}$, is the unique functor from $\mathbb{C}$ to $\mathbb{C}_{*\mathcal{M}}$ that maps each morphism $f : A \to B$ in $\mathbb{C}$ to the total map $\Gamma f = (\mathrm{id}_A, f) : A \to B$ in $\mathbb{C}_{*\mathcal{M}}$ (and thus satisfies $\Gamma(A) = A$ for all $A \in \mathbb{C}$).

We shall call arrows in $\mathbb{C}$ *morphisms* and reserve '*map*' for arrows of $\mathbb{C}_{*\mathcal{M}}$.

We conclude this section with the definition of inverse image functions between meet-semilattices of $\mathcal{M}$-subobjects and a review of direct image functions. For this, recall that each $\mathcal{M}$-morphism $m : M \hookrightarrow A$ is a representative of the subobject $[m]$, i.e. its isomorphism class in the slice category $\mathbb{C}/A$. Note that a $\mathcal{M}$ono-subobject in $\mathbb{SET}$ is essentially a subset and closed subgraphs correspond to $\mathcal{Cl}$-subobjects in $\mathbb{PG}$. We denote the poset of $\mathcal{M}$-subobjects over any object $A \in \mathbb{C}$ by $\mathrm{Sub}_{\mathcal{M}}A$; given $\mathcal{M}$-subobjects $[m], [n] \in \mathrm{Sub}_{\mathcal{M}}A$, the subobject $[m]$ is *included* in $[n]$, written $m \sqsubseteq n$, if there exists a morphism $i : m \to n$ in $\mathbb{C}/A$. For $A \in \mathbb{SET}$, $\mathrm{Sub}_{\mathcal{M}\mathrm{ono}}A$ is isomorphic to the powerset $\wp(A)$ and the relation $\sqsubseteq$ is just the appropriate generalisation of inclusions of subsets. The meet $[m] \sqcap [n]$ is given by the diagonal of the pullback of m along n. Finally, inverse images are obtained by pulling back representatives of subobjects along morphisms, and for a partial map (n, f), its *domain of definition* is the subobject $[n]$.

---

[2] This implies that the left leg of each representative partial map span is an $\mathcal{M}$-morphism.

[3] Arbitrary pullbacks suffice as they are unique up to isomorphism, thus avoiding unnecessary choices.

Definition 8 (Inverse Images). *Let $f\colon A \to B$ in $\mathbb{C}$ be a morphism. The* inverse *image function* $f^{-1}\colon \mathrm{Sub}_{\mathcal{M}}B \to \mathrm{Sub}_{\mathcal{M}}A$ *maps each* $[M \hookleftarrow^{m} B] \in \mathrm{Sub}_{\mathcal{M}}B$ *to the subobject* $f^{-1}([m])$ *such that for all pullbacks* $A \leftarrow^{m'} M' -f_m \to M$ *of* $A - f \to B \leftarrow^{m} M$ *we have* $f^{-1}([m]) = [m']$.

For each $\mathcal{M}$-morphism $m\colon Y \hookrightarrow X$, *post-composition with* m, which maps $[y] \in \mathrm{Sub}_{\mathcal{M}}Y$ to $[m \circ y]$, is a monotone function; it is denoted by $\exists_m\colon \mathrm{Sub}_{\mathcal{M}}Y \to \mathrm{Sub}_{\mathcal{M}}X$ (as it is the lower adjoint to m^{-1}).

2   Pushouts of Partial Maps: The State of the Art

Pushouts of partial maps are at the heart of SPO rewriting [3], one of the standard approaches to graph rewriting; rules of rewriting can be arbitrary partial maps and applying rewriting rules amounts to taking pushouts of rules along a class of *matching morphisms*, which often are assumed to be total. Thus, the existence of pushouts of partial maps (along total maps) is pivotal. In this section, we therefore discuss results on the existence of certain pushouts in $\mathbb{C}_{*\mathcal{M}}$ and their corresponding diagrams in our fixed category $\mathbb{C}$ with its admissible class of monos $\mathcal{M}$. We have not found the following results formulated anywhere in the literature (despite closely related work [14,15,7]).

2.1   A Necessary Condition for Pushouts of Partial Maps

We begin with a discussion of the crucial role of right adjoints to inverse image functors, which appears to have been neglected in the literature; we use terminology from Galois connections as subobjects form posets to make clear that we are not discussing right adjoints to pullback functors.

Definition 9 (Upper Adjoints to Inverse Images). *Let $f\colon A \to B$ in $\mathbb{C}$, and let $f^{-1}\colon \mathrm{Sub}_{\mathcal{M}}B \to \mathrm{Sub}_{\mathcal{M}}A$ be its inverse image function. A $\sqsubseteq$-monotone function $\mathcal{U}\colon \mathrm{Sub}_{\mathcal{M}}A \to \mathrm{Sub}_{\mathcal{M}}B$ is an* upper adjoint *of f^{-1} if for all $n \in \mathrm{Sub}_{\mathcal{M}}B$ and all $m \in \mathrm{Sub}_{\mathcal{M}}A$, we have $f^{-1}(n) \sqsubseteq m$ if and only if $n \sqsubseteq \mathcal{U}(m)$; if an upper adjoint of f^{-1} exists, it is denoted by $\forall_f$ and we write $f^{-1} \dashv \forall_f$ or $\forall_f \vdash f^{-1}$.*[4]

An example of how the upper adjoint of a morphism in $\mathbb{PG}$ can act on subobjects is given in Example 2.

Proposition 1 (Necessity of Upper Adjoints to Inverse Images). *If $\mathbb{C}_{*\mathcal{M}}$ has all pushouts (along total maps), i.e. if for every morphism $f\colon A \to B$ in $\mathbb{C}$ and every map $\phi\colon A \to C$ in $\mathbb{C}_{*\mathcal{M}}$, there is a pushout of $C \leftarrow\phi- A -(\mathrm{id},f)\to B$ in $\mathbb{C}_{*\mathcal{M}}$, then the upper adjoint $\forall_f \vdash f^{-1}$ exists for any morphism f in $\mathbb{C}$.*

Thus, if we want $\mathbb{C}_{*\mathcal{M}}$ to have pushouts along total maps, we need upper adjoints of inverse image functions of $\mathcal{M}$-subobjects in $\mathbb{C}$. It is typically easy to check whether the latter exist; it suffices to show that for all morphisms $f\colon A \to B$ in $\mathbb{C}$ and every subobject $[m] \subset \mathrm{Sub}_{\mathcal{M}}A$, the join $[m'] := \bigsqcup\{[n] \in \mathrm{Sub}_{\mathcal{M}}B \mid f^{-1}([n]) \sqsubseteq m\}$ exists and that setting $\forall_f([m]) := [m']$ yields $\forall_f \vdash f^{-1}$.

---

[4] Recall that upper adjoints are actually unique if they exist.

Lemma 2 (Upper Adjoint for Inverse Images of Closed Monos). *In* $\mathbb{PG}$, *for all* $f: A \to B$, *the upper adjoint* $\forall_f: \mathrm{Sub}_{\mathrm{cl}}A \to \mathrm{Sub}_{\mathrm{cl}}B$ *exists.*

Example 2 (Implicit Deletion in SPO Rewriting). In (2), we have a closed subgraph $m: K \hookrightarrow L$, a morphism $f: L \to G$, and $\forall_f([m])$ yields the result of applying the rule (m, id_K) at f using the SPO approach, i.e. the pushout of $G \leftarrow\!(f,\mathrm{id})\!- L -\!(m,\mathrm{id}_K)\!\to K$ in $\mathbb{PG}_{*\mathrm{cl}}$ is $G - (m', \mathrm{id}_D) \to D \leftarrow\!(f',\mathrm{id})\!- K$ where $[m'] = \forall_f([m])$ and $f': K \to D$ in $\mathbb{PG}$ is the unique morphism satisfying $f \circ m = m' \circ f'$ (cf. Proposition 1, Theorem 2). Roughly, to obtain $\forall_f([m]) \in \mathrm{Sub}_{\mathrm{cl}}G$, we remove from G everything that is in L but not in K. Due to the choice of closed monos, removal of the node ⑤ forces the removal of the node ⓪, which would leave the "dangling edge" ⊸⊸⑤, which is therefore also removed.

$$
\tag{2}
$$

We now turn to our second condition for the existence of pushouts in $\mathbb{C}_{*\mathrm{M}}$, which is necessary if $\mathbb{C}$ is a slice category $\mathbb{C} = \mathbb{D}/T$.

2.2   A Locally Necessary Condition

One might expect that taking a pushout of a span of total maps in $\mathbb{C}_{*\mathrm{M}}$ yields a cospan of total maps; however, this is only true if spans in $\mathbb{C}$ have cocones, as implicitly assumed in [7]. This assumption implies that all pushouts in $\mathbb{C}$ are hereditary if $\mathbb{C}_{*\mathrm{M}}$ has pushouts.

Definition 10 (Hereditary Pushouts). *A pushout* $B -b\to D \leftarrow c- C$ *of a span* $B \leftarrow f- A -g\to C$ *in* $\mathbb{C}$ *is* hereditary *if* $B -\Gamma b\to D \leftarrow\Gamma c- C$ *is a pushout of the span* $B \leftarrow\Gamma f- A -\Gamma g\to C$ *in* $\mathbb{C}_{*\mathrm{M}}$.

Proposition 2 (Pushouts of Total Maps). *Suppose the category* $\mathbb{C}$ *has cocones of spans, i.e. for each span* $C \leftarrow g- A -f\to B$, *there exists a cospan* $C -f'\to D \leftarrow g'- B$ *such that* $g' \circ f = f' \circ g$. *If* $\mathbb{C}_{*\mathrm{M}}$ *has pushouts of partial maps (along total maps), then* $\mathbb{C}$ *has pushouts and the latter are hereditary.*

Proof. Let $C \leftarrow g- A -f\to B$ be a span in $\mathbb{C}$ with cocone $C -f'\to D \leftarrow g'- B$; moreover let $C -(m,M,h)\to E \leftarrow(k,N,n)- B$ be a pushout of $C \leftarrow\Gamma g- A -\Gamma f\to B$. By the universal property of the pushout in $\mathbb{C}_{*\mathrm{M}}$, there is a unique map $\phi: E \to D$ such that $\Gamma(f') = \phi \circ (m, h)$ and $\Gamma(g') = \phi \circ (n, k)$. The latter implies that $\mathrm{id}_C \sqsubseteq m$ and $\mathrm{id}_B \sqsubseteq n$ and thus both of n and m are isomorphisms. Now one can show that $C -\mathrm{hom}^{-1}\to E \leftarrow k \circ n^{-1}- B$ is a pushout of $C \leftarrow g- A -f\to B$ in $\mathbb{C}$ and that it is hereditary follows from $\Gamma(h \circ m^{-1}) = (m, h)$ and $\Gamma(k \circ n^{-1}) = (n, k)$. $\square$

Remark 1. As is well-known, pushouts are not hereditary, in general. The category of jungles [7] is one example; a very similar example occurs naturally for pattern graphs, namely $\Gamma: \mathbb{PG} \to \mathbb{PG}_{*\mathrm{Mono}}$ does not preserve pushouts. To see why, consider the span ⊙⟨a⟩⊙ ←ı− ⊙⊙ −ı→ ⊙⟨a⟩⊙ where the morphism $\imath$ is the inclusion; this span has the pushout ⊙⟨a⟩⊙ −id→ ⊙⟨a⟩⊙ ←id− ⊙⟨a⟩⊙ in $\mathbb{PG}$. However, the embedding of this pushout into $\mathbb{PG}_{*\mathrm{Mono}}$ is not a pushout. To see this, note that the cospan ⊙⟨a⟩⊙ −(ı,ı)→ ⊙⟨a⟩⊙ ←id− ⊙⟨a⟩⊙ is a cocone in $\mathbb{PG}_{*\mathrm{Mono}}$; moreover it is easy to show that there is no mediating morphism, making a case distinction on whether the edge is in the domain of definition or not. Thus, even if all pushouts exist, they need not be hereditary; the class of monos is crucial.

Thus, under mild assumptions on $\mathbb{C}$, having pushouts of partial maps (along total ones) implies that $\mathbb{C}$ has hereditary pushouts. The latter condition is often easy to check using the theorem that left adjoint functors preserve *all* colimits. Thus, to show that all pushouts (that exist) are hereditary, it suffices to establish a right adjoint to the covariant embedding $\Gamma\colon \mathbb{C}\to\mathbb{C}_{*\mathrm{M}}$.

Proposition 3 (Hereditary Pushouts of Pattern Graphs). *Pushouts of spans in the category* $\mathbb{PG}$ *are hereditary w.r.t.* $\Gamma\colon \mathbb{PG}\to\mathbb{PG}_{*\mathrm{el}}$.

Proof. Spelling out the definition of a right adjoint to Γ leads to the fact that it is enough to give, for each PG G, a closed inclusion $\tilde{g}\colon G \hookrightarrow G'$ such that for each partial map $(n, H', f)\colon H \rightharpoonup G$ there is a unique morphism $f'\colon H \to G'$ satisfying $[n] = f'^{-1}([\tilde{g}])$. In fact, taking $G' = \left(V_G\cup\{\bot\}, E_G\cup\{\bot\}\times\Lambda\times(V_G\cup\{\bot\}\cup\wp_\le(\Lambda^*))\right)$ we obtain the desired inclusion (cf. [8, Section 3.3]).   □

Our main result will show that the discussed two conditions for the existence of pushouts of partial maps (which are necessary in the presence of cocones of spans) are in fact sufficient. To understand the main difficulty of this result, we discuss a peculiar fact about pushouts in $\mathbb{C}_{*\mathrm{M}}$ in terms of the underlying diagrams in $\mathbb{C}$.

2.3   Challenge for a Sufficient Condition

Our main theorem will establish that upper adjoints of inverse image functions and hereditary pushouts together are sufficient to obtain pushouts of all spans of partial maps. The crucial point in the proof is the construction of the domain of definition of the diagonal of a pushout candidate. The main difficulty is showing the existence of the join of subobjects illustrated in Figure 1 and spelled-out in the next proposition (following the proof idea of Theorem 3.2 of [7]).

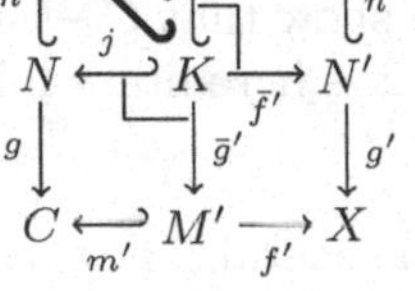

$$
\begin{array}{ccc}
A \xleftarrow{m} M \xrightarrow{f} B \\
\end{array}
\qquad
[k] = \bigsqcup \left\{ x \in \mathrm{Sub}_{\mathrm{M}}A \;\middle|\; \begin{array}{l} \exists b \in \mathrm{Sub}_{\mathrm{M}}B.\, \exists c \in \mathrm{Sub}_{\mathrm{M}}C. \\ \exists m(f^{-1}(b)) = x = \exists n(g^{-1}(c)) \end{array} \right\}
$$

Fig. 1. The domain of definition of the diagonal of a pushout of partial maps

Proposition 4 (Pushout Diagonal). *Assuming that $\mathbb{C}$ has cocones of spans, let $C \leftarrow\!(g,N,n)\!-\ A\ -(m,M,f)\!\rightarrow B$ be a span in $\mathbb{C}_{*\mathcal{M}}$, let $C\ -(m',M',f')\!\rightarrow X\ \leftarrow\!(g',N',n')\!-\ B$ be the pushout of the latter span in $\mathbb{C}_{*\mathcal{M}}$, and let $(k,h) = (n',g') \circ (m,f)$ be the diagonal of the resulting pushout square. Then $[k] \in \mathrm{Sub}_{\mathcal{M}}A$ is the join of all those subobjects $[x] \in \mathrm{Sub}_{\mathcal{M}}A$ for which there are morphisms $i'\colon x \to m$ and $j'\colon x \to n$ in $\mathbb{C}/A$ that are representatives of inverse images of subobjects of B and C, respectively, i.e. i' and j' are subject to the additional condition that there exist $[\tilde{n}] \in \mathrm{Sub}_{\mathcal{M}}B$ and $[\tilde{m}] \in \mathrm{Sub}_{\mathcal{M}}C$ satisfying $[i'] = f^{-1}([\tilde{n}])$ and $[j'] = g^{-1}([\tilde{m}])$ (cf. Figure 1).*

In previous work, the existence of the join $[k]$ in Figure 1 was either trivial [1] or assumed implicitly [7]; related assumptions are used for span-based rewriting, namely limits of small diagrams in [15] and the rather unwieldy final *triple diagrams* in [14]. In contrast, we shall *show* how existence of $[k]$ follows from upper adjoints to inverse image functions and hereditary pushouts. Interestingly, this will involve the following characterisation of hereditary pushouts from [8] (see also Theorem B.4 of [16]).

Theorem 1 (Hereditary Pushout Characterisation [8]). *Let $\mathbb{C}$ be a category with pushouts and let $\mathcal{M}$ be an admissible class of monos in $\mathbb{C}$; let $B \leftarrow A \to C$ be a span with pushout $B \to D \leftarrow C$.*

The pushout is hereditary if and only if for every completion to a commutative cube as shown to the right, where the morphisms $B' \hookrightarrow B$ and $C' \hookrightarrow C$ are $\mathcal{M}$-morphisms and the back faces are pullback squares, the top face is a pushout if and only if the front faces are pullbacks and $d\colon D' \to D$ is an $\mathcal{M}$-morphism.

In the proof of our main theorem, we also shall use the following consequence from [8], generalising Lemma 2.3 of [5].

Lemma 3 (Pushouts along $\mathcal{M}$-morphisms [5,8]). *Let $C \leftarrow\!m\!\rightarrow A\ -f\!\rightarrow B$ be a span with $m \in \mathcal{M}$ and let $C\ -g\!\rightarrow D \leftarrow\!n\!-\ B$ be a pushout that is hereditary and assume $\mathbb{C}$ has pushouts. Then n is an $\mathcal{M}$-morphism, $[m] = f^{-1}[n]$, and $[n] = \forall_f([m])$.*

In particular, $\mathcal{M}$ is pushout stable and pushouts along $\mathcal{M}$ yield pullback squares.

3   Partial Map Pushouts by Inheritance

We now present our main contribution: a construction of pushouts of partial maps that uses only hereditary pushouts and upper adjoints of inverse image functions. Thus, the conditions from the previous section, which are necessary locally, turn out to be sufficient. As a direct consequence, our construction of partial map pushouts directly transfers to slice categories, which turns out to be surprisingly useful in practice [9].

Theorem 2 (Existence of Pushouts of Partial Maps). *Let $\mathbb{C}$ be a category with cocones of spans with an admissible class of monos $\mathcal{M}$. The partial map category $\mathbb{C}_{*\mathcal{M}}$ has pushouts if and only if $\mathbb{C}$ has hereditary pushouts and inverse image functions between $\mathcal{M}$-subobject posets have upper adjoints.*

Proof. The *only if*-part follows from Proposition 1 and Proposition 2.

For the converse, let $C \leftarrow (g,N,n) - A - (m,M,f) \rightarrow B$ be a span in $\mathbb{C}_{*\mathcal{M}}$, assume that $\mathbb{C}$ has hereditary pushouts, and that for any morphism $h \colon Y \to Z$ in $\mathbb{C}$, the upper adjoint $\forall_h$ of the inverse image function $h^{-1} \colon \mathrm{Sub}_{\mathcal{M}} Z \to \mathrm{Sub}_{\mathcal{M}} Y$ exists. We shall first construct a suitable subobject $[k] \in \mathrm{Sub}_{\mathcal{M}} A$ (cf. Figure 1).

The Domain of Definition of the Diagonal Working in $\mathbb{C}$, we start by constructing the diagram on the left in (3).

$$[k] = l^{-1}\big(\forall_l(m \sqcap n)\big) \qquad\qquad [q] = \forall_g([j]) \qquad [p] = \forall_f([i]) \tag{3}$$

Thus, $C \hookleftarrow^{\bar{n}} \to G \leftarrow \bar{g} - A$ is the pushout of $C \leftarrow g - N \hookrightarrow^{n} A$ and $A - \bar{f} \to F \leftarrow \bar{m} \to B$ is the pushout of $A \leftarrow m \to M - f \to B$; moreover, $G - v \to W \leftarrow u - F$ is the pushout of $G \leftarrow \bar{g} - A - \bar{f} \to F$ and $l = v \circ \bar{g} = u \circ \bar{f}$, which is the diagonal of the latter pushout in $\mathbb{C}$. Finally, we put $[k] := l^{-1}\big(\forall_l(m \sqcap n)\big)$.

Note that $[k] = l^{-1}\big(\forall_l(m \sqcap n)\big) \sqsubseteq (n \sqcap m)$. Hence, there are unique morphisms $j \colon k \to n$ and $i \colon k \to m$ (in $\mathbb{C}/A$), witnessing the respective inclusions $k \sqsubseteq n$ and $k \sqsubseteq m$ (which follow from $k \sqsubseteq (m \sqcap n) \sqsubseteq n$ and $k \sqsubseteq (m \sqcap n) \sqsubseteq m$).

The Construction of a Pushout Candidate. Let $[q] = \forall_g([j])$ and $[p] = \forall_f([i])$. As illustrated in (3) on the right, we claim that there exist arrows $g' \colon (g \circ j) \to q$ in $\mathbb{C}/C$ and $f' \colon (f \circ i) \to p$ in $\mathbb{C}/B$, which then let us construct a pushout $Q - v' \to X \leftarrow u' - P$ of $Q \leftarrow g' - K - f' \to P$ (in $\mathbb{C}$) to obtain $C - (q,v') \to X \leftarrow (u',p) - B$ as a pushout candidate, i.e. $(q, v') \circ (n, g) = (p, u') \circ (m, f)$ in $\mathbb{C}_{*\mathcal{M}}$. Thus, we first have to prove the following claim.

Claim. The equations $g^{-1} \circ \forall_g([j]) = [j]$ and $f^{-1} \circ \forall_f([i]) = [i]$ hold.

The relevant steps are two: first, we verify that $(\bar{g})^{-1}\big(\forall_{\bar{g}}([k])\big) = [k]$, and thus $\forall_{\bar{g}}([k]) \sqsubseteq [\bar{n}]$ (using Lemma 3); second, we show that $\forall_{\bar{g}}([k]) = [\bar{n} \circ q]$, whence the desired result follows.

Finally, one can verify the universal property of the pushout candidate. $\qquad\square$

Corollary 1. *If a category has cocones of spans of morphisms and pushouts of partial maps, the same is true for all of its slice categories.*

We give a name to categories that "inherit" partial map pushouts.

Definition 11 (Inherited Partial Map Pushouts). *A category $\mathbb{C}$ with an admissible class of monos $\mathcal{M}$ has inherited $\mathcal{M}$-partial map pushouts or is a MIPMAP category if $\mathbb{C}$ has hereditary pushouts and upper adjoints to inverse image functions.*

Note that MIPMAP-categories are in particular vertical weak adhesive High Level Replacement Categories (cf. [4]) and partial map adhesive [8]. The category $\mathbb{PG}$ belongs to this class as does every (quasi-)topos (which directly follows from the definition of quasi-topos given in [17]).

4   On Pushouts in Full Subcategories

MIPMAP-categories share many properties with adhesive categories [5], are a development of recent generalisations [16,8], and fit well with the theory of categorical frameworks for rewriting, surveyed in Ref. [4]. In particular, they allow the development of standard results of graph rewriting [3] that can be applied to a wide range of graph-like structures. However, some applications require restriction to a full subcategory of a MIPMAP-category: the case of *coherent* pattern graphs [9] is the motivation for the present section, but term graphs (being a full subcategory of hypergraphs [11,12]) are another important example.

The approach taken in [9] to reason about existence of pushouts in a full subcategory $\mathbb{D} \subseteq \mathbb{C}_{*\mathcal{M}}$ of the partial map category of a MIPMAP-category $\mathbb{C}$ amounts to characterising the largest full subcategory $\mathbb{X} \subseteq \mathbb{C}_{*\mathcal{M}}$ that has $\mathbb{D}$ as reflective subcategory; then, all pushouts that exist in $\mathbb{D}$ can be *lifted* from $\mathbb{X}$ using the reflection. Finding a concrete description for the objects of $\mathbb{X}$ is usually non-trivial, and the full details for the case of *coherent* pattern graphs are quite involved (see [9]). We use a simplified example to illustrate the type of phenomena that have to be taken care of in the encoding of Kappa [9].

Example 3 (Branching-Free Graphs I). Let $\mathbb{B} \subseteq \mathbb{BG}$ be the full subcategory of all basic graphs that have at most one outgoing edge per node, i.e. in every graph $G \in \mathbb{B}$, any two edges (v, λ, u) and (v, λ', u') that share the same source node v are identical, i.e. $\lambda = \lambda'$ and $u = u'$. In this full subcategory $\mathbb{B} \subseteq \mathbb{BG}$, we have the following example of a span without cocone.

Note that if a cocone of this span would exist in $\mathbb{B}$, the image of node ① in the "tip" of the cocone would be the source of two different edges, namely one labelled a and one labelled b – a contradiction to branching-freeness.

In contrast, the embedding of this span into $\mathbb{B}_{*el}$ has not only a cocone but we even have the pushout that is shown to the right. Note that both partial maps of the pushout cocone have $\{②\}$ as domain of definition and thus are properly partial (cf. Proposition 2). To see that this square actually is a pushout square, we first observe that the maps of any cocone cannot have node ① in the domain of definition as then both maps would also have the outgoing edge in the domain of definition, which in turn would imply that the "tip" of the cocone is not branching-free. The only remaining choice for a cocone is to either not contain node ① in the domains of definition or that it is mapped to the same node by both morphisms. There is an obvious unique mediating morphism for both cases.

This example shows that pushouts in partial map categories are even more intricate if the category of total maps is not a MIPMAP-category. The concrete details of conditions for spans of partial maps that ensure the existence of a pushout can be rather complex; the motivating example is the situation of [9], but the same issues arise for term graphs [11,12].[5] In general, we can show (non-constructively) that all pushouts that do exist in a full subcategory $\mathbb{D} \subseteq \mathbb{C}_{*\mathcal{M}}$ can be *lifted* from a canonical subcategory $\mathbb{X} \subseteq \mathbb{C}_{*\mathcal{M}}$.

Lemma 4 (Pushout via Reflection). *Let $\mathbb{D} \subseteq \mathbb{E}$ be a full subcategory of an arbitrary category $\mathbb{E}$. There exists a greatest full subcategory $\mathbb{X} \subseteq \mathbb{E}$ such that $\mathbb{D} \subseteq \mathbb{X}$ is a reflective subcategory.*

Proof. Clearly, $\mathbb{D}$ is a reflective subcategory of itself. Moreover, a subcategory $\mathbb{X} \subseteq \mathbb{E}$ contains $\mathbb{D}$ as reflective subcategory if and only if for each object $X \in \mathbb{X}$ there exists a morphism $\eta_X \colon X \to \bar{X}$ in $\mathbb{E}$ with $\bar{X} \in \mathbb{D}$ such that for every other morphism $f \colon X \to D$ in $\mathbb{E}$ with $D \in \mathbb{D}$, there is a unique arrow $f^{\sharp} \colon \bar{X} \to D$ in $\mathbb{E}$ satisfying $f = f^{\sharp} \circ \eta_X$. Now, $\mathbb{X}$ is just the category that contains all objects $X \in \mathbb{X}$ for which there exist η_X as above, because these η_X define the unit of the reflection $\mathbb{D} \subseteq \mathbb{X}$. $\square$

This result allows to characterise when pushouts in $\mathbb{D}$ exist: a span $B \leftarrow f - A - g \rightarrow C$ in $\mathbb{D}$ has a pushout in $\mathbb{D}$ if, and only if, it has a pushout $B - g' \rightarrow X \leftarrow f' - C$ in $\mathbb{E}$ such that $X \in \mathbb{X}$. If such a pushout exists, then it can be lifted from $\mathbb{X}$ to $\mathbb{D}$, using the left adjoint $\mathcal{L}$ to the inclusion $\mathbb{D} \subseteq \mathbb{X}$, namely $B - \mathcal{L}(g') \rightarrow \mathcal{L}(X) \leftarrow \mathcal{L}(f') - C$ is the pushout of $B \leftarrow f - A - g \rightarrow C$; finally we have $\mathcal{L}(g') = \eta_X \circ g$ and $\mathcal{L}(f') = \eta_X \circ f$.

The category $\mathbb{X}$ of Lemma 4 can be non-trivial, i.e. $\mathbb{D} \neq \mathbb{X} \neq \mathbb{E}$, as in the example of branching-free graphs.

Example 4 (Branching-Free Graphs II). The greatest subcategory of $\mathbb{X} \subseteq \mathbb{BG}$ that contains the category of branching-free graphs $\mathbb{B}$ as reflective category is non-trivial. To see this, we first consider the *fork* graph F, below on the left.

$$F = \qquad\qquad\qquad \eta_F :$$

While F is clearly branching and $F \notin \mathbb{B}$, it is easy to verify that the map η_F above on the right is the universal way to make F branching-free, i.e. for any other $f \colon F \to F'$ such that F' is branching-free, there exists a unique $f^{\sharp} \colon \bar{F} \to F'$ such that $f = f^{\sharp} \circ \eta_F$.

In contrast, consider the situation in the *lollipop* L, below one the left.

$$L = \qquad\qquad g_1 : \qquad\qquad\qquad g_2 :$$

There are essentially two ways to remedy the branching at node ①: either ① is in the domain of definition, or not; the above partial maps g_1 and g_2 are examples for the

---

[5] The illustration in Example 3 could equally well be seen as a pushout of partial maps of term graphs (requiring regular monos for the left legs of partial map spans).

respective cases. Now, suppose there was a universal arrow $\eta_L \colon L \to \bar{L}$ with $L \in \mathbb{B}$. If ⓞ is in the domain of definition, then $\eta_L(\text{ⓞ}) = \eta_L(\text{ⓢ})$ by branching-freeness and closure of domains of definition; as a consequence, there does not exist any $g_2^\sharp$ such that $g_2 = g_2^\sharp \circ \eta_L$. Thus, the only possibility would be that ⓞ is not in the domain of definition. However, in the latter case, there is no $g_1^\sharp$ such that $g_1 = g_1^\sharp \circ \eta_L$. In the end, we see that also $L \notin \mathbb{X}$, and thus $\mathbb{B} \neq \mathbb{X} \neq \mathbb{BG}$.

The encoding of the Kappa calculus into pattern graphs from Ref. [9] fits the situation of Lemma 4, using a full subcategory of a suitable slice category of pattern graphs (as discussed further in the next section). Similar situations arise for the category of term graphs (cf. Example 3 and Footnote 5).

5   Related and Future Work

The reference article for SPO rewriting using the algebraic approach is Ref. [1], which gives set-theoretic characterisations of pushouts; the idea of a categorical characterisation of pushouts of partial maps was first given in [7]. The present article gives a streamlined and rigorous account of (consequences of) results from [7], fixing minor omissions of the latter (see Footnote 1). Most importantly, our pushout construction in (3) does not involve any assumptions about existence of joins in subobject lattices (which again are assumed implicitly in [7]), and it only uses pushouts, pullbacks, and upper adjoints of inverse images in $\mathbb{C}$. This can be useful for applications as we can develop algorithms to construct pushouts in $\mathbb{C}_{*\mathcal{M}}$ using well-understood constructions in $\mathbb{C}$. Even in the case of algebras over a signature [1], our main results sheds new light on pushouts of partial maps.

The restriction to full subcategories in applications has an elegant theoretical solution (Lemma 4), even if the complexity of the details of the encoding of Kappa [9] as a full subcategory of $(\mathbb{PG}/T_\kappa)_{*\mathcal{M}}$ for a suitable type graph T_κ are daunting. Another example of a subcategory of an adhesive category has been used in [18] in combination with the double pushout approach (DPO) [2], which is a special case of SPO in the presence of hereditary pushouts (by Lemma 3).

For DPO rewriting, the literature contains a variety of categorical frameworks and here we comment only on those of the last decade that are surveyed in Ref. [4]. In proposing $\mathcal{M}$IPMAP-categories as a framework for SPO rewriting, we do not intend to replace any of these; $\mathcal{M}$IPMAP-categories are also not the most modest strengthening, as partial map adhesive categories with relatively pseudo-complemented subobject posets have already pushouts along monos in $\Gamma(\mathcal{M})$ (cf. [6]). $\mathcal{M}$IPMAP-categories are based on our main theorem, can be instantiated to many examples (including all quasi-topoi), and have additional properties that are relevant for double pushout rewriting, e.g. the so-called Twisted-Triple-Pushout property (reusing the proof of Lemma 8.5 of [5]) without additional assumptions.

As future work, it remains to explore whether the results of the present paper can shed new light on term graph rewriting [11], making use of the categorical framework of $\mathcal{M}$IPMAP-categories and complementing the study of term graphs as a (quasi-)adhesive category [12]. Moreover, guided by the idea that partial map adhesive categories are the natural weakening of adhesive categories when moving "down" from bi-categories of spans to categories of partial maps [8], it is natural to go "up" and study existence conditions for bi-pushouts of spans; a related goal is the characterisation of sesqui-pushout rewriting with monic matches [19] as a single bi-pushout, complementing existing work on span-based rewriting [15,14].

6   Conclusion

The main result is a theorem of category theory that shows that upper adjoints of inverse images are necessary and sufficient for the existence of pushouts of partial maps, provided that spans have cocones. Based on this theorem, we propose MIPMAP-categories as a uniform framework for SPO and DPO rewriting. They are a natural strengthening of partial map adhesive categories [8], and even though there is scope for further generalisation, MIPMAP-categories are the first categorical framework that is relevant to both single and double pushout rewriting. A subtle point is the restriction to full subcategories. While it does not pose any theoretical problems (cf. Lemma 4), it adds an extra level of complexity to the pushout construction which can require substantial additional work in practice [9].

In summary, Theorem 2 justifies the categorical framework of MIPMAP-categories, distilling central ideas of [7]; moreover, Lemma 4 isolates the problems that one has to solve to characterise pushouts of (partial maps of) a full subcategory of a MIPMAP-category. The motivating example is the encoding of the rule-based modelling language Kappa of [9]; however, very similar problems arise in SPO rewriting of term graphs and jungle rewriting [7].

Acknowledgements. We wish to thank the reviewers for their useful comments.

References

1. Löwe, M.: Algebraic Approach to Single-Pushout Graph Transformation. Theoretical Computer Science 109(1&2), 181–224 (1993)
2. Ehrig, H., Pfender, M., Schneider, H.J.: Graph-Grammars: An Algebraic Approach. In: IEEE Computer Society SWAT (FOCS), pp. 167–180 (1973)
3. Ehrig, H., Heckel, R., Korff, M., Löwe, M., Ribeiro, L., Wagner, A., Corradini, A.: Algebraic approaches to graph transformation – part II: Single pushout approach and comparison with double pushout approach. In: Handbook of Graph Grammars, pp. 247–312 (1997)
4. Ehrig, H., Golas, U., Hermann, F.: Categorical Frameworks for Graph Transformation and HLR Systems Based on the DPO Approach. Bulletin of the EATCS 102, 111–121 (2010)
5. Lack, S., Sobociński, P.: Adhesive and quasiadhesive categories. Theoretical Informatics and Applications 39(2), 511–546 (2005)
6. Baldan, P., Corradini, A., Heindel, T., König, B., Sobociński, P.: Unfolding grammars in adhesive categories. In: Kurz, A., Lenisa, M., Tarlecki, A. (eds.) CALCO 2009. LNCS, vol. 5728, pp. 350–366. Springer, Heidelberg (2009)

7. Kennaway, R.: Graph rewriting in some categories of partial morphisms. In: Ehrig, H., Kreowski, H.-J., Rozenberg, G. (eds.) Graph Grammars 1990. LNCS, vol. 532, pp. 490–504. Springer, Heidelberg (1991)
8. Heindel, T.: Adhesivity with Partial Maps instead of Spans. Fundama Informaticae 118(1-2), 1–33 (2012)
9. Hayman, J., Heindel, T.: Pattern graphs and rule-based models: The semantics of kappa. In: Pfenning, F. (ed.) FOSSACS 2013. LNCS, vol. 7794, pp. 1–16. Springer, Heidelberg (2013)
10. Danos, V., Feret, J., Fontana, W., Harmer, R., Hayman, J., Krivine, J., Thompson-Walsh, C., Winskel, G.: Graphs, Rewriting and Pathway Reconstruction for Rule-Based Models. In: D'Souza, D., Radhakrishnan, J., Telikepalli, K. (eds.) FSTTCS 2012. LIPIcs (2012)
11. Plump, D.: Term graph rewriting. In: Ehrig, H., Engels, G., Kreowski, H.-J., Rozenberg, G. (eds.) Handbook of Graph Grammars and Computing by Graph Transformation, II: Applications, Languages and Tools, pp. 3–61. World Scientific (1999)
12. Corradini, A., Gadducci, F.: On term graphs as an adhesive category. Electronic Notes Theorertic Computer Science 127(5), 43–56 (2005)
13. Robinson, E., Rosolini, G.: Categories of Partial Maps. Information and Computation 79(2), 95–130 (1988)
14. Monserrat, M., Rosselló, F., Torrens, J., Valiente, G.: Single-pushout rewriting in categories of spans I: The general setting. Informe d'investigació, Department of Software (LSI) Universitat Politècnica de Catalunya (May 1997)
15. Löwe, M.: Refined graph rewriting in span-categories. In: Ehrig, H., Engels, G., Kreowski, H.-J., Rozenberg, G. (eds.) ICGT 2012. LNCS, vol. 7562, pp. 111–125. Springer, Heidelberg (2012)
16. Heindel, T.: A Category Theoretical Approach to the Concurrent Semantics of Rewriting – Adhesive Categories and Related Concepts. PhD thesis. Universität Duiburg-Essen (2009)
17. Adámek, J., Herrlich, H., Strecker, G.E.: Abstract and Concrete Categories – The Joy of Cats. Dover Publications (2009)
18. Dixon, L., Kissinger, A.: Open graphs and monoidal theories. CoRR abs/1011.4114 (2010)
19. Corradini, A., Heindel, T., Hermann, F., König, B.: Sesqui-Pushout Rewriting. In: Corradini, A., Ehrig, H., Montanari, U., Ribeiro, L., Rozenberg, G. (eds.) ICGT 2006. LNCS, vol. 4178, pp. 30–45. Springer, Heidelberg (2006)

The Subgraph Isomorphism Problem on a Class of Hyperedge Replacement Languages

H.N. de Ridder[1,*] and N. de Ridder[2,**]

[1] University of Konstanz, Department of Computer and Information Science,
78457 Konstanz, Germany
ernst.de-ridder@uni-konstanz.de
[2] Department of Computer Science, University of Rostock, 18051 Rostock, Germany

Abstract. A graph class is called A-free if every graph in the class has no graph in the set A as an induced subgraph. Such characterisations by forbidden induced subgraphs are (among other purposes) very useful for determining whether A-free is a subclass of B-free, by determining whether every graph in B has some graph in A as an induced subgraph. This requires solving the Subgraph Isomorphism Problem, which is NP-complete in general, but for which effective practical algorithms for general and specific purposes exist. However, if B is infinite, these algorithms cannot be used. We introduce *Head-Mid-Tail grammars* (a special case of hyperedge replacement grammars) which have the property that if an infinite set B can be defined by a Head-Mid-Tail grammar then it is decidable whether every graph in B contains some graph from a finite set A of graphs as an induced subgraph, thereby solving the A-free $\subseteq$ B-free problem. Moreover, our algorithm is both simple and efficient enough to be practical.

1 Introduction

1.1 Notation

In this article all graphs are simple and undirected, with edges written as unordered pairs (u,v) of vertices. Let $G = (V,E)$ be a graph and $S \subseteq V$. The subgraph *induced* by S is $G[S] = (S, \{(u,v) \mid (u,v) \in E \wedge u,v \in S\})$. Two graphs $G = (V,E)$ and $H = (W,F)$ are *isomorphic* if a bijection $\pi : V \rightarrow W$ exists, such that $(v,w) \in E$ iff $(\pi(v), \pi(w)) \in F$. A graph G' is an *induced subgraph* of G, written $G' \sqsubseteq G$, iff there is some $S \subseteq V$ such that G' is isomorphic to $G[S]$. The *complement* $\overline{G}$ of G has precisely those edges that are not in G: $\overline{G} = (V, V \times V \setminus E)$ and the complement co-$\mathscr{C}$ of a graph class $\mathscr{C}$ consists of the complements of the graphs in $\mathscr{C}$: $\mathscr{C} = \{\overline{G} \mid G \in \mathscr{C}\}$. A graph class $\mathscr{C}$ is *induced-hereditary* if for every graph G in $\mathscr{C}$, every induced subgraph of G belongs to $\mathscr{C}$. For such classes a unique set A of graphs exists, such that no graph $A' \in A$ is in $\mathscr{C}$, but every proper induced subgraph of A' is. $\mathscr{C}$ then equals the class A-*free* of graphs that have no induced subgraph from the set A.

---

* H.N. de Ridder: Part of the work done at the Department of Computer Science, University of Rostock, D-18051 Rostock, Germany.
** N. de Ridder: Supported by Landesgraduiertenförderung Mecklenburg-Vorpommern, Germany.

H. Giese and B. König (Eds.): ICGT 2014, LNCS 8571, pp. 192–206, 2014.

1.2 Background

Such forbidden subgraph characterisations have many uses:

1. They provide insight into the structure of graphs in the class.
2. Since a characterisation by minimal forbidden induced subgraphs is unique, it can be used for easily proving relations between graph classes.
3. They can be extremely helpful in the design of algorithms for problems like Independent Set or Colourability.
4. If a recognition algorithm for a class decides that the input does not belong to the class, it can return a forbidden subgraph in the input as an easily checkable certificate for the validity of the decision.
5. If the set of forbidden subgraphs is finite, then the graph class can be recognised in polynomial time by brute force comparison of all induced subgraphs of the input graph against the forbidden subgraphs.

For these reasons, finding forbidden subgraph characterisations is one of the evergreen topics in graph theory, with results ranging from König's classical theorem [12] that the bipartite graphs are precisely the odd cycle-free graphs, to the long open Strong Perfect Graph Theorem (SPGT[1]) [2] that the perfect graphs are precisely the (odd hole, odd anti-hole)-free graphs, where a *hole* is a chordless cycle with at least five vertices, an *anti-hole* the complement of a hole, an *odd* (anti-)hole is an (anti-)hole with an odd number of vertices.

It is the second aspect of forbidden subgraph characterisations, the determination of relations between graph classes, that interests us here in particular. Over the years, the mathematics and computer science communities have described many special graph classes in an effort to enlarge both our understanding of fundamental properties of graphs, and our ability to solve practical problems efficiently. This work has been — and still is — so fruitful that very many classes have been defined, whose relations are difficult to overview even for the initiated. In an effort to make this field more accessible the book [1] documents over 300 classes. Out of this book sprang the online database ISGCI, the Information System on Graph Classes and their Inclusions [7]. The first version of ISGCI was released in 1999 with 300 graph classes and 10,000 inclusions and over the years it has grown to its current contents of about 1500 classes and over 170,000 inclusions, plus other relations. It provides the user with the ability to

- find the definition of graph classes;
- check the relation between graph classes and get a witness for these relations;
- find common super/subclasses of given graph classes;
- find the complexity of selected algorithmic problems on specific graph classes;
- find graph classes which are open with respect to the complexity of selected algorithmic problems;
- print inclusion diagrams of graph classes, optionally coloured according to the complexity of a specific problem;
- and provide literature references on graph classes, inclusions and algorithms.

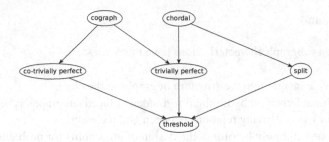

Fig. 1. Inclusion relations between some famous graph classes. There is an arrow $A \rightarrow B$ if $A \supseteq B$ is known.

Currently, the focus of development on ISGCI is on the user interface, wikification and graph parameters (cliquewidth etc.).

For a system like ISGCI completeness and soundness are two major goals. This not only refers to getting classes *into* the system, but especially to relations between classes that *are* in the system. For any two classes $\mathscr{A}, \mathscr{B}$ in the database the objective is to know as precisely as possible the relation between them: $\mathscr{A} = \mathscr{B}$, $\mathscr{A} \subseteq \mathscr{B}$, $\mathscr{A} \subsetneq \mathscr{B}$, $\mathscr{A} \cap \mathscr{B} = \emptyset$, or $\mathscr{A}, \mathscr{B}$ are incomparable (i.e. $\mathscr{A} \cap \mathscr{B}$, $\mathscr{A} - \mathscr{B}$, $\mathscr{B} - \mathscr{A}$ are all non-empty). To this end ISGCI incorporates a rule engine that derives new relations and checks them. As a simple example, if graph classes *split, cograph, cograph $\cap$ split* are present in the database, then from the rule $\mathscr{C}_1 \cap \mathscr{C}_2 \cap \ldots \mathscr{C}_n \subseteq \mathscr{C}_i$ we can deduce without further knowledge about the definition of *split* and *cograph* that *cograph $\cap$ split $\subseteq$ split*. But from this data we cannot deduce whether the inclusion is proper. Take a look at Fig. 1. It shows 6 classes[2] with (taking transitivity into account) 9 out of 15 relations.

All of these classes have in fact known characterisations by forbidden subgraphs, see Fig. 3 and Fig. 2, and from the forbidden subgraphs we can derive much more information.

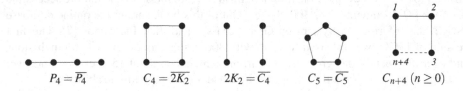

Fig. 2. The forbidden subgraphs for the classes in Fig. 3

---

[1] Formerly known as SPGC — Strong Perfect Graph Conjecture.

[2] These classes are well-known and well-studied. In the discussion that follows we aim to demonstrate the power of forbidden subgraph characterisations. Many of the relations we derive are also obtained in the literature using structural characterisations of these graph classes. This is a consequence of how well-studied these classes are and does not detract from the power and generality of the derivations sketched.

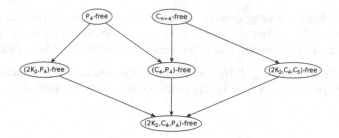

Fig. 3. Inclusion relations between the graph classes from Fig. 1, now characterised by forbidden subgraphs depictured in Fig. 2. Every class is drawn in the same position as in Fig. 1.

Definition 1. *Let A and B be sets of graphs. A forbids B iff for every graph $B' \in B$ there is a graph $A' \in A$ such that $A' \sqsubseteq B'$.*

So A-free $\subseteq B$-free iff A forbids B. For convenience, we use the same term for single graphs: A forbids B iff A is an induced subgraph of B.

Applying this together with some basic set theory to our example we arrive at the following conclusions:

- First of all, the forbidden subgraphs allow us to derive $\subseteq$-relations, e.g. threshold $\subseteq$ split follows from $\{2K_2, C_4, P_4\}$ forbidding $\{2K_2, C_4, C_5\}$.
- Then, because $\{2K_2, C_4, C_5\}$ does not forbid $\{2K_2, C_4, P_4\}$ (none of $2K_2, C_4, C_5$ forbids P_4), it follows that split $\not\subseteq$ threshold and therefore that the inclusion is proper: threshold $\subsetneq$ split.
- For the properness of this relation we also get a witness: P_4 is a split graph but not a threshold graph.
- The forbidden subgraphs of the intersection of cograph and split are $2K_2, C_4, P_4, C_5$. Because $P_4 \sqsubseteq C_5$, we get cograph $\cap$ split $= (2K_2, C_4, P_4)$-free, which are the threshold graphs. Hence, cograph $\cap$ split = threshold.
- The complement of a $(2K_2, C_4, P_4)$-free graph is $(\overline{2K_2}, \overline{C_4}, \overline{P_4})$-free, which is to say $(C_4, 2K_2, P_4)$-free. Hence threshold = co-threshold, or threshold is *self-complementary*. Note that this does not imply that every threshold graph is isomorphic to its complement, but rather that every threshold graph is the complement of some threshold graph. Similar statements apply to cographs and split graphs.
- If we take any two classes in the diagram that have no inclusion relation drawn between them (possibly over multiple edges), then we can verify by the forbidden subgraphs that there indeed is no inclusion between them. For example, cograph $\not\subseteq$ split and split $\not\subseteq$ cograph with witnesses P_4 and $2K_2$, respectively. Our diagram is therefore complete: every inclusion relation is indeed depicted.

Thus, using forbidden subgraphs improved our knowledge from 9 $\subseteq$-relations and 6 unknowns to 9 $\subset$-relations, 6 incomparables and 0 unknowns. Moreover, we have witnesses for everything. As a further indication of the importance of forbidden subgraphs for deducing relations, in ISGCI more than half of the inclusions alone (so not counting incomparabilities) are due to forbidden subgraph characterisations.

The uses of forbidden subgraphs don't even end here. Because not only does A forbid B imply A-free $\subseteq$ B-free, but also A-free $\subseteq$ B-free implies that A forbids B, these characterisations are of great importance to the verification of the soundness of the system.

– If the inclusion A-free $\subseteq$ B-free is present in the database, then it must be that A forbids B, or there is an error (either in the input or in the derivation).

1.3   Goal

All of this hinges on being able to determine by computer program whether a set of graphs A forbids a set of graphs B and, if not, giving a witness.

If A and B are both single graphs, then this asks whether $A \sqsubseteq B$, which is known as the *Subgraph Isomorphism Problem*. When $A \not\sqsubseteq B$, the witness is B itself. The Subgraph Isomorphism Problem is a generalisation of the Hamiltonian Cycle and Maximum Clique problems and is NP-complete in general [10]. As it has applications in pattern recognition, computer vision, biocomputing and databases, among others, practical solutions are in much demand. The classical algorithm for the general case is due to Ullman [15], and [9,13] are surveys comparing the performance of current state of the art algorithms.

If A and B are finite sets, then we can loop through them according to Def. 1 and solve the Subgraph Isomorphism Problem for each pair A', B'. In fact, this is precisely what ISGCI does. If we find some B' that is not forbidden by A, then this B' is our witness.

However, when (one of) these sets is infinite, this doesn't work anymore. And infinite sets of forbidden subgraphs are not at all rare. A set of graphs with a common structure that is used in a forbidden subgraph characterisation is called a *family*. Infinite families we have seen are the odd cycles for bipartite graphs; odd holes and odd anti-holes for perfect graphs; and C_{n+4} for chordal graphs[3].

In this article we let B be infinite and study the problem of deciding whether every member of B contains an induced subgraph in the finite set A. We shall have more to say on the case of infinite A in Sec. 3. We introduce *Head-Mid-Tail grammars*, or HMT-grammars for short, which are a special case of *hyperedge replacement grammars* [11] and use these to solve the following specialization of the "forbidding" problem:

Problem 2. Given a finite set of graphs A and a Head-Mid-Tail grammar $\mathcal{G}$, does A forbid the graphs generated by $\mathcal{G}$?

In case of a negative answer, a good solution should also provide a witness.

It is important that A is a set and not just a single graph, because often multiple graphs are needed to forbid a family. For example, in Sec. 1.2, the (C_4, P_4)-free graphs are a subclass of C_{n+4}-free graphs, because C_4 forbids C_4 and P_4 forbids C_5 and larger cycles.

We show that to solve Problem 2 it suffices to examine only a finite subset of the graphs generated by $\mathcal{G}$ and give an efficient procedure to calculate this subset. This results in an algorithm that is simple and efficient enough to be practical and can also return a witness when A docs not forbid the graphs of $\mathcal{G}$.

---

[3] Although these families are all cycles of some constrained length, or complements thereof, many families with another structure exist, as well.

2 Head-Mid-Tail Grammars

2.1 Definition

Definition 3. *The graphs we consider in this paper are tuples* $X = (V_X, E_X, att_X, ext_X)$, *with* V_X *(or* $V(X)$*) and* E_X *(or* $E(X)$*) the vertex and edge sets of* X, *respectively, and* att_X, ext_X *are bijective functions numbering a subset of vertices, with domains* $\underline{att}_X$ *and* $\underline{ext}_X$, *respectively:*

$$att_X : \underline{att}_X \to [1 \dots |\underline{att}_X|] \ , \qquad \underline{att}_X \subseteq V_X$$
$$ext_X : \underline{ext}_X \to [1 \dots |\underline{ext}_X|] \ , \qquad \underline{ext}_X \subseteq V_X$$

For a set of vertices S, $att_X(S) = \{att_X(v) \mid v \in S\}$, where implicitly $S \subseteq \underline{att}_X$. Analogously for $ext_X(S)$.

If a graph has no attachment (extension), then we consider its attachment (extension) empty. We will number the vertices in $\underline{att}_X$ with $1a, 2a, \dots$ where $att_X(1a) = 1$, $att_X(2a) = 2, \dots$ and similarly the vertices in $\underline{ext}_Y$ with $1e, 2e, \dots$.

We call an ordered pair (X, Y) *compatible* if $|\underline{att}_X| = |\underline{ext}_Y|$. Let X, Y be disjoint graphs such that the pair (X, Y) is compatible and let $k := |\underline{att}_X| = |\underline{ext}_Y|$. The graph XY (the juxtaposition of X and Y) is the *composition* of X and Y and is defined as the union of X and Y such that vertices with the same number in $\underline{att}_X, \underline{ext}_Y$ get identified ($1a \in \underline{att}_X$ with $1e \in \underline{ext}_Y$, $2a \in \underline{att}_X$ with $2e \in \underline{ext}_Y$ and so on) and all multiple edges are made simple. The formal definition is as follows:

Definition 4. *Let* X, Y *be graphs such that the pair* (X, Y) *is compatible.*

- $\lambda_{XY} : V(Y) \to V(XY)$ *is defined by*

$$\lambda_{XY}(y) = \begin{cases} y & \text{if } y \notin \underline{ext}_Y \\ att_X^{-1}(ext_Y(y)) & \text{if } y \in \underline{ext}_Y \end{cases} .$$

When there is no danger of confusion we shall leave out the subscript $_{XY}$ *and write simply* λ.

- *Without loss of generality, assume* X, Y *are disjoint (otherwise work with isomorphic copies of* X, Y*). The composition* XY *is defined up to isomorphism by*
  - $V(XY) = V(X) \cup V(Y) \setminus \underline{ext}_Y$,
  - $E(XY) = E(X) \cup \{(\lambda(y'), \lambda(y'')) \mid (y', y'') \in E(Y)\}$,
  - $att_{XY} = att_Y \circ \lambda^{-1}$,
  - $ext_{XY} = ext_X$.

Note that composition never deletes vertices nor edges. According to this definition edge-preserving functions $\chi : V(X) \to V(XY)$ and $\upsilon : V(Y) \to V(XY)$ exist. To keep the notation light, we shall leave out these functions and write in XY simply S instead of $\chi(S)$ or $\upsilon(S)$, for $S \subseteq V(X)$ and $S \subseteq V(Y)$, respectively. For the i-repeated composition of a graph X with itself we write X^i, with the individual copies of X numbered with a subscript: $X^i = X_1 X_2 \dots X_i$.

Proposition 5. *Composition is associative:* $X(YZ) = (XY)Z$.

Definition 6. *A* Head-Mid-Tail grammar $\mathscr{G}$ *is a tuple* (H,M,T) *where*

- *H is a graph with* $att_H \neq \varnothing$ *and* $ext_H = \varnothing$;
- *M is a graph with* $att_M \neq \varnothing$ *and* $ext_M \neq \varnothing$;
- *T is a graph with* $att_T = \varnothing$ *and* $ext_T \neq \varnothing$;
- $|\underline{att_H}| = |\underline{ext_M}| = |\underline{att_M}| = |\underline{ext_T}| > 0$. *That is, the pairs* (H,M), (H,T), (M,M), (M,T) *are compatible.*

A Head-Mid-Tail grammar $\mathscr{G} = (H,M,T)$ is called *growing* if $\underline{ext_M} \neq V(M)$. The *language* $L(\mathscr{G})$ generated by a given Head-Mid-Tail grammar $\mathscr{G} = (H,M,T)$ is the family of all graphs that can be composed from H, i copies of M ($i \geq 0, i \in \mathbb{N}$) and T:

$$HT(=HM^0T), HMT(=HM^1T), HMMT(=HM^2T), \ldots$$

Proposition 7. $L(\mathscr{G})$ *is infinite iff* $\mathscr{G}$ *is growing.*

Note that for every graph in $L(\mathscr{G})$ both attachment and extension are empty. As an example, Fig. 4 gives a Head-Mid-Tail grammar for the family holes. We remark that the holes, like many other families, cannot be generated by iterated composition of only two graphs.

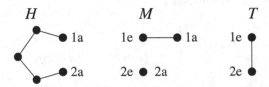

Fig. 4. $\mathscr{G} = (H,M,T)$ generating the family Holes

Head-Mid-Tail grammars are a special case of hyperedge replacement grammars, where *att*, *ext* and composition have their origin. We'll have more to say on hyperedge replacement grammars in Sec. 3.

2.2 Normalised Head-Mid-Tail Grammars

Infinite families of graphs that can be represented with a Head-Mid-Tail grammar may have several different such representations. Our algorithm for solving Problem 2 is applied to *normalised* Head-Mid-Tail grammars. Before explaining what this is we need some auxiliary definitions.

Let $\mathscr{G} = (H, M, T)$ be a Head-Mid-Tail grammar. Define $Q(M) := \underline{ext_M} \cap \underline{att_M}$ and $q := |Q(M)|$. A *transitive cycle* of M is a minimal set of vertices $C \subseteq Q(M)$ such that $ext_M(C) = att_M(C)$. The *transitive set* of M is the maximal set of vertices $M_{tr} \subseteq Q(M)$ such that $ext_M(M_{tr}) = att_M(M_{tr})$. We remark that M_{tr} is the maximal set of vertices on which λ_{MM} is a permutation and that a transitive cycle is precisely an orbit of λ_{MM}. Clearly, $M_{tr} = \bigcup_C C$ and $M_{tr} \subseteq Q(M)$. See Fig. 5 for an example of transitive cycles and the transitive set of M.

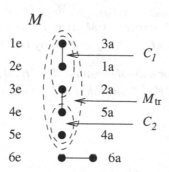

Fig. 5. Transitive cycles C_1, C_2, transitive set M_{tr}. Here $Q(M) = M_{tr}$.

If the pair (X, M) is compatible, then we define $X_{tr} := att_X^{-1}(ext_M(M_{tr}))$ and if (M, X) is compatible, then $X_{tr} := ext_X^{-1}(att_M(M_{tr}))$. Well-definedness requires that if both (M, X) and (X, M) are compatible then these definitions must be equal and for $X = M$, the solution to this equality is given by the definition of M_{tr} above.

Definition 8. *A Head-Mid-Tail grammar* $\mathscr{G} = (H, M, T)$ *is called* normalised *if*

1) $Q(M) = M_{tr}$;
2) $\forall v \in M_{tr} : ext_M(v) = att_M(v)$;
3) $M[M_{tr}]$ *is edgeless.*

Intuitively, in all graphs HM^i generated by a normalised grammar, H_{tr} induces the same subgraph: By 8.1 its set of vertices doesn't change; by 8.2 the attachment/extension numbering doesn't change; and by 8.3 its set of edges doesn't change. The consequences and use of being normalised will be established in section 2.3.

Every Head-Mid-Tail grammar can be reformulated as a set of normalised grammars.

Theorem 9. *For any Head-Mid-Tail grammar* $\mathscr{G} = (H, M, T)$ *we can construct a finite set* S *of graphs and a finite set of normalised grammars* $\mathscr{G}'_i$ *such that* $S \cup \bigcup_i L(\mathscr{G}'_i) = L(\mathscr{G})$.

Proof. See [6].

2.3 Properties of Normalised Grammars

We next prove some properties of normalised grammars.

Lemma 10. *Let $\mathcal{G} = (H,M,T)$ be normalised. Then in graphs $HM^iT, \forall i > 0$ the sets $\underline{att}_H \setminus H_{\mathrm{tr}}$, $\underline{ext}_T \setminus T_{\mathrm{tr}}$ and $\underline{ext}_{M_j} \setminus M_{\mathrm{tr}}, \forall j, 2 \le j \le i$ are pairwise disjoint.*

Proof. By definition of transitive set, in HM^iT: $\underline{att}_H \setminus H_{\mathrm{tr}} = \underline{ext}_{M_1} \setminus M_{\mathrm{tr}}$ and $\underline{ext}_T \setminus T_{\mathrm{tr}} = \underline{att}_{M_i} \setminus M_{\mathrm{tr}}$. If we define $\underline{ext}_{M_{i+1}} := \underline{ext}_T$, then $\forall j : 1 \le j \le i, \forall k : 1 \le k \le i - j + 1$ $\underline{ext}_{M_j} \setminus M_{\mathrm{tr}} \cap \underline{ext}_{M_{j+k}} \setminus M_{\mathrm{tr}} = (\underline{ext}_{M_j} \cap \underline{ext}_{M_{j+k}}) \setminus M_{\mathrm{tr}} \subseteq (\underline{ext}_{M_j} \cap \underline{att}_{M_j}) \setminus M_{\mathrm{tr}}$, because by definition of composition all vertices of M_j that are in $M_{j+1} \ldots M_i$ are in $\underline{att}_{M_j}$. By definition of $Q(M)$, this equals $Q(M) \setminus M_{\mathrm{tr}} = \varnothing$, as by Def. 8.1, $Q(M) = M_{\mathrm{tr}}$. $\quad\square$

Definition 11. *Consider a normalised grammar $\mathcal{G} = (H,M,T)$. We say that a graph X fits easily into HM^iT if $X \sqsubseteq HM^iT$ such that at least one of the following conditions holds:*

a) $V(X) \cap \underline{att}_H \setminus H_{\mathrm{tr}} = \varnothing$ or
b) $V(X) \cap \underline{ext}_T \setminus T_{\mathrm{tr}} = \varnothing$ or
c) $\exists j, 2 \le j \le i : V(X) \cap \underline{ext}_{M_j} \setminus M_{\mathrm{tr}} = \varnothing$.

In particular, if X fits easily into HT then both a) and b) hold. See Fig. 6: if X fits easily into HM^iT, then X has no vertices in at least one of the hatched areas.

Theorem 12. *Let $\mathcal{G} = (H,M,T)$ be growing and normalised and let X be a graph. If $|V(X)| \le i$, then $X \sqsubseteq HM^iT$ iff X fits easily into HM^iT.*

Proof. $\Longleftarrow$: follows from Def. 11.
$\Longrightarrow$: Let $X \sqsubseteq HM^iT, |V(X)| \le i$. Suppose X does not fit easily into HM^iT. Then:

– $|V(X) \cap \underline{att}_H \setminus H_{\mathrm{tr}}| \ge 1$,
– $|V(X) \cap \underline{ext}_T \setminus T_{\mathrm{tr}}| \ge 1$,
– $\forall j, 2 \le j \le i : |V(X) \cap \underline{ext}_{M_j} \setminus M_{\mathrm{tr}}| \ge 1$.

By Def. 8.1, Lemma 10 holds, thus $|V(X)| \ge i + 1$ — contradiction. $\quad\square$

Theorem 13. *Let $\mathcal{G} = (H,M,T)$ be growing and normalised and let X be a graph. Let i be the smallest exponent such that X fits easily into HM^iT, then $i \le |V(X)|$.*

Proof. Let i be the smallest exponent such that X fits easily into HM^iT. Assume $i > |V(X)|$, then X does not fit easily into $HM^{|V(X)|}T$. It follows that

– $|V(X) \cap \underline{att}_H \setminus H_{\mathrm{tr}}| \ge 1$,
– $|V(X) \cap \underline{ext}_T \setminus T_{\mathrm{tr}}| \ge 1$,
– $\forall j, 2 \le j \le |V(X)| : |V(X) \cap \underline{ext}_{M_j} \setminus M_{\mathrm{tr}}| \ge 1$.

By Def. 8.1, Lemma 10 holds, thus $|V(X)| \ge |V(X)| + 1$ — contradiction. $\quad\square$

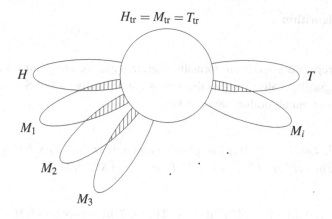

Fig. 6. $\mathscr{G} = (H,M,T)$ is normalised. Illustration for Def. 11.

The next theorem formalizes the following intuition: If a graph X fits easily into HM^i, then X has no vertices in one of the disjoint hatched areas of Fig. 6. As discussed after Def. 9, the graph induced by H_{tr} doesn't change by pumping. Hence, we can insert new copies of M precisely in a hatched area without changing any vertex of X. Thereby X is also an induced subgraph of all HM^lT, with $l > i$.

Theorem 14. *Let* $\mathscr{G} = (H,M,T)$ *be growing and normalised and let* X *be a graph. If* X *fits easily into* HM^iT, *then* X *fits easily into* HM^lT, *for all* $l > i$.

Proof. See [6].

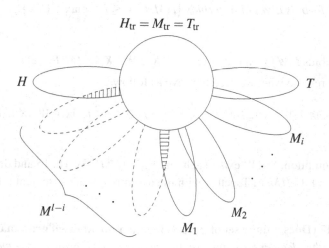

Fig. 7. $\mathscr{G} = (H,M,T)$ is normalised. Illustration for Th. 14

2.4   The Algorithm

After this preparatory work on normalised grammars, we are going to exploit their properties algorithmically. The first theorem is conceptually part of Sec. 2.3, but since it is the basis of our algorithm, we list it here.

Theorem 15. *Let $\mathscr{G} = (H, M, T)$ be growing and normalised, let X be a graph and let $i = |V(X)|$. Then either $\forall l \geq i : X \sqsubseteq HM^l T$ or $\forall l \geq i : X \not\sqsubseteq HM^l T$.*

Proof. If X fits easily into $HM^i T$, then, by Th. 14, X fits easily into $HM^l T$ for all $l > i$, as well, and it follows that $\forall l \geq i : X \sqsubseteq HM^l T$.

If X does not fit easily into $HM^i T$, then, by Th. 14, X does not fit easily into $HM^k T, \forall k \leq i$. By Th. 13, there is no j such that X fits easily into $HM^j T$. Thus, by Th. 12, $\forall l \geq i : X \not\sqsubseteq HM^l T$.   □

We remark that this theorem implies that (for a normalised grammar!) situations where, for example, X forbids all the odd members of $L(\mathscr{G})$, but not the even members, and Y forbids the even members, but not the odd members, are not possible. For general Head-Mid-Tail grammars such situations are possible. For X and a normalised grammar, either the set of forbidden members or the set of non-forbidden members is finite and a subset of the $|V(X)|$ smallest members.

Theorem 16. *Let $\mathscr{G} = (H, M, T)$ be growing and normalised and let W be a finite set of graphs. W forbids $L(\mathscr{G})$ iff W forbids $\{HM^i T \mid 0 \leq i \leq \max_{X \in W} |V(X)|\}$.*

Proof. W forbids $L(\mathscr{G})$ means that $\forall i \geq 0 \; \exists X \in W : X \sqsubseteq HM^i T$. Let $t := \max_{X \in W} |V(X)|$. Then we can rewrite what we need to prove as follows:

$$\forall i \geq 0 \; \exists X \in W : X \sqsubseteq HM^i T \quad \Longleftrightarrow \quad \forall i : 0 \leq i \leq t \; \exists X \in W : X \sqsubseteq HM^i T$$

$\Longrightarrow$: Obvious.

$\Longleftarrow$: By assumption, $Y \in W$ exists such that $Y \sqsubseteq HM^t T$. By Th. 15 and definition of t, we get $\forall i \geq t : Y \sqsubseteq HM^i T$. Together with the assumption, the statement follows.   □

Algorithm 17 (Does a finite set of graphs forbid a normalised grammar?) *The following algorithm decides for a finite set of graphs W and a growing, normalised Head-Mid-Tail grammar $\mathscr{G} = (H, M, T)$ whether W forbids $L(\mathscr{G})$.*

$$t = \max_{X \in W} |V(X)|$$

```
loop:   for i = 0...t
            for X ∈ W  such that  i ≤ |V(X)|           (∗)
                if X ⊑ HM^iT                            (∗∗)
                    if i = |V(X)|                       (∗∗∗)
                        return true
                    continue loop
            return false
        return true
```

Proof. The algorithm is a straightforward implementation of Th. 16, with two optimizations: By Th. 15 we don't need to check whether $X \sqsubseteq HM^iT$ for $i > |V(X)|$ (∗) and can return true immediately when $X \sqsubseteq HM^{|V(X)|}T$ (∗∗∗).   □

If at (∗∗) a subgraph isomorphism algorithm is used that not only returns true/false, but also returns the actual subgraph, if one exists, then we can add a further optimization at (∗∗∗) by returning true as soon as either $i = |V(X)|$ or X fits easily into HM^iT. If a witness is required, then we can return the witness HM^iT together with the result false.

Theorem 18. *There is a simple algorithm that for a finite set of graphs W and a Head-Mid-Tail grammar $\mathscr{G}$ decides whether W forbids $L(\mathscr{G})$.*

Proof. If $\mathscr{G}$ is not normalised, then by Th. 9 we can construct S and $\mathscr{G}'_i$ such that $S \cup \bigcup_i L(\mathscr{G}'_i) = L(\mathscr{G})$, with $\mathscr{G}'_i$ normalised. Check S. Check all $\mathscr{G}'_i$ (Alg. 17).   □

By Th. 15, we can formulate similar algorithms to check whether W forbids any graph in $L(\mathscr{G})$, or no graph in $L(\mathscr{G})$.

3 Discussion

By Alg. 17 we have that Problem 2 is at most as hard as the subgraph isomorphism problem, which is NP-complete [10]. In the other direction, by using a Head-Mid-Tail grammar that generates only a single graph, we can reduce the subgraph isomorphism problem to Problem 2. Hence Problem 2 is NP-complete, as well. Our algorithm makes a linear number of calls to a subgraph isomorphism algorithm and performance heavily depends on the subgraph algorithm used. Comparisons of different subgraph isomorphism algorithms are [9,13]; ISGCI uses VFLib [16] which implements the VF2 algorithm [4].

Although it is straightforward [6] to implement Th. 9 in order to normalise any given Head-Mid-Tail grammar, in practice it has turned out to be better to skip this step and instead require that grammars be presented normalised: Because normalised grammars are far easier to understand for humans than not-normalised ones, this reduces the number of input mistakes.

Head-Mid-Tail grammars are a special case of hyperedge replacement grammars [11,14] limited to producing simple graphs: Consider a hyperedge replacement grammar $G = (N,T,P,S)$, such that $N = \{S,A\}$ and $P = \{S := Head, A := Mid \mid Tail\}$, where *Head* is an A-handled hypergraph, *Mid* is an A-handled $type(A)$-hypergraph and *Tail* is a $type(A)$-hypergraph, then we can define G by the tuple (*Head*, *Mid*, *Tail*). The theory of hyperedge replacement grammars has a mechanism for deciding whether every graph generated by a grammar satisfies a certain property, if that property is so-called *compatible*. Roughly speaking, a compatible property can be decided for a graph by combining the property (or a related one) on the components used in the derivation of the graph. Unfortunately, the induced subgraph property is not compatible and therefore this mechanism doesn't help us in solving Problem 2. Another approach is expressing the subgraphs in monadic second order logic, $\mathscr{G}$ as an equational set, and verifying that the expression is universally valid on this set. This can be done algorithmically [5], but the algorithm is complicated and prohibitively resource-hungry even for simple families. Our algorithm, on the other hand, is not only very easy to implement, but also profits directly from any improvements to the subgraph isomorphism algorithm without the need to change a byte in our code.

As Head-Mid-Tail grammars are a special case of hyperedge replacement grammars, it follows that if an infinite family of graphs cannot be generated by hyperedge replacement grammars, then it cannot be generated by Head-Mid-Tail grammars. Because the number of edges in a hyperedge replacement grammar exhibits linear growth, hyperedge replacement families cannot contain cliques of arbitrary size. An example of graphs that contain arbitrary large cliques and therefore cannot be represented by Head-Mid-Tail grammars are the anti-holes. All is not lost, however. Since A is an induced subgraph of B iff the complement of A is an induced subgraph of the complement of B, and since holes can be represented by Head-Mid-Tail grammars, we can switch to the complement and still solve Problem 2. A prominent family where this trick does not work are the suns [8,7].

Another way to enhance the reach of our algorithm is taking unions of families. Fig. 8 shows the Dumbbells, which can be represented by a hyperedge replacement grammar, but not by a Head-Mid-Tail grammar. They cannot be represented by a finite union of Head-Mid-Tail grammars either. But when the cycles $x_0, \ldots, x_i$ and $z_0, \ldots, z_k$ are limited to length either 3 or 4, we get the bicycles [3,7], which can be represented by a union of three Head-Mid-Tail grammars, with cycles of length 3 and 3; 3 and 4; and 4 and 4, respectively.

Considering the practical purpose of deciding relations between sets of forbidden subgraphs, we can say that surprisingly[4] many families used in forbidden subgraph characterisations of existing graph classes are in fact representable by Head-Mid-Tail grammars. Of the roughly 100 infinite families currently used[5] by ISGCI less than 10 cannot be represented by Head-Mid-Tail-Grammars and these cannot be represented by hyperedge replacement grammars, either. This implies that an improvement to our

---

[4] Actually not so very surprising, because Head-Mid-Tail grammars were developed with the goal to have a formalism that is as simple as possible and yet powerful enough to handle as many infinite families used by ISGCI as possible.

[5] Many more await being used; see Fig. 8 for some examples.

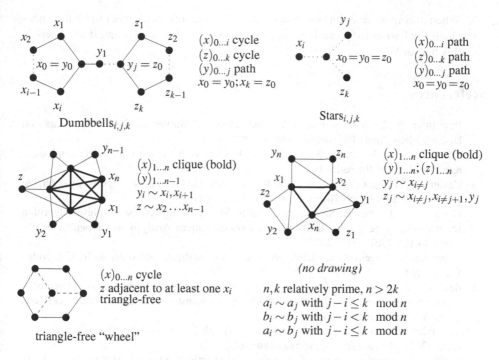

$(x)_{0...i}$ cycle
$(z)_{0...k}$ cycle
$(y)_{0...j}$ path
$x_0 = y_0; x_k = z_0$

Dumbbells$_{i,j,k}$

$(x)_{0...i}$ path
$(z)_{0...k}$ path
$(y)_{0...j}$ path
$x_0 = y_0 = z_0$

Stars$_{i,j,k}$

$(x)_{1...n}$ clique (bold)
$(y)_{1...n-1}$
$y_i \sim x_i, x_{i+1}$
$z \sim x_2 ... x_{n-1}$

$(x)_{1...n}$ clique (bold)
$(y)_{1...n}; (z)_{1...n}$
$y_j \sim x_{i \neq j}$
$z_j \sim x_{i \neq j}, x_{i \neq j+1}, y_j$

$(x)_{0...n}$ cycle
z adjacent to at least one x_i
triangle-free

triangle-free "wheel"

(no drawing)

n, k relatively prime, $n > 2k$
$a_i \sim a_j$ with $j - i \leq k \mod n$
$b_i \sim b_j$ with $j - i < k \mod n$
$a_i \sim b_j$ with $j - i \leq k \mod n$

Fig. 8. Typical examples of infinite families used in forbidden subgraph characterisations that cannot be generated by Head-Mid-Tail grammars. $(x)_{1...n}$ is shorthand for $x_0 ... x_n$ and $x \sim y$ means x ist adjacent to y.

algorithm should preferably act on a grammar more general than hyperedge replacement grammars.

Then there is the issue of deciding whether an infinite set A forbids a finite or infinite set B. If B is finite, Head-Mid-Tail grammars in A work fine: Since only graphs that have at most as many vertices as the graphs in B can forbid B, we can generate this finite set from the grammar and again check finite A and B. But we currently cannot decide programmatically the case where both A and B are infinite.

This leaves us with the following open questions:

1. How far can Alg. 17 be extended *easily*? For example, a variation that results in derivation stars instead of a path seems straightforward and would already be powerful enough to represent dumbbells.
2. We think it is possible to extend the algorithm to work on hyperedge replacement grammars, but a naive extension would check an exponential number of graphs and therefore be impractical. What would a smarter extension look like?
3. In light of the families used by ISGCI, is there a practical algorithm for Problem 2 on grammars more powerful than hyperedge replacement, like pushout grammars?
4. The most burning open question of all is how to determine whether one Head-Mid-Tail family forbids another one.

5. When asking whether one grammar forbids another one, we expect to hit the undecidability barrier for powerful enough grammars. Is this correct and if so, where is this barrier?

References

1. Brandstädt, A., Le, V.B., Spinrad, J.P.: Graph Classes: A Survey. In: SIAM Monographs on Discrete Math. Appl., Philadelphia, vol. 3 (1999)
2. Chudnovsky, M., Robertson, N., Seymour, P.D., Thomas, R.: The strong perfect graph theorem. Annals of Mathematics 164, 51–229 (2006)
3. Chvátal, V., Hoàng, C.T., Mahadev, N.V.R., De Werra, D.: Four classes of perfectly orderable graphs. Journal of Graph Theory 11(4), 481–495 (1987)
4. Cordella, L., Foggia, P., Sansone, C., Vento, M.: A (sub)graph isomorphism algorithm for matching large graphs. IEEE Transactions on Pattern Analysis and Machine Intelligence 26(10), 1367–1372 (2004)
5. Courcelle, B., Engelfriet, J.: Graph structure and monadic second-order logic. Cambridge University Press (2012)
6. de Ridder, H.N., de Ridder, N.: The subgraph isomorphism problem on a class of hyperedge replacement languages. Technical report. Universität Konstanz, Fachbereich Informatik und Informationswissenschaft (2014)
7. de Ridder, H.N., et al.: Information System on Graph Classes and their Inclusions (ISGCI) (2001-2014), http://www.graphclasses.org
8. Farber, M.: Characterizations of strongly chordal graphs. Discrete Mathematics 43(23), 173–189 (1983)
9. Foggia, P., Sansone, C., Vento, M.: A performance comparison of five algorithms for graph isomorphism. In: Proceedings of the 3rd IAPR TC-15 Workshop on Graph-based Representations in Pattern Recognition, pp. 188–199 (2001)
10. Garey, M.R., Johnson, D.S.: Computers and Intractability: A Guide to the Theory of NP-Completeness. W. H. Freeman & Co., New York (1979)
11. Habel, A.: Hyperedge Replacement: Grammars and Languages. LNCS, vol. 643. Springer, Heidelberg (1992)
12. König, D.: Theorie der endlichen und unendlichen Graphen, Leipzig (1936)
13. Lee, J., Han, W.-S., Kasperovics, R., Lee, J.-H.: An in-depth comparison of subgraph isomorphism algorithms in graph databases. In: Proceedings of the 39th International Conference on Very Large Data Bases, pp. 133–144 (2013)
14. Rozenberg, G. (ed.): Handbook of Graph Grammars and Computing by Graph Transformation. World Scientific Publishing Co., Inc. (1997)
15. Ullmann, J.R.: An algorithm for subgraph isomorphism. J. ACM 23(1), 31–42 (1976)
16. Vento, M., Foggia, P.: The vflib graph matching library, http://mivia.unisa.it/datasets/graph-database/vflib

Canonical Derivations
with Negative Application Conditions

Andrea Corradini[1] and Reiko Heckel[2]

[1] Dipartimento di Informatica, Università di Pisa, Italy
andrea@di.unipi.it
[2] University of Leicester, UK
reiko@mcs.le.ac.uk

Abstract. Using graph transformations to specify the dynamics of distributed systems and networks, we require a precise understanding of concurrency. Negative application conditions (NACs) are an essential means for controlling the application of rules, extending our ability to model complex systems. A classical notion of concurrency in graph transformation is based on shift equivalence and its representation by canonical derivations, i.e., normal forms of the shift operation anticipating independent steps. These concepts are lifted to graph transformation systems with NACs and it is shown that canonical derivations exist for so-called incremental NACs.

Keywords: graph transformation, canonical derivation, incremental NACs.

1 Introduction

Graph Transformation Systems (GTS) provide a visual formal specification technique and computational model for concurrent and distributed systems, where graphs modelling system states evolve through the application of rules with local effects. A significant body of literature is dedicated to the study of parallelism and concurrency of graph transformation systems [7,14,5,1,2].

The classical theory includes, among others, the definition of the parallel composition of several rules (by coproduct, i.e. disjoint union) and its application to a graph, and of sequential independence between two consecutive rule applications [7]. These notions are exploited in the Church-Rosser theorem that shows that sequentially independent rule applications can be switched obtaining the same resulting graph. Furthermore, as stated by the Parallelism theorem, the same effect can be obtained by applying the parallel composition of the two rules to the start graph. This leads to the definition of a natural equivalence on the set of parallel derivations, i.e. on sequences of possibly parallel rule applications, called the *shift equivalence*: parallel derivations that differ only by the order in which independent rule applications appear are considered to be equivalent.

Kreowski showed in [14] that shift-equivalence classes of parallel derivations have canonical representatives, obtained by anticipating as much as possible the

H. Giese and B. König (Eds.): ICGT 2014, LNCS 8571, pp. 207–221, 2014.

rule applications. Such representatives are called *canonical derivations*, and they feature an "early maximal parallelism": each rule application depends on at least one rule application in the preceding parallel rule application, if one exists.

Using graph transformations to specify the dynamics of distributed systems and networks, we require a precise understanding of concurrency. The notions of shift equivalence and canonical derivation are the original expression of such an understanding. They have been fundamental to more recent research on concurrency of graph transformation systems (including graph processes [5] and unfolding [3]), as well as their generalisation to transformation systems based on ($\mathcal{M}$-)adhesive categories [6,2].

Negative application conditions (NACs) [9] are an essential means for controlling the application of rules, extending our ability to model complex systems. However, a corresponding notion of concurrency has been missing so far. This paper addresses the generalisation of the results on canonical derivations to rules with NACs.

A NAC allows one to describe a "forbidden context", whose presence around a match inhibits the application of the rule. NACs introduce new kinds of dependencies among rule applications, as stressed in [11,4], due to the fact that a forbidden context for a rule can be generated by two sequentially independent rule applications. As a consequence, unlike the case without NACs, sequential independence of rule applications with NACs (as defined in [16]) is not stable under switching independent rule applications: as recalled in Sect. 3 two independent consecutive rule applications may become dependent if both are switched with a third rule application, independent of both. As shown in [4], this problem does not occur if the NACs are *incremental*, that is, if each morphism embedding the left-hand side of a rule into a forbidden context can be decomposed in an essentially unique way.

After presenting basic definitions in Sect. 2, Sect. 3 recalls concepts and results on independence of rule applications, including the definition of incremental NACs and some of their properties. The original contribution starts with Sect. 4 where we first introduce an operational semantics for transformation systems. Given a set of pairwise independent rule matches in a graph, we compose them via an *amalgamation* construction obtaining a *step*, whose application to a graph has the same effect of the parallel application of the given rules. In the presence of NACs, such a step is not always serializable. We call it *safe* if any interleaving of its constituent rule applications satisfies all the NACs, that is, if it is serializable in all possible ways. This might be expensive to check, but we show that if NACs are incremental then safety of a step made of several rules can be checked by a pairwise analysis of the rules.

Next in Sect. 5 we discuss existence of canonical derivations made of steps. We show that they do not exist in general for rules with arbitrary NACs, but are guaranteed to exist if all NACs are incremental. Finally, Sect. 6 provides a conclusion and sketches future developments.

2   Basic Definitions

In this paper we use the double-pushout approach [7] to graph transformation with negative application conditions [9]. However, we will state all definitions and results at the level of adhesive categories [15]. We recall that a category is *adhesive* if (1) it has pullbacks, (2) it has pushouts along monomorphisms (hereafter *monos*), and (3) all pushouts along a mono are *Van Kampen squares*. That means, when such a pushout is the bottom face of a commutative cube such as in the diagram to the right, whose rear faces are pullbacks, the top face is a pushout if and only if the front faces are pullbacks. In any adhesive category pushouts along monos are also pullbacks, and if a pushout complement of two composable arrows $K \xrightarrow{l} L \xrightarrow{m} G$ with l mono exists, then it is unique (up to iso).

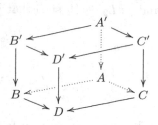

As an example, the category of typed graphs for a fixed type graph TG, defined as the slice category (**Graph** $\downarrow TG$), is adhesive. In the rest of the paper, all objects and arrows live in an arbitrary but fixed adhesive category **C**.

A *rule* $p = (L \xleftarrow{l} K \xrightarrow{r} R)$ consists of a span of two monos l and r. A *redex* in a graph G is a pair (p, m), where p is a rule and $m : L \to G$ is a mono, called a *match*. Given a redex (p, m), a transformation $G \xRightarrow{p,m} H$ from G to H exists if the *double-pushout (DPO) diagram* to the right can be constructed, where (1) and (2) are both pushouts.[1]

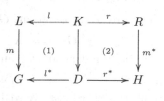

The applicability of rules can be restricted by imposing some negative constraints, defined as follows. A *(negative) constraint* over an object L is a mono $n : L \to N$. A mono $m : L \to G$ satisfies n (written $m \models n$) iff there is no mono $q : N \to G$ such that $n; q = m$. A negative application condition (NAC) is a set of constraints. A mono $m : L \to G$ *satisfies* a NAC ϕ over L (written $m \models \phi$) if and only if m satisfies every constraint, i.e., $\forall n \in \phi, m \models n$. In this paper we shall consider only monic matches and monic constraints.

A *graph transformation system (GTS)* $\mathcal{G} = (P, \pi)$ consists of a set of rule names P and a function π assigning to each name p a rule $\pi(p) = L \xleftarrow{l} K \xrightarrow{r} R$. A *conditional GTS* (P, π, Φ) consists of an *underlying GTS* (P, π) and a function Φ providing for each $p \in P$ a NAC $\Phi(p)$ over $\pi(p)$'s left-hand side. A *derivation* in a GTS $\mathcal{G} = (P, \pi)$ is a finite sequence of transformations $s = (G_0 \xRightarrow{p_1,m_1} G_1 \xRightarrow{p_2,m_2} \cdots \xRightarrow{p_n,m_n} G_n)$ with $p_i \in P$. A *conditional derivation* in a conditional GTS (P, π, Φ) is a derivation in its underlying GTS such that each $G_{i-1} \xRightarrow{p_i,m_i} G_i$ is a *conditional transformation*, that is, it is a DPO diagram where match m_i satisfies the NAC $\Phi(p_i)$.

We write $s : G_0 \xRightarrow{*} G_n$ for a generic derivation and, given $s' : G_k \xRightarrow{*} G_m$ with $G_n = G_k$, denote their sequential composition by $s; s' : G_0 \xRightarrow{*} G_m$.

---

[1] Note that we stick to DPO rewriting with monic matches only (see [10]).

3   Independence and Switch Equivalence

This section recalls the notions of parallel and sequential independence and switch equivalence, and illustrates the problem that sequential independence with NACs is not stable under switching. In the DPO approach, two transformations from the same graph $G_1 \overset{p_1,m_1}{\Longleftarrow} G_0 \overset{p_2,m_2}{\Longrightarrow} G_2$ as in the left of the diagram below are *parallel independent* iff there exist morphisms $i : L_1 \to D_2$ and $j : L_2 \to D_1$ such that $j; l_1^* = m_2$ and $i; l_2^* = m_1$.

$$R_1 \overset{r_1}{\prec} K_1 \overset{l_1}{\succ} L_1 \cdots\cdots L_2 \overset{l_2}{\prec} K_2 \overset{r_2}{\succ} R_2 \qquad L_1 \overset{l_1}{\prec} K_1 \overset{r_1}{\succ} R_1 \cdots\cdots L_2 \overset{l_2}{\prec} K_2 \overset{r_2}{\succ} R_2$$

$$\downarrow m_1^* \quad \downarrow \quad j \overset{}{\searrow} m_1 \quad m_2' \overset{i}{\swarrow} \downarrow \quad \downarrow m_2^* \qquad \downarrow m_1 \quad \downarrow \quad j \overset{}{\searrow} m_1^* \quad m_2' \overset{i}{\swarrow} \downarrow \quad \downarrow m_2^*$$

$$G_1 \underset{r_1^*}{\prec} D_1 \xrightarrow{l_1^*} G \xleftarrow{} D_2 \underset{r_2^*}{\succ} G_2 \qquad G_0 \underset{l_1^*}{\prec} D_1 \xrightarrow{r_1^*} G_1 \xleftarrow{l_2^*} D_2 \underset{r_2^*}{\succ} G_2$$

Two consecutive transformations $G_0 \overset{p_1,m_1}{\Longrightarrow} G_1 \overset{p_2,m_2}{\Longrightarrow} G_2$ as in the right of the diagram above are *sequentially independent* iff there exist morphisms $i : R_1 \to D_2$ and $j : L_2 \to D_1$ such that $j; r_1^* = m_2$ and $i; l_2^* = m_1^*$.

The Local Church-Rosser theorem (Thm. 3.20 of [6], called "LCR" in the following) states that (1) given the two parallel independent transformations, there are transformations $G_1 \overset{p_2,m_2'}{\Longrightarrow} X \overset{p_1,m_1'}{\Longleftarrow} G_2$ such that $G_0 \overset{p_1,m_1}{\Longrightarrow} G_1 \overset{p_2,m_2'}{\Longrightarrow} X$ (and $G_0 \overset{p_2,m_2}{\Longrightarrow} G_2 \overset{p_1,m_1'}{\Longrightarrow} X$) are sequentially independent and that (2) given sequentially independent transformations $\sigma = G_0 \overset{p_1,m_1}{\Longrightarrow} G_1 \overset{p_2,m_2}{\Longrightarrow} G_2$, there exists $\sigma' = G_0 \overset{p_2,m_2'}{\Longrightarrow} G_2' \overset{p_1,m_1'}{\Longrightarrow} G_2$ such that $G_1 \overset{p_1,m_1}{\Longleftarrow} G_0 \overset{p_2,m_2'}{\Longrightarrow} G_2'$ are parallel independent. In the last case we write $\sigma \sim_{sw} \sigma'$ to denote that derivations σ and σ' are in the *switch relation*. This relation extends to an equivalence on derivations of arbitrary finite length, called *switch equivalence*, by setting $s ; s_1 ; s' \sim_{sw} s ; s_2 ; s'$ if $s_1 \sim_{sw} s_2$, for arbitrary derivations s, s', and closing under transitivity.

The definition of parallel independence carries over to conditional transformations [17] by requiring that the matches $j; r_1^*$ for p_2 in G_1 and $i; r_2^*$ for p_1 in G_2 satisfy their respective NACs. Similarly, for sequential independence, the match for p_2 in G_0 given by $m_2' = j; l_1^*$ must satisfy the NAC of p_2 and the induced match of p_1 into graph G_1' obtained by $G_0 \overset{p_2,m_2'}{\Longrightarrow} G_1'$ must satisfy the NAC of p_1. With these definitions, the statements of LCR carry over verbatim to the conditional case.

However, as mentioned in the introduction and as shown in the next example, in the conditional case the sequential independence of two transformations is not preserved by switch equivalence, a property that is known to hold for derivations without NACs by the results in [14]. The same problem also holds if the coarser *permutation equivalence* is considered on conditional derivations, as defined in [12].

Example 1 (independence with NACs). Along the paper we use the rules of Fig. 1 to show the problems that arise using rules with arbitrary NACs. Rule p_1 has a negative constraint $n_1 : L_1 \to N_1$: it creates a new node only if the node it is

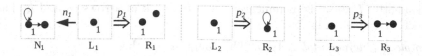

Fig. 1. Three simple graph transformation rules

applied to does not have both a loop and an edge to another node; p_2 adds a loop to a node; p_3 adds to a node an edge connecting a new node. Recall that rules are spans: the mid graph is omitted as it coincides with the left-hand side.

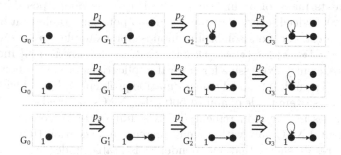

Fig. 2. Independence of p_1 and p_2 is not preserved by switching with p_3

Fig. 2 shows three conditional derivations from G_0 applying p_1, p_2 and p_3 in different orders. All rules are applied to node labeled 1. In the first derivation $s = G_0 \xrightarrow{p_1} G_1 \xrightarrow{p_2} G_2 \xrightarrow{p_3} G_3 = (s_1; s_2; s_3)$ (top), s_1 and s_2 are sequentially independent, and so are s_2 and s_3. After switching s_2 and s_3 by point (2) of LCR we obtain $s' = (s_1; s_3'; s_2')$ (middle of Fig. 2), thus we have $s \sim_{sw} s'$. Since s_1, s_3' are independent, we can perform a further switch obtaining $s'' = (s_3''; s_1'; s_2')$ (bottom). However, s_1' and s_2' are sequentially dependent in s'', because the match for p_1 into G_2'', obtained as $G_1' \xrightarrow{p_2} G_2''$, does not satisfy its NAC. Hence, sequential independence may not be preserved in equivalent derivations.

Essentially, the problem identified in Ex. 1 is due to the fact that the NAC of rule p_1 is made of two parts that can be created, in any order and independently, by p_2 and p_3. In fact, as shown in [4], the problem disappears if all NACs are *incremental* in the following sense.

Definition 1 (incremental monos and NACs). *A mono $f: A \to B$ is called* incremental, *if for any pair of decompositions $g_1; g_2 = f = h_1; h_2$ as in the diagram to the right where all morphisms are monos, there is either a mediating morphism $o : O \to O'$ or $o' : O' \to O$, such that the resulting triangles commute. A monic NAC ϕ over L is* incremental *if each constraint $(n : L \to N) \in \phi$ is incremental.*

Equivalently, f above is incremental if the pullback of arrows g_2 and h_2 has a projection that is an iso.[2]

Example 2 (Incremental NACs). The diagram to the right shows that the negative constraint $n_1 : L_1 \to N_1$ of rule p_1 of Ex. 1 is not incremental, because N_1 extends L_1 in two independent ways: by the loop on 1 in O_1, and by the outgoing edge with an additional node 2 in O'_1. Indeed, there is no mediating arrow from O_1 to O'_1 or vice versa relating these two decompositions.

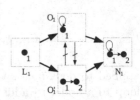

Incremental monos in the category of directed graphs can only represent forbidden contexts made of a single edge or of a single node (possibly with an incident edge). Indeed, every incremental mono between graphs can be decomposed in at most two non-isomorphic monos: for example, $\{\bullet\} \to \{\bullet \to \bullet\} = \{\bullet\} \to \{\bullet \quad \bullet\}; \{\bullet \quad \bullet\} \to \{\bullet \to \bullet\}$. Despite that, as discussed in [4] incremental NACs are sufficiently expressive for several applications of GTSs. For example, [13] describes a significant model transformation problem addressed with a graph transformation system made of almost 500 rules. NACs are used extensively, and only a few rules have non-incremental NACs, but could be converted to rules with incremental NAC only with a small increase in execution time.

The next result shows that independence is stable under switching in the case of incremental NACs. It follows easily from Thm. 1 of [4], where only the sequential case was considered.

Theorem 1 (indep. stable under switch). *In the cube on the right, edges represent conditional transformations $t_i^j = (p_i, m_i^j)$ for $1 \le i, j \le 3$, where all rules have incremental NACs only. Furthermore, each face $A \lessdot_a= B \lessdot_b= C =c\gtrdot D =d\gtrdot A$ is a switch operation iff either b and c are parallel independent and a, d are obtained by point (1) of LCR, or b; a (resp. c; d) are sequentially independent, and c; d (resp. b; a) are obtained by point (2) of LCR. Then we have:*

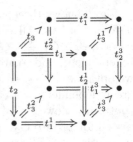

1. *If the top and right faces are switch operations, the front face is a switch operation iff the back face is.*
2. *If the top and front faces are switch operations, the right face is a switch operation iff the left face is.*

Another way of reading item 1 above is that $t_1; t_2^1; t_3^3 \sim_{sw} t_3; t_1^2; t_2^3$ implies that t_1 and t_2^1 are sequentially independent iff t_1^2 and t_3^3 are, i.e., as shown in [4] sequential independence is stable under switching (the property that did not hold for Ex. 1 because of the non-incremental NAC of p_1). By symmetry, this implies analogous statements assuming that the faces in the top and back, front and bottom, front and right, left and back, left and bottom, respectively, represent switch operations.

---

[2] We are grateful to an anonymous referee for this observation.

Item 2 states stability of parallel independence under switch, so statements symmetrical to 2 hold by assuming that the faces in the front and left, left and top, respectively, represent switch operations. This can be shown by exploiting the duality between parallel and sequential independency, i.e., t_1 and t_2 are sequentially independent iff the inverse t_1^{-1} of t_1 and t_2 are parallel independent.

4   Conditional Step Derivations

We investigate conditional step derivations as a computational model where, in each state, a number of rules with NACs can be applied in parallel as long as they are not in conflict, i.e., the same effect can be obtained by applying them sequentially. We face two challenges. First, to define a notion of parallel transformation the traditional approach is to construct a parallel rule as the disjoint union of all the rules to be applied. If the matches of these rules are not disjoint, this results in a non-monic match for the parallel rule. However, we have restricted our matches to monos because they work more naturally with the notion of satisfaction employed by NACs. That means, rather than parallel rules based on disjoint union, we consider amalgamated rules which merge individual rules wherever their matches overlap. We will call the application of such amalgamated rules to a graph a *step*.

The notion of amalgamated transformation and derivation has been introduced for graphs and HLR systems [18], and it is adapted here to adhesive categories and extended by NACs. A similar construction is also presented in [10], but for two rules only. Paper [8] considers amalgamation of several rules with *nested* application conditions, which generalize negative ones, in arbitrary ($\mathcal{M}$-)adhesive categories. However, rules are amalgamated along *a single* common subrule, thus this approach is not applicable to our setting.

The second challenge lies in ensuring that conditional steps do not contain conflicting application conditions. We call these conditional steps *safe* in the sense that they can be serialised in any order with the same effect, so that conditional derivations and conditional step derivations induce the same reachability relation on graphs. While for the unconditional case this follows from the Parallelism theorem [7,6], in the conditional case checking serializability of a step requires to check that the negative constraints of a rule cannot be produced by applying an arbitrary subset of the remaining rules, which is computationally very costly. We show that if all NACs are incremental, in order to check that a conditional step is safe a pairwise analysis of the involved rules is sufficient.

A different approach is taken in [17], where to ensure serialisation binary parallel rules are equipped with additional NACs, besides those of the component rules. The process can be iterated, but the high number of constraints generated would make the approach difficult to apply.

We will define steps with a two-level construction. For a multiset of redexes enabled individually, first an amalgamated rule is constructed, then it is applied at a match obtained by combining the individual matches. The amalgamated rule is obtained using a colimit construction called *pushout star*.

Proposition 1 (existence of pushout stars). *Given a family of spans* $(O_i \leftarrow O_{ij} \rightarrow O_j)_{i<j\in I}$ *connecting a finite collection of objects* $(O_i)_{i\in I}$, *its colimit, denoted* $POS(O_i \rightarrow O \leftarrow O_j)_{i<j\in I}$, *is called a* pushout star *(see [7,18]).*

Let $I = \{1, \ldots, n\}$, $\{m_i : K_i \rightarrow G\}_{i\in I}$ *be a set of monos, and* $(K_i \leftarrow K_{ij} \rightarrow K_j)_{i<j\in I}$ *be a family of spans in an adhesive category, where* $K_{ij}K_iK_jG$ *is a pullback for each* $i < j \in I$. *Then the colimit of this family, i.e.* $POS(K_i \rightarrow K \leftarrow K_j)_{i<j\in I}$, *exists.*

Definition 2 ((conditional) steps and step derivations). *Given a graph transformation system* (P, π), *assume a graph* G *and a finite multiset of redexes* $S = [(p_i, m_i)]_{i\in I}$ *with* $p_i \in P$, $\pi(p_i) = L_i \xleftarrow{l_i} K_i \xrightarrow{r_i} R_i$, *and* $m_i : L_i \rightarrow G$ *for* $i \in I = \{1, \ldots, n\}$.[3] *The multiset of redexes* S *is* enabled *in* G *if*

(1) there exist transformations $G \overset{p_i,m_i}{\Longrightarrow} H_i$ *for all* $i \in I$;
(2) for all $i \neq j \in I$, *transformations* $G \overset{p_i,m_i}{\Longrightarrow} H_i$ *and* $G \overset{p_j,m_j}{\Longrightarrow} H_j$ *are parallel independent.*

Then a step $G \overset{S}{\Longrightarrow} H$ *in* (P, π) *is obtained as shown in the left of Fig. 3 by*

1. *constructing pullbacks* $K_i \leftarrow K_{ij} \rightarrow K_j$ *of* $K_i \xrightarrow{l_i} L_i \xrightarrow{m_i} G \xleftarrow{m_j} L_j \xrightarrow{l_j} K_j$ *for all* $i < j \in I$;
2. *building (thanks to Prop. 1) the amalgamated rule* p_S *with* $\pi(p_S) = L \xleftarrow{l} K \xrightarrow{r} R$ *by*
   - $L = POS(L_i \rightarrow L \leftarrow L_j)_{i<j\in I}$ *of* $(L_i \xleftarrow{l_i} K_i \leftarrow K_{ij} \rightarrow K_j \xrightarrow{l_j} L_j)_{i<j\in I}$
   - $K = POS(K_i \rightarrow K \leftarrow K_j)_{i<j\in I}$ *of* $(K_i \leftarrow K_{ij} \rightarrow K_j)_{i<j\in I}$
   - $R = POS(R_i \rightarrow R \leftarrow R_j)_{i<j\in I}$ *of* $(R_i \xleftarrow{r_i} K_i \leftarrow K_{ij} \rightarrow K_j \xrightarrow{r_j} R_j)_{i<j\in I}$
   with l, r *induced by the universal property of the POS forming* K;
3. *applying* p_S *at the match* $m_S : L \rightarrow G$ *induced by the universal property of the POS forming* L. *This is possible because it can be shown that under the above hypotheses the pushout complement of* l *and* m_S *exists.*

A conditional step *in a conditional GTS* (P, π, Φ) *is a step* $G \overset{S}{\Longrightarrow} H$ *in* (P, π) *such that for each* $i \in I$, *match* m_i *satisfies* $\Phi(p_i)$. *(Conditional) step derivations[4] are sequences of (conditional) steps, defined as expected.*

If S is a singleton, step $G \overset{S}{\Longrightarrow} H$ specialises to a transformation. If S is empty, the amalgamated rule is empty and we assume $G = H$. Note that steps are based on *multisets* of redexes. This allows auto-concurrency, i.e., applying a redex twice in the same step is possible if it is not in conflict with itself. In the

---

[3] We denote a finite multiset over a set X as $[x_1, \ldots, x_n]$, where $x_i \in X$ for all $0 < i \leq n$ and the order is irrelevant. Multiset union is denoted by juxtaposition, and a singleton multiset $[x]$ is sometimes represented simply as x.

[4] Based on this, the title of this paper should really refer to Canonical Step Derivations with NACs. We stick to the more traditional title for consistency with the literature.

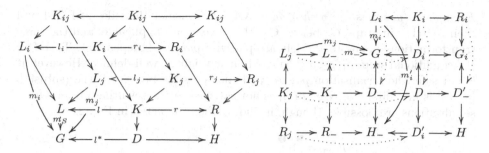

Fig. 3. A step $G \overset{S}{\Longrightarrow} H$ and a serialisation

unconditional case, a step is a special case of amalgamated transformation with constant interface rules [18].

A step is called *safe* if it can be serialised in any order.

Definition 3 (serialisation, parallelisation, safe step, residual). *A (conditional) step $G \overset{S}{\Longrightarrow} H$ has a serialisation $G \overset{S_-}{\Longrightarrow} H_- \overset{s_i'}{\Longrightarrow} H$ or $G \overset{s_i}{\Longrightarrow} G_i \overset{S'}{\Longrightarrow} H$ if such derivations follow the diagram on the right of Fig. 3, where*

- *$s_i = (p_i, m_i)$ and $S = [s_i]S_-$.*
- *$L_i \to D_-$ exists such that $L_i \to D_- \to G = m_i$, allowing to define m_i' as the residual match of m_i by $L_i \to D_- \to H_-$, and $s_i' = (p_i, m_i')$*
- *for all $(p_j, m_j) \in S_-$, an $L_j \to D_i$ exists such that $L_j \to D_i \to G = m_j$ allowing to define $m_j' = L_j \to D_i \to G_i$ as the residual match of m_j, and S_-' from S_- by replacing, in each (p_j, m_j), match m_j by its residual m_j'.*

Vice versa, $G \overset{S}{\Longrightarrow} H$ is a parallelisation of either of the two derivations. A step $G \overset{S}{\Longrightarrow} H$ is safe if either S is empty or a singleton, or for all $s_i \in S$ it has serialisations $G \overset{S_-}{\Longrightarrow} H_- \overset{s_i'}{\Longrightarrow} H$ and $G \overset{s_i}{\Longrightarrow} G_i \overset{S'}{\Longrightarrow} H$ where $G \overset{S_-}{\Longrightarrow} H_-$ and $G_i \overset{S'}{\Longrightarrow} H$ are safe.

In the unconditional case, due to pairwise parallel independence of the redexes all steps are safe (see also [18]). Instead, spelling out the definition, a conditional step $G \overset{S}{\Longrightarrow} H$ is safe if for all $i \in I$, the residual match $m_i' : L_i \to H_-$ satisfies $\Phi(p_i)$ and for all $j \in I \setminus \{i\}$ the residual match $m_j' : L_j \to G_i$ satisfies $\Phi(p_j)$. That means, to decide if the step is safe each individual transformation has to be checked for parallel independence against the amalgamation of the rest of the redexes in the step. Because this applies recursively, there are $\mathcal{O}(n!)$ independence checks. Each of these requires the construction of the individual or amalgamated transformations in order to check if NACs are satisfied in the resulting graphs. That means, even if we assume an efficient implementation of individual transformations, an operational semantics of conditional GTSs based on conditional step derivations would not be efficient, as it requires to check safety of each step.

Example 3 (unsafe steps with general NACs). Consider the rules of Fig. 1 and their matches in graph G_0 below. Clearly all rules are applicable, and the three transformations below to the left are pairwise parallel independent. Therefore the conditional step $[p_1, p_2, p_3]$ depicted on the right is well-defined. However, it is not safe: the serialisation $p_2 ; p_3 ; p_1$ is not possible because if the (sub)step $[p_2, p_3]$ is fired first the NAC of p_1 is not satisfied in the resulting graph. Other serialisations are possible, though, including those depicted in Fig. 2.

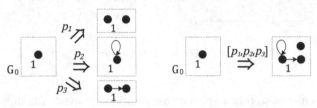

Instead, using incremental NACs only, the situation is comparable to the unconditional case. In fact, as shown in Thm. 2, it is sufficient to consider pairwise parallel independence.

Theorem 2 (safety with incremental NACs). *A conditional step $G \overset{S}{\Longrightarrow} H$ with incremental NACs only is safe if for all redexes $s \neq t \in S$, the corresponding conditional transformations are parallel independent.*

Proof. Following Def. 3 we work by induction. For steps based on 1 or 0 redexes, the statement holds by definition. Assume that all conditional steps are safe if they are based on no more than n redexes, and consider $G \overset{S}{\Longrightarrow} H$ with $S = [t_1, \dots, t_n, t_{n+1}]$. Since t_{n+1} is parallel independent of each of the $t_1, \dots, t_n$ we can use the Local Church-Rosser theorem to form a derivation $t_{n+1}; [t_1', \dots, t_n']$ such that t_{n+1} is sequentially independent of each $t_i, 1 \leq i \leq n$. This provides one of the two sequentialisations required, since item 2 of Thm. 1 implies that the $t_1' \dots t_n'$ are still pairwise parallel independent.

Vice versa, we can transform $[t_1, \dots, t_n, t_{n+1}]$ into $t_1; \dots; t_n'; t_{n+1}'$ by successively delaying independent steps and using Local Church Rosser and item 2 of Thm. 1 to show that this operation preserves independence of the remaining steps. Finally, the resulting conditional derivation $t_1; \dots; t_n'; t_{n+1}'$ can be transformed into the conditional step derivation $[t_1, \dots, t_n'']; t_{n+1}$ again preserving pairwise independence due to Local Church Rosser and Thm. 1. □

Given conditional transformations $G_0 \overset{p_1, m_1}{\Longrightarrow} G_1$ and $G_0 \overset{p_2, m_2}{\Longrightarrow} G_2$, as recalled in Sect. 3, they are parallel independent if the underlying unconditional transformations are, and the residual match $j; r_1^*$ for rule p_2 in G_1 satisfies the NACs $\Phi(p_2)$ (and symmetrically for match $i; r_2^*$). The next result shows that if NACs are incremental, then it is sufficient to check for each negative constraint $n : L_2 \to N$ of p_2, if the negative part $N \setminus n(L_2)$ of n already occurs in the right-hand side of p_1. More abstractly, this negative part of N is characterised by the initial pushout over constraint n, defined as follows.

Definition 4 (initial pushout). *The rectangle with vertices $ABNM$ to the right is an initial pushout over $f : M \to N$ if it is a pushout and for every pushout $IJNM$ there exist unique morphisms $A \to I$ and $B \to J$ such that the triangles commute and $ABJI$ is a pushout.*

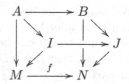

Theorem 3 (satisfaction of incremental NACs by residual match). *In an adhesive category $\mathbf{C}$ with initial pushouts, assume a GTS (P, π, Φ) with incremental NACs, and conditional transformations $s_i = (G_0 \stackrel{p_i, m_i}{\Longrightarrow} G_i)$ for $i = 1, 2$ as in the left of Fig. 4, such that there exists $j : L_2 \to D_1$ with $j; l_1^* = m_2$. Furthermore, let $(n_2 : L_2 \to N_2) \in \Phi(p_2)$ be a negative constraint of p_2, and let $L_2^0 N_2^0 L_2 N_2$ be an initial pushout over n_2. Then match $j; r_1^*$ satisfies n_2 if and only if there exists no monic $k : N_2^0 \to R_1$ such that $L_2^0 N_2^0 G_1 L_2$ is a pullback.*

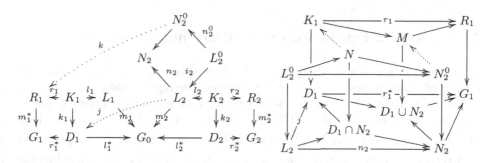

Fig. 4. Satisfaction of incremental NACs (left) and Proof of Thm. 3 (right)

Proof. "⇒": We show that, if k as above exists, then match $j; r_1^*$ does not satisfy constraint n_2. Indeed, given k and j as in the left diagram of Fig. 4, since $L_2^0 N_2^0 L_2 N_2$ is a pushout there exists a morphism $i : N_2 \to G_1$ commuting with n_2 and $j; r_1^*$. Since $\mathbf{C}$ is adhesive, $\mathbf{C} \downarrow G_1$ has *effective unions* [15], that is, the coproduct of two objects (arrows of $\mathbf{C}$ with target G_1) is computed as the pushout over their pullback. Since $L_2^0 N_2^0 G_1 L_2$ is a pullback, i is the coproduct of $k; m_1^*$ and $j; r_1^*$ in $\mathbf{C} \downarrow G_1$, and thus i is monic. Therefore constraint n_2 is violated.

"⇐": We show that, if mono $N_2 \to G_1$ exists commuting with n_2 and $j; r_1^*$, we can find k as required such that $L_2^0 N_2^0 G_1 L_2$ is a pullback. The back and bottom faces of the diagram in the right of Fig. 4 represent the second pushout of s_1 and the non-satisfaction of the constraint by the match $j; r_1^*$. The initial pushout over n_2 is shown in the front of the diagram.

Let $D_1 \leftarrow D_1 \cap N_2 \to N_2$ be a pullback of $D_1 \to G_1 \leftarrow N_2$ and $D_1 \to D_1 \cup N_2 \leftarrow N_2$ a pushout over $D_1 \leftarrow D_1 \cap N_2 \to N_2$, with $L_2 \to D_1 \cap N_2$ and $D_1 \cup N_2 \to G_1$ given by universality. Both arrows are monic, the second one because $\mathbf{C} \downarrow G_1$ has effective unions. All other arrows in the front, bottom, and back faces are also monic, so in particular the front initial pushout is also a pullback.

We can thus decompose it into pullbacks $L_2^0 N L_2 (D_1 \cap N_2)$ and $N N_2^0 (D_1 \cap N_2) N_2$ and then infer by pushout-pullback decomposition [15] that both diagrams are pushouts, too. It can be shown, by a general result about decomposition of initial pushouts, that $N N_2^0 (D_1 \cap N_2) N_2$ is an initial pushout over arrow $D_1 \cap N_2 \to N_2$. In particular, $D_1 \cap N_2 \to N_2$ is not iso because otherwise we would find $N_2 \to D_1 \to G_0$, i.e., m_2 would not satisfy n_2, contradicting the existence of s_2.

Like the front face, also the pushout in the back decomposes into pushouts along $D_1 \to D_1 \cup N_2 \to G_1$, in particular $K_1 M D_1 (D_1 \cup N_2)$. It is easy to check that the composition of an initial pushout and a pushout is again initial if the second is all monic. Hence, since $N N_2^0 (D_1 \cap N_2) N_2$ is initial pushout over $D_1 \cap N_2 \to N_2$ and $(D_1 \cap N_2) N_2 D_1 (D_1 \cup N_2)$ is a pushout, this implies that $N N_2^0 D_1 (D_1 \cup N_2)$ is initial pushout over $D_1 \to D_1 \cup N_2$. Therefore there exists a pushout $N N_2^0 K_1 M$, providing us with the desired $N_2^0 \to M \to R_1$. Finally, $L_2^0 N_2^0 G_1 L_2$ is a pullback because $L_2^0 N_2^0 L_2 N_2$ is a pushout over monos and therefore a pullback also, and $N_2 \to G_1$ is mono.   □

5   Canonical Derivations with Incremental NACs

As recalled in the Introduction, Kreowski showed in [14] that *shift*-equivalence classes of *parallel derivations* of a GTS have unique representatives (up to iso), called canonical derivations. Parallel derivations are sequences of parallel transformations, defined as follows. Given a multiset of redexes $S = [(p_i, m_i)]_{i \in I}$ enabled in a graph G (see Def. 2), a *parallel transformation* $G \overset{\Sigma S}{\Longrightarrow} H$ is defined as a transformation $G \overset{p,m}{\Longrightarrow} H$, where rule $p = L \overset{l}{\longleftarrow} K \overset{r}{\longrightarrow} R$ is obtained as the coproduct of rules $[p_i]_{i \in I}$, and match $m : L \to G$ (which is not mono, in general) is induced by the universal property of the coproduct forming L. The *shift* equivalence is the generalization of the *switch* equivalence on *sequential* derivations (see Sect. 3): two parallel derivations are shift equivalent if any of their serializations, which exist by the Parallelism theorem [6], are switch equivalent.

Existence of canonical representatives is shown by Kreowski by defining a *shift operation* on parallel derivations, that anticipates a redex from a parallel transformation to the previous one, if it is sequentially independent. This operation transforms a derivation into an equivalent one, and it is shown to be confluent and terminating, from which uniqueness of normal forms follows (up to iso, because so is the shift operation). The *canonical derivations* obtained this way feature maximal parallelism by applying each redex as early as possible.

We generalize here definitions and constructions by Kreowski to conditional step derivations, showing that canonical derivations exist if NACs are incremental. First we formalise sequential independence of redexes in steps and introduce the shift operation.

Definition 5 (sequential independence of redexes). *Given two consecutive safe steps* $\sigma = (G_0 \overset{S_1}{\Longrightarrow} G_1 \overset{S_2}{\Longrightarrow} G_2)$ *in a (conditional) GTS, redexes* $s_1 \in S_1$ *and* $s_2 \in S_2$ *are sequentially independent if for a serialisation* $(G_0 \overset{S_1^-}{\Longrightarrow} G_1' \overset{s_1'}{\Longrightarrow}$

$G_1 \overset{s_2}{\Rightarrow} G_2' \overset{S_2^-}{\Rightarrow} G_2)$ of σ, transformations $G_1' \overset{s_1'}{\Rightarrow} G_1 \overset{s_2}{\Rightarrow} G_2'$ are independent. Steps $G_0 \overset{S_1}{\Rightarrow} G_1 \overset{S_2}{\Rightarrow} G_2$ are sequentially independent if for all $s_1 \in S_1$ and $s_2 \in S_2$, s_1 and s_2 are.

Definition 6 (shift operation). *Given two consecutive safe steps $\sigma = (G_0 \overset{S_1}{\Rightarrow} G_1 \overset{S_2}{\Rightarrow} G_2)$ in (P, π, Φ), let $s_2 = (p_2, m_2) \in S_2$ be such that for all $s_1 = (p_1, m_1) \in S_1$, s_1 and s_2 are sequentially independent, and let $G_1 \overset{s_2}{\Rightarrow} G_2' \overset{S_2'}{\Rightarrow} G_2$ be a serialization of $G_1 \overset{S_2}{\Rightarrow} G_2$. Furthermore, let $\sigma' = (G_0 \overset{S_1'}{\Rightarrow} G_2' \overset{S_2'}{\Rightarrow} G_2)$ be the step derivation where S_1' is the parallelisation of S_1 and s_2. Then we say that σ' is a shift of σ, denoted $\sigma \overset{sh(s_2)}{\longrightarrow} \sigma'$. The shift relation over step derivations in a (conditional) GTS is defined as the smallest relation including $\overset{sh(\_)}{\longrightarrow}$ and such that $\sigma \overset{sh(s)}{\longrightarrow} \sigma'$ implies $\sigma_1; \sigma; \sigma_2 \overset{sh(s)}{\longrightarrow} \sigma_1; \sigma'; \sigma_2$ for any derivations σ_1, σ_2. Shift equivalence $\equiv_{sh}$ is the smallest equivalence containing $\overset{sh(\_)}{\longrightarrow}$.*

It is easy to see that if $\sigma \overset{sh(s_2)}{\longrightarrow} \sigma'$ according to the above definition, then in σ' all redexes satisfy their NACs and that redexes are pairwise parallel independent in both steps. Therefore the result is a legal step derivation. It remains to analyse under which conditions it is safe. The following example provides some intuition.

Example 4 (unsafe shift). Using again the rules of Fig. 1, we can build the following safe step derivation. Notice that p_3 is independent of p_1 and p_2 in isolation, but if it is shifted using the construction of Def. 6 this would result in the unsafe step of Ex. 3.

As expected, this problem can be avoided by restricting to incremental NACs, as shown in the theorem below. First we need the following technical result.

Proposition 2 (parallelisation). *Given two consecutive safe steps $G_0 \overset{S_1}{\Rightarrow} G_1 \overset{s_2}{\Rightarrow} G_2$ using incremental NACs only, they are sequential independent if and only if there is a parallelisation $G_0 \overset{S_1[s_2']}{\Rightarrow} G_2$ that is safe (see Def. 3).*

Theorem 4 (safe and confluent shift with incremental NACs). *Let σ be a safe step derivation over a GTS with incremental NACs only, and let $\sigma \overset{sh(s)}{\longrightarrow} \sigma'$. Then σ' is a safe step derivation. Moreover, shift is confluent, that is, given $\sigma \overset{sh(s_1)}{\longrightarrow} \sigma_1$ and $\sigma \overset{sh(s_2)}{\longrightarrow} \sigma_2$ we also have $\sigma_1 \overset{sh(s_2'')}{\longrightarrow} \sigma_3$ and $\sigma_2 \overset{sh(s_1'')}{\longrightarrow} \sigma_3$ where s_1'', s_2'' are the residuals of s_1 and s_2, respectively.*

Proof. The first point follows easily from Prop. 2. To show confluence of shift, assume a two step derivation $\sigma = (G_0 \overset{S_1}{\Rightarrow} G_1 \overset{S_2}{\Rightarrow} G_2)$ and let $s_1, s_2 \in S_2$ such that

$\sigma \xrightarrow{sh(s_1)} \sigma_1$ and $\sigma \xrightarrow{sh(s_2)} \sigma_2$ exist. Therefore, σ is safe and s_1, s_2 are independent of all $s \in S_1$. Then, let $\sigma_1 = (G_0 \xRightarrow{S_1 s_1'} G_1^1 \xRightarrow{S_2^1} G_2)$ and $\sigma_2 = (G_0 \xRightarrow{S_1 s_2'} G_1^2 \xRightarrow{S_2^2} G_2)$.

By Prop. 2 all steps are safe, so in order to have shifts $\sigma_1 \xrightarrow{sh(s_2'')} \sigma_3$ and $\sigma_2 \xrightarrow{sh(s_1'')} \sigma_3$ we have to show that $s_2'' \in S_2^1$ (resp. $s_1'' \in S_2^2$) is independent of all transformations in $S_1 s_1'$ (resp. in $S_1 s_2'$). Then, by confluence of shift on the underlying step derivations, which can be proved essentially as in [14] for parallel derivations, we know that $sh(s_1'')$ and $sh(s_2'')$ lead to isomorphic results.

This is similar, and in fact a consequence, of the stability of independence under switching in item 1 of Thm. 1. More generally we show that the shift operation does not affect the independence of the remaining transformations in S_1 and S_2, i.e., for any $s \in S_1$ and $t \in S_2 - s_1$ with residual $t' \in S_2'$, s, t are independent iff s, t' are. Consider the cube in Thm. 1, and let s be t_1 and t be t_2^1, with t_3^1 being the transformation s_1 anticipated by the shift to $t_3 = s_1'$ and t_2^3 playing the role of t'. By item 1 of Thm. 1, t_1, t_2^1 are independent iff t_1^2, t_2^3 are, while the latter is equivalent to independence of s, t' by Def. 5.   □

Thus the shift relation on safe step derivations with incremental NACs is confluent. Since it is also terminating, as it can be proved along the lines of [14], it has normal forms characterising shift equivalence, called *canonical derivations*. That is, two safe step derivations with incremental NACs are shift equivalent if and only if their canonical derivations are isomorphic.

6 Conclusion

We have investigated the computational model of safe step derivations for conditional graph transformation systems. We have proved that, unlike the general case, if NACs are restricted to be incremental then the safety of a step, which ensures its serializability, follows from the pairwise independence of the component match-rule pairs. In turn, Thm. 3 showed that parallel independence of conditional transformations can be checked efficiently if NACs are incremental: we think that this result can be exploited for the efficient computation of critical pairs. Finally Thm. 4 showed the existence of canonical representatives for the shift-equivalence classes of step derivations with incremental NACs. This result holds for any system satisfying Thm. 1: the stability of independence under switching provides a semantic characterisation for well-behaved concurrency, for which incremental NACs are a sufficient structural condition on rules.

There are several possible developments of the present work, that we intend to explore in the future. First, we aim at investigating graph processes with incremental NACs as a more compact representation of shift equivalence classes of derivations, and how far this can be generalised to arbitrary NACs. Next we intend to look for weaker structural conditions able to ensure that a GTS, even if including non-incremental NACs, still satisfies Thm. 1 (and therefore has canonical derivations). Finally, the extension of the proposed step-based computational model to non-monic matches and non-monic NACs looks like an interesting challenge to face.

References

1. Baldan, P., Corradini, A., Montanari, U., Rossi, F., Ehrig, H., Löwe, M.: Concurrent Semantics of Algebraic Graph Transformations. In: Rozenberg, G. (ed.) The Handbook of Graph Grammars and Computing by Graph Transformations, Concurrency, Parallelism and Distribution, vol. 3, pp. 107–188. World Scientific (1999)
2. Baldan, P., Corradini, A., Heindel, T., König, B., Sobociński, P.: Unfolding grammars in adhesive categories. In: Kurz, A., Lenisa, M., Tarlecki, A. (eds.) CALCO 2009. LNCS, vol. 5728, pp. 350–366. Springer, Heidelberg (2009)
3. Baldan, P., Corradini, A., Montanari, U., Ribeiro, L.: Unfolding semantics of graph transformation. Inf. Comput. 205(5), 733–782 (2007)
4. Corradini, A., Heckel, R., Hermann, F., Gottmann, S., Nachtigall, N.: Transformation systems with incremental negative application conditions. In: Martí-Oliet, N., Palomino, M. (eds.) WADT 2012. LNCS, vol. 7841, pp. 127–142. Springer, Heidelberg (2013)
5. Corradini, A., Montanari, U., Rossi, F.: Graph processes. Fundamenta Informaticae 26(3/4), 241–265 (1996)
6. Ehrig, H., Ehrig, K., Prange, U., Taentzer, G.: Fundamentals of Algebraic Graph Transformation. EATCS Monographs in Theor. Comp. Science. Springer (2006)
7. Ehrig, H.: Introduction to the algebraic theory of graph grammars (a survey). In: Claus, V., Ehrig, H., Rozenberg, G. (eds.) Graph Grammars 1978. LNCS, vol. 73, pp. 1–69. Springer, Heidelberg (1978)
8. Golas, U., Ehrig, H., Habel, A.: Multi-amalgamation in adhesive categories. In: Ehrig, H., Rensink, A., Rozenberg, G., Schürr, A. (eds.) ICGT 2010. LNCS, vol. 6372, pp. 346–361. Springer, Heidelberg (2010)
9. Habel, A., Heckel, R., Taentzer, G.: Graph Grammars with Negative Application Conditions. Fundamenta Informaticae 26(3,4), 287–313 (1996)
10. Habel, A., Müller, J., Plump, D.: Double-pushout graph transformation revisited. Mathematical Structures in Computer Science 11(5), 637–688 (2001)
11. Heckel, R.: DPO Transformation with Open Maps. In: Ehrig, H., Engels, G., Kreowski, H.-J., Rozenberg, G. (eds.) ICGT 2012. LNCS, vol. 7562, pp. 203–217. Springer, Heidelberg (2012)
12. Hermann, F.: Permutation equivalence of DPO derivations with negative application conditions based on Subobject Transformation Systems. ECEASST 16 (2008)
13. Hermann, F., Gottmann, S., Nachtigall, N., Braatz, B., Morelli, G., Pierre, A., Engel, T.: On an automated translation of satellite procedures using triple graph grammars. In: Duddy, K., Kappel, G. (eds.) ICMT 2013. LNCS, vol. 7909, pp. 50–51. Springer, Heidelberg (2013)
14. Kreowski, H.J.: Is parallelism already concurrency? Part 1: Derivations in graph grammars. In: Ehrig, H., Nagl, M., Rosenfeld, A., Rozenberg, G. (eds.) Graph Grammars 1986. LNCS, vol. 291, pp. 343–360. Springer, Heidelberg (1987)
15. Lack, S., Sobocinski, P.: Adhesive and quasiadhesive categories. ITA 39(3) (2005)
16. Lambers, L., Ehrig, H., Orejas, F., Prange, U.: Parallelism and Concurrency in Adhesive High-Level Replacement Systems with Negative Application Conditions. In: Proceedings of the ACCAT workshop at ETAPS 2007. ENTCS, vol. 203 / 6, pp. 43–66. Elsevier (2008)
17. Lambers, L.: Certifying Rule-Based Models using Graph Transformation. Ph.D. thesis. Technische Universität, Berlin (2009)
18. Taentzer, G.: Parallel high-level replacement systems. Theor. Comput. Sci. 186(1-2), 43–81 (1997)

Van Kampen Squares for Graph Transformation

Harald König[1], Michael Löwe[1], Christoph Schulz[1], and Uwe Wolter[2]

[1] University of Applied Sciences, FHDW Hannover,
Freundallee 15, 30173 Hannover, Germany
{harald.koenig,michael.loewe,christoph.schulz}@fhdw.de
[2] Department of Informatics, University of Bergen,
P.O.Box 7803, 5020 Bergen, Norway
uwe.wolter@ii.uib.no

Abstract. This paper demonstrates the benefits of a recent result by the authors, proving a necessary and sufficient condition for a pushout of two morphisms to be a Van Kampen Square, *even if both morphisms are not monomorphisms*. The theorem can be applied in categories that are based on graph structure signatures. We discuss its value in the context of general views on co-transformations and illustrate an application in a software co-evolution scenario.

Keywords: Van Kampen Square, Co-Transformation, Co-Evolution.

1 Introduction

Adhesive categories are a commonly used framework for the theory of Graph Transformation. The Van Kampen property for pushouts is the most influential ingredient on which the definition of adhesiveness is based. In particular, there is the axiom that all pushouts along $\mathcal{M}$-morphisms are Van Kampen Squares, where $\mathcal{M}$ is a distinguished subclass of all monomorphisms. This property and its derived features provide the ability to generalize all important results for graph rewriting to high-level structures such as typed graphs, attributed graphs, elementary Petri nets, hypergraphs, etc.[6]

The pushout of m and f is called a *Van Kampen Square*, if in every commutative cube as in Fig.1 with the pushout under consideration on the top and with back and left faces being pullbacks, the following equivalence holds: The bottom face is a pushout if and only if the front and right faces are pullbacks.

In many involved categories pushouts may enjoy the Van Kampen property, even if in Fig.1 neither m nor f are monomorphisms. Furthermore, there are practical relevant problems, in which theoretical conclusions rely on the fact that certain pushouts with non-injective m and f possess the Van Kampen property. The fact, however, that the property cannot be detected in this case, restricts analysis in an unnecessary way.

The objectives of this paper are to present a necessary and sufficient criterion for a pushout to have the Van Kampen property (especially if both morphisms are not in $\mathcal{M}$), and to demonstrate the use of this criterion in a software co-evolution scenario.

H. Giese and B. König (Eds.): ICGT 2014, LNCS 8571, pp. 222–236, 2014.
© Springer International Publishing Switzerland 2014

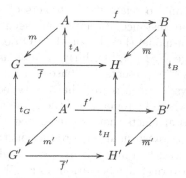

Fig. 1. Van Kampen Property

In software engineering, many artefacts (e.g. class models) can be represented as graphs or derived structures, e.g. attributed graphs. Nodes denote classes or primitive types, edges formalize associations or attributes. Thus, a suitable framework for our considerations are structures that are based on a signature Σ with unary operation symbols only. Depending on Σ, the category of all Σ-algebras can be e.g. directed graphs, E-graphs (for attributed graph transformations), or elementary Petri nets. It is well-known that these categories admit all pullbacks and pushouts. Moreover, pushouts along monomorphisms are Van Kampen Squares [9], [14].

We will show that applications of the presented criterion can be found in research branches in which consistency maintenance of two ore more related software artefacts is formalized. In this case, pushouts are constructed simultaneously on different levels of abstraction which often requires an exact interplay with pullbacks, i.e. the validity of the Van Kampen condition.

We summarize this topic in Section 2 in order to motivate the announced criterion which is presented in Section 3. Afterwards, we apply the main theorem to a software co-evolution scenario using the formalization of model and meta-model co-evolutions suggested in [21]. Finally, related work is discussed and some open research questions are stated.

2 Coupled Transformations

Recent research approaches deal with the challenge of maintaining consistency of two related sources of information. There are the following prominent examples: (1) In the context of model driven software development, (semi-)automatic model transformations have to be induced, if meta-models evolve [3], [18], [21]. (2) Conversion by means of a relation between old and new database schemas requires adjustment of database contents [1], [13]. (3) Different models on similar abstraction levels have to be kept synchronized in evolving information systems, e.g. in the theory of bidirectional transformations [12] or triple-graph-grammars (TGG) [20]. In all scenarios two or more *coupled* transformations on interdependent artefacts have to produce consistent output.

Formally, coupling means to relate two artefacts G and G' by a morphism $t_G : G' \to G$. t_G may be an instance-of-relation between a class model G' which conforms either to a meta-model (Level 2 of Meta Object Facility (MOF)) or a Domain Specific Language (DSL) G. Similarly, t_G may assign to each element of G' a data type, if a graph G' represents database entities (nodes are table entries and edges are foreign keys between them) that are typed in a database schema G. There are also cases where t_G remains on the same abstraction level: E.g. if G and G' are interrelated artefacts that are generated alongside by TGG-rules.

In the theory of graph transformation, production rules p are spans or co-spans that usually consist of two morphisms. Let $f : A \to B$ be any of these two morphisms, then p governs transformations of graphs G by constructing the pushout of f and an occurrence of A in G, i.e. a match $m : A \to G$, cf. the top face in Fig.1. Model co-evolution approaches [21], however, seek to maintain consistency of *two* coupled artefacts during co-transformation. The match m gets traded for a *pair* of morphisms

$$(m : A \to G, m' : A' \to G')$$

indicating the occurence of $t_A : A' \to A$ in t_G. Since elements of the matched region in G' must be typed by elements of the matched region of G, the left face in Fig.1 has to be commutative. Moreover, the match m must *completely* and *uniquely* be reflected on the bottom level, i.e. A' must be the preimage of the coupling t_G restricted to the occurrence of A in G (this is dubbed "complete match" in [21]), i.e. the left face in Fig.1 is claimed to be a pullback.

Clearly, not only the synchronously transformed pair (G', G) is coupled, but also the rule parts, which control transformation, must be compatible with coupling, i.e. the back face in Fig.1 shall at least be commutative. However, if we require a *complete* propagation of the action of the rule part f to the bottom level, each t_B-preimage of $x \in B$ shall be treated w.r.t. f' to the same extent than x, i.e. for all $x \in B$: $x = t_B(x') \Rightarrow |(f')^{-1}(\{x'\})| = |f^{-1}(\{x\})|$, which is true if the back face is also a pullback square. This concept is called "action-reflecting" in [21]. An example with this property is presented in Section 4.2.

There are, however, applications where one abandons this requirement [18] in order to achieve retyping effects. Then one defines $f' := id_{A'}$, $t_B := f \circ t_A$ which shifts effects of f to t_B and may not completely reflect the action of f in f'. In this paper, we will restrict to action-reflecting co-transformations.

Uniform action on the coupled structures G and G' means that the pushout on the top of the cube in Fig.1 has to invoke the same universal construction for the bottom span (m', f'). But then a subsequent application of another co-transformation at match $(\overline{m}, \overline{m}')$ can only be carried out if this match is complete, i.e. the right face has to be a pullback. This can be guaranteed if the top pushout has the Van Kampen property.

Another important application of the Van Kampen property in the context of coupled transformations arises from the fact that the bottom pushout in Fig.1 has the Van Kampen property if the top face has (this can easily be proved by pullback composition and decomposition). One can thus infer that an evolution on MOF level n not only correctly extends to level $n - 1$ but to *all* model levels k with $0 \leq k < n$, if we start with a Van Kampen square on level n.

In adhesive categories pushouts along monomorphisms are Van Kampen Squares, such that there are no problems with co-transformations, if f or m is a monomorphism. But already the classical DPO-approach [6] does not require m to be a monomorphism and there are many research lines where parts of the rule, i.e. f, are not claimed to be monomorphisms, e.g. sesqui-pushout rewriting with cloning [4,5], graph-rewriting in span categories [16], co-transformations with merging [18], and investigations on computing pushout complements [11], [19].

With the current knowledge, all approaches are restricted to either a monic match or a monic rule part if conclusions rely on the Van Kampen property. These limitation can be relaxed if the class of squares known to possess the Van Kampen property can be extended. This will be done in the next section, where we exactly characterize this class by means of an easily verifiable criterion.

3   Characterizing the Van Kampen Property

Since we want to cover common software engineering structures, the following theory is formulated in the context of algebras that are based on graph structure signatures (see [6], Section 11.1), i.e. signatures $\Sigma = (S, Op)$ with a collection S of sorts and a familiy of unary operation symbols

$$Op = (Op_{s,s'})_{s,s' \in S}.$$

Besides the category SET of sets and mappings (one sort, no operation symbols), further examples are signatures for *directed graphs*:

$$sorts : E, V$$
$$opns : src, tgt : E \to V,$$

E-graphs [7]:

$$sorts : E_1, E_2, E_3, V_1, V_2$$

$opns : src_1, tgt_1 : E_1 \to V_1$	(Associations)
$src_2 : E_2 \to V_1, tgt_2 : E_2 \to V_2$	(Attributes)
$src_3 : E_3 \to E_1, tgt_3 : E_3 \to V_2$	(Edge Attributes),

and *Petri nets*:

$$sorts : P, T, Pre, Post$$

$opns : src_{pre} : Pre \to P, tgt_{pre} : Pre \to T$	(Precondition Relation)
$src_{post} : Post \to T, tgt_{post} : Post \to P$	(Postcondition Relation).

We fix Σ and let $\mathcal{C}$ be the category of all Σ-algebras together with homomorphisms between them. For a Σ-algebra G, we denote with G_s the carrier set of sort s and for any operation symbol $op : s \to s'$ we write $op^G : G_s \to G_{s'}$ for the corresponding function of G. A homomorphism $f : G \to H$ is then a sort-indexed family of mappings $(f_s : G_s \to H_s)_{s \in S}$ compatible with operations. As pointed out in Section 1, C has all pushouts and pullbacks (by sortwise construction) and C is adhesive.

Let now $m : A \to G, f : A \to B$ be arbitrary C-morphisms and $\overline{m} : B \to H$, $\overline{f} : G \to H$ be its pushout, cf. the top face in Fig.1. The goal is to partition all spans (m, f) into those for which the pushout has the Van Kampen property and those for which this is not true. An example for $C = SET$ is the top square in Fig.2. The functions map according to the names of the elements (e.g. $f(y) = f(x) = yx$).

For a map $f : A \to B$ we let

$$ker(f) := \{(a, b) \in A \times A \mid f(a) = f(b)\}$$

be the *kernel relation*. Since these relations will be important in subsequent considerations, they are emphasized in Fig.2: $ker(f)$ is depicted by dashed lines in A, i.e. a and b are connected iff $f(a) = f(b)$. Analogously, dotted lines represent $ker(m)$. The pushout object H consists of the equivalence classes of the smallest equivalence relation over $\{(f(a), m(a)) \mid a \in A\} \subseteq (B+G) \times (B+G)$. It is easily verifiable that it consists of one element.

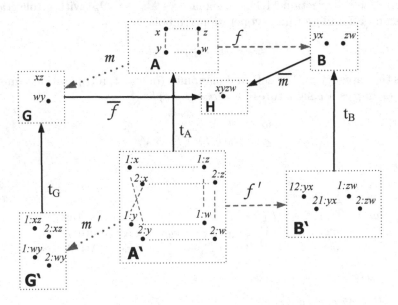

Fig. 2. Violated Van Kampen property

In this example, the Van Kampen property is violated: To see this, consider the "instances" t_G, t_A, and t_B which map according to the notation: $t_X(a{:}b) = b$ for all $X \in G, A, B$. Moreover, dashed lines in A' depict $ker(f')$, dotted lines represent $ker(m')$. It can easily be checked that the two rear faces are pullbacks.

One can now either enforce pullbacks in the front and right by creating pullback complements, or we can construct the pushout in the bottom and check whether its unique mediator into H yields two pullbacks. The first construction, however, fails to be commutative in the bottom, in the second construction the bottom pushout is a singleton set, pullback construction being unsuccessful.

Obviously, in this example, the kernels of f and m are intertwined too much and are thus not enough "separated". m' and f' "benefit" from this by interweaving t_A-fibres, such that the extent of identification in the construction of the bottom pushout is significantly enlarged.

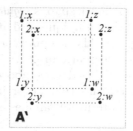

Fig. 3. Uniform kernel relations

We want to place emphasis on the fact that an *alternative* definition for f' does yield a cube with four pullbacks and a pushout in the bottom, namely if $f'(1:x) = f'(1:y)$ and $f'(2:x) = f'(2:y)$ with all other definitions preserved, see Fig.3 in which the kernel pairs in A' are depicted for this case. Obviously, t_A-fibres are now uniformly treated by m' and f'.

Returning to the general case of a category $\mathcal{C}$ based on signatures with unary operation symbols, the example has shown that the Van Kampen property may vanish in the presence of cyclic dependencies amongst the kernel pairs of m and f. In the example above this is the cycle $(x, y), (y, w), (w, z), (z, x)$. This leads to the following definition:

Definition 1 (Separated Kernels). *Let $s \in S$ be a sort and f and m be given as in Figure 1. A sequence $(x_i)_{i \in \{0,1,\ldots,2k+1\}}$ of elements in A_s is called an A_s-domain cycle of f and m (or just domain cycle, if it becomes clear from the context which carrier is considered), if $k \in \mathbb{N}$ and the following conditions hold:*

1. $\forall j \in \{0, 1, \ldots, 2k+1\} : x_j \neq x_{j+1}$
2. $\forall i \in \{0, \ldots, k\} : (x_{2i}, x_{2i+1}) \in ker(f)$
3. $\forall i \in \{0, \ldots, k\} : (x_{2i+1}, x_{2i+2}) \in ker(m)$

where the sums are understood modulo $2k+2$ (i.e. $2k+2 = 0$).

The pair f and m is said to have separated kernels, if it has no domain cycle.

In Fig.2, the sequences $[x, y, w, z]$ and $[w, z, x, y]$ are domain cycles.

If f or m is monomorphic, the definition immediately yields separated kernels for f and m. Clearly, the reverse statement is false in general. To see this, let $\tilde{A} := A - \{w\}$ in Fig.2 and furthermore $\tilde{f} := f|_{\tilde{A}}$, $\tilde{m} := m|_{\tilde{A}}$. Then $\tilde{f}$ and $\tilde{m}$ are both non-injective, but they have separated kernels.

We can now state the main theorem:

Theorem 1. *Let Σ be a graph structure signature, $\mathcal{C}$ be the category of Σ-algebras, and the top square in Figure 1 be a pushout in $\mathcal{C}$.*

The square is a Van Kampen square $\Longleftrightarrow$ f and m have separated kernels.

The theoretical background for the theorem is a categorical characterization of situations, in which a backface pullback span is *regular*, i.e the pushout construction in the bottom together with the unique mediator from the bottom to the top pushout object (t_H in Fig.1) yields two pullbacks as front and right face. It has been stated and proved as Theorem 1 in [23].

In the special case of a category $\mathcal{C}$ as defined above, the involved data reduces to kernel relations of homomorphisms that occur in pullbacks: e.g. the kernel relation of f' can be described in terms of families of bijections

$$(\xi^{f'}_{x,x'} : (t_A)_s^{-1}(x) \to (t_A)_s^{-1}(x'))_{(x,x')\in ker(f)}$$

on fibres of $(t_A)_s$ for each sort s which are compatible with operations (see the dashed lines in Fig.2). The family is *neutral*, i.e. $\xi^{f'}_{x,x} = id_{(t_A)_s^{-1}(x)}$, and *associative*, i.e. $\xi^{f'}_{x,x''} = \xi^{f'}_{x',x''} \circ \xi^{f'}_{x,x'}$ whenever $(x,x'),(x',x'') \in ker(f)$.

If d and d' are the diagonals of the top and bottom pushout, resp., in Fig.1, i.e. $d = \overline{m} \circ f = \overline{f} \circ m$ and analogously for d', then the general result translates as follows: Two backface pullbacks are regular, if and only if for each sort s there is a neutral and associative family of bijections $(\xi^{d'}_{x,x'} : (t_A)_s^{-1}(x) \to (t_A)_s^{-1}(x'))_{(x,x')\in ker(d)}$ compatible with operations such that

$$\forall(x,x') \in ker(f): \xi^{d'}_{x,x'} = \xi^{f'}_{x,x'} \text{ and } \forall(x,x') \in ker(m): \xi^{d'}_{x,x'} = \xi^{m'}_{x,x'} \quad (1)$$

(note that $ker(f) \subseteq ker(d)$ and $ker(m) \subseteq ker(d)$). Intuitively, the construction of two front pullbacks fails, if and only if the equivalence relation of the diagonal d' causes substantially more non-injectivity than m' and f' already provided. E.g. in Fig.2, $d'(1{:}x) = d'(2{:}x)$ which is neither specified by m' nor f'.

Proof sketch of Theorem 1:

"$\Rightarrow$": Assume to the contrary that there is an an s-domain cycle $(x_i)_{i\in\{0,1,...,2k+1\}}$ of f and m for some $k \in \mathbb{N}$. We seek a pullback span which yields a similar effect as in the example above, cf. Fig.2. This can be done by defining three "instances" $t_G : G' \to G$, $t_A : A' \to A$, and $t_B : B' \to B$. We explain the construction of t_A, the definitions of t_G and t_B are analogous. A' has the same carrier sets than A except for the carrier of sort s, which occurs three times (i.e. $A'_s = \bigcup_{i=1}^{3}\{(x,i) \mid x \in A_s\}$). For an operation symbol $op : s_1 \to s_2$ we define $op^{A'}$ distinguishing 4 cases:

- If $s_1 \neq s \neq s_2$, $op^{A'}$ coincides with op^A.
- If $s_1 = s \neq s_2$, $op^{A'}(x, i) = op^A(x)$.
- If $s_1 \neq s = s_2$, $op^{A'}(x) = (op^A(x), 3)$.
- If $s_1 = s = s_2$, $op^{A'}(x, i) = (op^A(x), 3)$.

$t_A : A' \to A$ is defined to be the identity on carrier sets of sort $s' \neq s$ and forgetting the index on the carrier of sort s. A straightforward calculation shows that t_A is a homomorphism.

Elementary arguments (i.e. sortwise pullback construction) show that defining $m' : A' \to G'$ by $m'_{s_1} = m_{s_1}$ if $s_1 \neq s$ and $m'_s(x, i) = (m_s(x), i)$ yields a homomorphism establishing a pullback square under m.

To establish the right rear square, we define $f' : A' \to B'$ by $f'_{s_1} := f_{s_1}$, if $s_1 \neq s$. In order to create the effect of Fig.2, the definition of f'_s introduces interweaving of t_A-fibres of the two distinct domain cycle elements x_0 and x_1:

$$f'_s(x, i) = \begin{cases} (f_s(x), i) & \text{if } x \notin \{x_0, x_1\} \text{ or } i = 3 \\ (f_s(x), i) & \text{if } i \neq 3 \text{ and } x = x_0 \\ (f_s(x), 3 - i) & \text{if } i \neq 3 \text{ and } x = x_1. \end{cases}$$

This means that f' maps according to f on the fibres but interchanges indices at x_0 and x_1:

$$f'_s(x_0, 1) = f'_s(x_1, 2) \quad \text{and} \quad f'_s(x_0, 2) = f'_s(x_1, 1). \tag{2}$$

The homomorphism property of f carries over to f' because (due to the definitions above) the compatibility condition must only be checked on the third copy of A_s which is not affected of interweaving. Moreover, it can be shown that this definition yields a pullback square under f.

The Van Kampen property of the top pushout yields regularity of the constructed pullback span and thus the existence of the neutral and associative family $(\xi^{d'}_{x, x'})_{(x, x') \in ker(d)}$ with property (1). The twist (2) yields $\xi^{d'}_{x_0, x_0}(x_0, 1) = (x_0, 2)$ by transitivity of $\xi^{d'}_{x_i, x_{i+1}}$ along the cycle. However, this violates neutrality.

"$\Leftarrow$": Let two arbitrary back face pullbacks be given under f and m, resp. Let $(x, x') \in ker(d)$, then there is a sequence $[x = x_0, \ldots, x_{2k+1} = x']$ with $(x_0, x_1) \in ker(f)$, $(x_1, x_2) \in ker(m)$, $(x_2, x_3) \in ker(f)$, $(x_3, x_4) \in ker(m)$ etc. We can mutually compose the respective components of the family of fibre bijections $\xi^{f'}_{x_{2i}, x_{2i+1}}$ and $\xi^{m'}_{x_{2i+1}, x_{2i+2}}$ along this sequence. Because we have separated kernels, this composition is unique and yields a family $(\xi^{d'}_{x, x'} : (t_A)^{-1}_s(x) \to (t_A)^{-1}_s(x'))_{(x, x') \in ker(d)}$ of bijections being compatible with operation symbols and still enjoying neutrality and associativity, such that the given back face pullback span is regular by (1).

This shows that constructing the pushout on top always yields four pullbacks as side faces. Vice versa, four pullback side faces always yield a pushout in the bottom of the cube, because $\mathcal{C}$ is a category in which pullback functors preserve colimits [9]. Thus the top face has the Van Kampen property. $\square$

4 An Application of the Van Kampen Characterization

4.1 Co-transformations

In order to motivate the use of Theorem 1, we shortly extract and repeat main ideas from [18] and [21]. In order to keep considerations simple, we let $\mathcal{M}$ be the collection of *all* monomorphisms.

Co-span transformation rules $L \xrightarrow{l} I \xleftarrow{r} R$ contain two jointly epimorphic morphisms l and r with $r \in \mathcal{M}$.[1]

Co-transfomation rules consist of an evolution rule tp (on the meta-model level, top row on the right hand side) and a migration rule p (model level, bottom row) where t_L, t_I, t_R provide model structure typing. As justified in Section 2, square (1) is a pullback. Action-reflection requires (2) to be a pullback, as well.

$$
\begin{array}{ccccc}
TL & \xrightarrow{tl} & TI & \xleftarrow{tr} & TR \\
t_L \uparrow & (1) & t_I \uparrow & (2) & \uparrow t_R \\
L & \xrightarrow{l} & I & \xleftarrow{r} & R
\end{array}
$$

Co-transformation: In Fig.4 (which is a copy of [21], Fig.7) the left face pullback is a complete match of the co-transformation rule. The evolution rule is applied at match tm yielding two pushouts on top of the double cube. The existence of the pushout complement (top right face) is guaranteed, if the classical gluing condition is satisfied. Constructing the pushout of m and l creates the unique mediator t_U. The right cube is completed by constructing the pullback of th and t_U and using the unique mediator m' into this pullback.

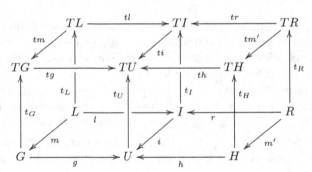

Fig. 4. Co-transformation: Applying rule (tp, p) at match (tm, m)

Co-transformations are only well-defined, if the following two properties can be inferred from this construction: To allow for sequential composition of co-transformations, the co-match (tm', m') must be complete. Moreover, as already mentioned in Section 2, the derivation $TG \overset{(tp,tm)}{\Longrightarrow} TH$ on the meta-model level must invoke the same constructions in the bottom, i.e. two pushout squares for the derivation $G \overset{(p,m)}{\Longrightarrow} H$.

---

[1] [21] prefers the co-span approach, because migrations can better be specified with an intermediate enclosing type graph instead of an intersection type graph.

Thus it is interesting to obtain the following conclusions from the above co-transformation construction:

(1) The right face is a pullback.
(2) The right bottom face is a pushout.

These conclusions can easily be drawn, if the assumption $tm \in \mathcal{M}$ holds: In this case the top face is a Van Kampen square and, because $\mathcal{C}$ is adhesive (see Section 3), the middle face is a pullback. Then (1) follows by pullback composition and decomposition. Thus the right cube has four pullbacks as side faces and (2) follows from the Van Kampen property of the top right pushout ($tr \in \mathcal{M}$ by definition).

The benefit of Theorem 1 is the ability to *give up the assumption "$tm \in \mathcal{M}$"*: If we only know that tm and tl have separated kernels, we can obviously perform the same deductions as above and obtain

Corollary 1. *If tm and tl have separated kernels, Conclusions (1) and (2) hold.*

$\square$

Thus, we may draw these conclusions even if $tm \notin \mathcal{M}$ and $tl \notin \mathcal{M}$. Obviously, the verification whether tm and tl have separated kernels is an easy task. Section 4.2 illustrates the use of Corollary 1 with the help of an example.

Likewise, one obtains the possibility to admit splitting ($tr \notin \mathcal{M}$), cf. the remarks in [18], still being able to draw the above conclusions, if tr and tm' have separated kernels and a unique pushout complement of ti and tr can be found.

4.2 A Well-Defined Co-transformation with Non-monic Match

In this section, we illustrate the use of Corollary 1 by an example where a meta-model evolution successfully invokes a model migration, although $tm, tl \notin \mathcal{M}$. Suppose in a web application, a DSL specifies different dialog models, cf. Fig.5. Hypertext Models 1 and 2 prescribe utilization of two classes typed in $http_1$, $http_2$ resp. to start HTML-dialogs which may use unsecure "http". For secure connections ("https"), Model 1 specifies *dynHttps*-instances starting JSP[2] dialogs, whereas Model 2 suggests *sttcHttps*-instances starting static HTML-dialogs. All dialog starting classes are *managers* which are responsible for preparing the dialog (they establish connections, read personalized configurations, etc). Moreover, there is a third simple model (Hypertext Model 3) which only admits a single page, either JSP or HTML, without a manager.

Since recent attacks spied confidential data during transmission, it was decided to avoid http-protocols wherever adequate: For this purpose, manager class pairs shall be merged in Model 1 and Model 2. A merge is preferred to a deletion, because deleting the *http*-manager would disable HTML-dialogs in Model 1 which is undesirable.

Moreover, the merge shall not be carried out for Model 3, because there is no distinguished manager. Thus, an evolution rule is created, whose left morphism

---

[2] Java Server Pages.

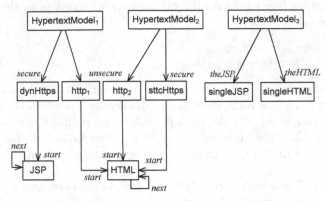

Fig. 5. Dialog model (the graph TG)

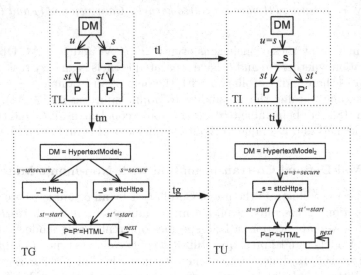

Fig. 6. Evolution Rule's left-hand side and Application: $tl, tm \notin \mathcal{M}$

is shown in the top row of Fig.6 ($tr = id$ is omitted). To avoid applicability of the rule to Model 3, TL and TI contain the start pages (P and P'), too.

This figure also shows the rule's application to Hypertext Model 2 restricted to the corresponding affected parts in TG (pushout of tm and tl). The square constitutes the left top of the double cube in Fig.4.

Let $t_G : G \to TG$ be any instance typed in TG, such that the resulting arrow $t_L : L \to TL$ (by pulling back along tm) admits a pullback complement with tl. An example of such a pullback span is shown in Fig.7: G is a TG-typed class model with mutual use of secure and unsecure connections where explicitly two different page classes $1{:}HTML$ and $2{:}HTML$ are modelled.

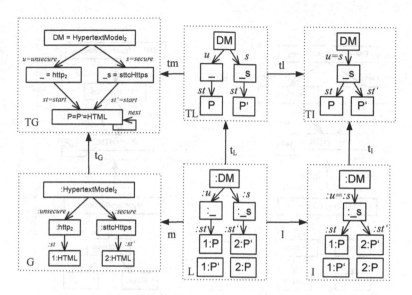

Fig. 7. Coupled match and rule's left-hand sides: 2 Pullbacks

As pointed out in Section 4.1, the co-transformation must produce a pullback as right face in Fig.4. Up to now, we could draw this conclusion only if we knew that the square in Fig.6 has the Van Kampen property, which is not obvious, since $tm \notin \mathcal{M}$ and $tl \notin \mathcal{M}$, cf. the reasoning for Conclusion (1) on page 231.

With Corollary 1 at hand, however, we can nevertheless conclude that the pushout in Fig.6 has the Van Kampen property: Non-injectivity on vertices appears as $tl(\_) = tl(\_s)$ and $tm(P) = tm(P')$, such that there is no domain cycle of vertices. Injectivity of tm on edges prevents domain cycles on edges, hence tl and tm have separated kernels. Consequently, without further effort, we can deduce that there is a well-defined co-transformation in the given example. The resulting instance $t_U : U \to TU$ is shown in Fig.8. Manager classes are merged to one instance of $sttcHttps$, which still maintains the two different HTML-classes as desired. Moreover, the co-transformation yields a complete co-match.

5   Related Work

An important characterization of the Van Kampen property is [10]. Pushout squares of a category $\mathcal{C}$ with chosen pullbacks are embedded into the bicategory of spans over $\mathcal{C}$. It is shown that a pushout in $\mathcal{C}$ has the Van Kampen property if and only if its embedding into the span category is a bipushout. In contrast to our work, this characterization is valid in a more general context, but it is not as easily verifiable as our criterion.

The question when a square is a Van Kampen square was first investigated in [15] which is also the origin of the technique in the first part of the proof sketch of Theorem 1, i.e. the interweaving of preimage structures.

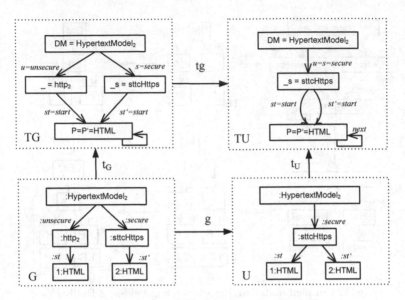

Fig. 8. Induced Model Migration $G \Rightarrow U$

A more special but closely related question was raised in [22] in the context of diagrammatic predicate frameworks: Is there a feasible condition for a back face pullback span (cf. Fig.1) to be regular? That paper uncovers the intrinsic cause of the non-regularity problem by examining the relationship between two semantic antipodes: *Indexed* semantics assigns to each type the set of all instances whereas *fibred* semantics assigns to each instance its type. Their categorical abstractions reveal an old concept: Indexed categories and fibred categories. The relation between these two concepts is established via the *Grothendieck Construction* [8]. The involved functor, however, is not an equivalence of categories: Additional fibred categories represent the cases in which unregular pullback spans occur.

An important research line deals with categorical views on general Van Kampen theorems [2], where the question is investigated under which circumstances a pushout (of $\mathcal{M}$-morphisms) has the Van Kampen property, if one restricts oneself to certain subsets of admissible fibres. This work may become important, if one analyzes coupled transformations with constrained instance structures.

6   Outlook

We presented a criterion which characterizes the Van Kampen property in categories based on signatures with unary operation symbols.

It is a goal to find out whether the criterion extends to more general settings: We conjecture that a "typed version" for comma categories based on presheaves can be formulated. Moreover, there should be corresponding versions for typed attributed graphs, and hopefully also for typed graphs with inheritance, which – for a certain formalization – has already been shown to be an adhesive HLR category [17].

In the example of Section 4, we had to restrict ourselves to situations, in which the left back face in Fig.4 is a pullback. A natural question is to investigate situations where this claim is omitted.

The scenario extends the theory of coupled graph transformations by investigating only pushouts. It is, however, also necessary to investigate the dual situation, namely under which conditions are pullbacks perserved when constructing *pushout complements*.

Since some non-adhesive categories play a role in computer science, another but more theoretical research line is to provide conditions for the Van Kampen property in a more abstract context (e.g. in the category of partially ordered sets or the category of categories).

References

1. Ambler, S.: Refactoring Databases: Evolutionary Database Design. Addison-Wesley (2006)
2. Brown, R., Janelidze, G.: Van Kampen Theorems for Categories of Covering Morphisms in Lextensive Categories. Journal of Pure and Applied Algebra 119, 255–263 (1997)
3. Cicchetti, A., Ruscio, D.D., Eramo, R., Pierantonio, A.: Automating Co-Evolution in Model-Driven Engineering. In: Proceedings of the 2008 12th International IEEE Enterprise Distributed Object Computing Conference, EDOC 2008, pp. 222–231. IEEE Computer Society, Washington, DC (2008), http://dx.doi.org/10.1109/EDOC.2008.44
4. Corradini, A., Heindel, T., Hermann, F., König, B.: Sesqui-Pushout Rewriting. In: Corradini, A., Ehrig, H., Montanari, U., Ribeiro, L., Rozenberg, G. (eds.) ICGT 2006. LNCS, vol. 4178, pp. 30–45. Springer, Heidelberg (2006)
5. Duval, D., Echahed, R., Prost, F., Ribeiro, L.: Transformation of Attributed Structures with Cloning (long version). CoRR abs/1401.2751 (2014)
6. Ehrig, H., Ehrig, K., Prange, U., Taentzer, G.: Fundamentals of Algebraic Graph Transformations. Springer (2006)
7. Ehrig, H., Prange, U., Taentzer, G.: Fundamental Theory for Typed Attributed Graph Transformation. In: Ehrig, H., Engels, G., Parisi-Presicce, F., Rozenberg, G. (eds.) ICGT 2004. LNCS, vol. 3256, pp. 161–177. Springer, Heidelberg (2004)
8. Fiadeiro, J.L.: Categories for Software Engineering. Springer (2005)
9. Goldblatt, R.: Topoi: The Categorial Analysis of Logic. Dover Publications (1984)
10. Heindel, T., Sobociński, P.: Van Kampen Colimits as Bicolimits in Span. In: Kurz, A., Lenisa, M., Tarlecki, A. (eds.) CALCO 2009. LNCS, vol. 5728, pp. 335–349. Springer, Heidelberg (2009), http://dx.doi.org/10.1007/978-3-642-03741-2_23
11. Heumüller, M., Joshi, S., König, B., Stückrath, J.: Construction of Pushout Complements in the Category of Hypergraphs. ECEASST 39 (2011)
12. Hu, Z., Schürr, A., Stevens, P., Terwilliger, J.: Bidirectional Transformation "bx" (Dagstuhl Seminar 11031). Dagstuhl Reports 1(1), 42–67 (2011), http://drops.dagstuhl.de/opus/volltexte/2011/3144

13. König, H., Löwe, M., Schulz, C.: Model Transformation and Induced Instance Migration: A Universal Framework. In: Simao, A., Morgan, C. (eds.) SBMF 2011. LNCS, vol. 7021, pp. 1–15. Springer, Heidelberg (2011)
14. Lack, S., Sobociński, P.: Toposes are Adhesive. In: Corradini, A., Ehrig, H., Montanari, U., Ribeiro, L., Rozenberg, G. (eds.) ICGT 2006. LNCS, vol. 4178, pp. 184–198. Springer, Heidelberg (2006)
15. Löwe, M.: Van-Kampen Pushouts for Sets and Graphs. Technical Report. FHDW Hannover 4 (2010)
16. Löwe, M.: Refined Graph Rewriting in Span-Categories - A Framework for Algebraic Graph Transformation. In: Ehrig, H., Engels, G., Kreowski, H.-J., Rozenberg, G. (eds.) ICGT 2012. LNCS, vol. 7562, pp. 111–125. Springer, Heidelberg (2012)
17. Löwe, M., König, H., Schulz, C., Schultchen, M.: Algebraic Graph Transformations with Inheritance. In: Iyoda, J., de Moura, L. (eds.) SBMF 2013. LNCS, vol. 8195, pp. 211–226. Springer, Heidelberg (2013)
18. Mantz, F., Taentzer, G., Lamo, Y.: Co-Transformation of Type and Instance Graphs Supporting Merging of Types and Retyping. ECEASST 61 (2013)
19. Müller, J.: Shifting Derivations of non-injective Rules in the Algebraic Graph Rewriting Approaches. Techn. Report. TU, Berlin 16 (1997)
20. Schuerr, A.: Specification of Graph Translators with Triple Graph Grammars. In: Mayr, E.W., Schmidt, G., Tinhofer, G. (eds.) WG 1994. LNCS, vol. 903, pp. 151–163. Springer, Heidelberg (1995)
21. Taentzer, G., Mantz, F., Lamo, Y.: Co-Transformation of Graphs and Type Graphs with Application to Model Co-Evolution. In: Ehrig, H., Engels, G., Kreowski, H.-J., Rozenberg, G. (eds.) ICGT 2012. LNCS, vol. 7562, pp. 326–340. Springer, Heidelberg (2012), http://dx.doi.org/10.1007/978-3-642-33654-6_22
22. Wolter, U., Diskin, Z.: From Indexed to Fibred Semantics – The Generalized Sketch File –. Reports in Informatics 361. Dep. of Informatics, University of Bergen (2007)
23. Wolter, U., König, H.: Fibred Amalgamation, Descent Data, and Van Kampen Squares in Topoi. Applied Categorical Structures, 1–40 (2013), http://dx.doi.org/10.1007/s10485-013-9339-2

Graph Transformation Meets Reversible Circuits: Generation, Evaluation, and Synthesis

Hans-Jörg Kreowski, Sabine Kuske, Aaron Lye, and Melanie Luderer

University of Bremen, Department of Computer Science
P.O. Box 33 04 40, 28334 Bremen, Germany
{kreo,kuske,lye,melu}@informatik.uni-bremen.de

Abstract. Reversible circuits are intensively studied for some years as a promising alternative to conventional circuits. Mainly for illustrative purposes and in a rather informal way, they are often visually represented. This inspired us to a graph-transformational approach to reversible circuits. The first steps are documented in this paper with emphasis on generation, evaluation, and synthesis.

1 Introduction

In this paper, we start a graph-transformational study of reversible circuits. According to Moores law, the integration density of circuits doubles every 18 months. While the development of the last 50 years supports this thesis, the process cannot go on much further because of physical limits in scale and power dissipation. For low power electronics [1,19] reversible computation is an alternative and hence a very promising research area, due to the fact that energy dissipation is significantly reduced or even eliminated in reversible circuits [2]. But also especially for its applications to a variety of emerging technologies, such as quantum computation [13,8], superconducting quantum interference devices [15], nanoelectromechanical systems [10,7], optical computing [4] and DNA computing [17]. Reversible circuits are cascades of reversible gates that compute inverse functions on Boolean vectors and are often represented in a graphical way.

The graphical representations of reversible circuits serve illustrative purposes and are used in a rather informal and intuitive way while the technical treatment is usually done on a set-theoretical and logical level or in a programming context. In contrast to this, we propose a formalization of the graphical description of reversible circuits by means of graph transformation (cf., e.g., [16]). In this way, we provide a novel rule-based framework for the investigation of many typical problems of reversible circuits. In particular, we model the generation, evaluation, and synthesis of (graph representations of) reversible circuits as graph transformation units [9] and show their correctness. The synthesis is considered in two variants adapted from the literature: The first one translates the truth tables of reversible functions into reversible circuits (cf. [12]), and the second one transforms binary decision trees into reversible circuits (cf. [21]). This new rule-based approach to reversible circuits on the level of graphs allows formal

H. Giese and B. König (Eds.): ICGT 2014, LNCS 8571, pp. 237–252, 2014.

visual modeling and formal proofs of correctness within the graph transformational setting. The hope is that the graph transformation theory and tools can be employed advantageously. Our first steps look promising in this respect. In particular, the two modeled synthesis procedures are significant examples of model transformation which are only meaningful if they are correct.

The paper is organized as follows. After the graph-transformational preliminaries in Section 2, Section 3 presents reversible circuits. Section 4 introduces lines graphs for representing reversible circuits. Section 5 presents graph-transformation units for generating reversible circuits as graphs. Section 6 concentrates on the correct evaluation of these graphs and Sections 7 and 8 focus on the mentioned synthesis variants. Section 9 contains a short conclusion.

2   Graph-Transformational Preliminaries

A *graph* over a set Σ is a system $G = (V, E, s, t, l)$ where V is a finite set of *nodes*, E is a finite set of *edges*, $s, t \colon E \to V$ are mappings assigning a *source* $s(e)$ and a *target* $t(e)$ to every edge in E, and $l \colon E \to \Sigma$ is a mapping assigning a *label* to every edge in E. An edge e with $l(e) = x$ is an *x-edge*; if $s(e) = t(e)$, it is also called an *x-loop* or a *loop*. The components V, E, s, t, and l of G are also denoted by V_G, E_G, s_G, t_G, and l_G, respectively. The set of all graphs over Σ is denoted by $\mathcal{G}_\Sigma$. We reserve a specific label $*$ which is omitted in drawings of graphs.

For graphs $G, H \in \mathcal{G}_\Sigma$, G is a *subgraph* of H, denoted by $G \subseteq H$, if $V_G \subseteq V_H$, $E_G \subseteq E_H$, $s_G(e) = s_H(e)$, $t_G(e) = t_H(e)$, and $l_G(e) = l_H(e)$, for each $e \in E_G$. A *graph morphism* $g \colon G \to H$ is a pair of mappings $g_V \colon V_G \to V_H$ and $g_E \colon E_G \to E_H$ such that $g_V(s_G(e)) = s_H(g_E(e))$, $g_V(t_G(e)) = t_H(g_E(e))$, and $l_H(g_E(e)) = l_G(e)$ for all $e \in E_G$. If the mappings g_V and g_E are bijective, then G and H are *isomorphic*, denoted by $G \cong H$. For a graph morphism $g \colon G \to H$, the image $g(G) \subseteq H$ of G in H is called a *match* of G in H.

A *rule* $r = (L \supseteq K \subseteq R)$ consists of three graphs $L, K, R \in \mathcal{G}_\Sigma$ such that K is a subgraph of L and R and $E_K = \emptyset$. The components L, K, and R of r are called *left-hand side*, *gluing graph*, and *right-hand side*, respectively. A rule r is drawn in the form $L \to R$ where the nodes of K are numbered.

The application of $r = (L \supseteq K \subseteq R)$ to a graph $G = (V, E, s, t, l)$ yields a directly derived graph H and consists of the following three steps.

1. A match $g(L)$ of L in G is chosen subject to the following conditions.
   - *dangling condition:* $v \in g_V(V_L)$ with $s_G(e) = v$ or $t_G(e) = v$ for some $e \in E_G - g_E(E_L)$ implies $v \in g_V(V_K)$.
   - *identification condition:* $g_V(v) = g_V(v')$ for $v, v' \in V_L$ implies $v = v'$ or $v, v' \in V_K$ as well as $g_E(e) = g_E(e')$ for $e, e' \in E_L$ implies $e = e'$ or $e, e' \in E_K$.
2. Now the nodes of $g_V(V_L - V_K)$ and the edges of $g_E(E_L)$ are removed yielding the *intermediate graph* $Z \subseteq G$.
3. Let $d \colon K \to Z$ be the restriction of g to K and Z. Then H is constructed as the disjoint union of Z and $R - K$ where all edges $e \in E_Z + (E_R - E_K)$ keep their labels and their sources and targets except for $s_R(e) = v \in V_K$ or $t_R(e) = v \in V_K$ which is replaced by $d_V(v)$.

A *rule with positive (negative) context* $r = (C \supseteq L \supseteq K \subseteq R)$ consists of a rule $(L \supseteq K \subseteq R)$ and a graph C with $C \supseteq L$ called *positive (negative) context*. It is applied by applying $(L \supseteq K \subseteq R)$ provided that the corresponding match $g(L)$ can be extended to a match $g'(C)$ (i.e., $g'|L = g$) in the case of positive context. If C is negative, g must not be extendable to a match of C. A rule with positive context is depicted in the form $C \underset{\text{PAC}}{\supseteq} L \to R$ and a rule with negative context as $C \to R$ where the items of $C - L$ are dashed.

The application of a rule r to a graph G is denoted by $G \underset{r}{\Longrightarrow} H$ and called a *direct derivation*. The sequential composition of direct derivations

$$d = G_0 \underset{r_1}{\Longrightarrow} G_1 \underset{r_2}{\Longrightarrow} \cdots \underset{r_n}{\Longrightarrow} G_n \quad (n \in \mathbb{N})$$

is called a *derivation* from G_0 to G_n. The string $r_1 \cdots r_n$ is the *application sequence* of the derivation d. If $r_1, \ldots, r_n \in P$ (for some set P of rules), d can be denoted as $G_0 \underset{P}{\overset{*}{\Longrightarrow}} G_n$.

A *graph class expression* may be any syntactic entity X that specifies a class $SEM(X) \subseteq \mathcal{G}_\Sigma$. A typical example is a forbidden structure $F \in \mathcal{G}_\Sigma$ where $SEM(forbidden(F))$ consists of all graphs G such that there is no subgraph in G that is isomorphic to F. For $M \subseteq \mathcal{G}_\Sigma$, $SEM(forbidden(M)) = \bigcap_{F \in M} SEM(forbidden(F))$.

A *control condition* may be any syntactic entity that restricts the non-determinism of the derivation process. A typical example is a regular expression over a set of rules. Let C be a regular expression specifying the language $L(C)$. Then a derivation with application sequence s is *permitted* by C if $s \in L(C)$. We use the operator ; for sequential composition, | for alternatives and $*$ for iteration. Alternatively to $(r|\cdots|r_n)^*$ with $n \geq 1$ we allow $(r_1|\cdots|r_n)!$ which requires to apply the rules $r_1, \ldots, r_n$ as long as possible. The *priority* $r_1 > r_2$ means that rule r_2 may only be applied if r_1 is not applicable. For control conditions C_1 and C_2, the condition $C_1 \& C_2$ specifies all derivations allowed by both C_1 and C_2. The fact that a derivation $G \underset{P}{\overset{*}{\Longrightarrow}} H$ is permitted by C is denoted by $G \underset{P,C}{\overset{*}{\Longrightarrow}} H$.

A *graph transformation unit* is a system $gtu = (I, P, C, T)$, where I and T are graph class expressions which specify initial and terminal graphs respectively. P is a finite set of rules, and C is a control condition. The operational semantics of gtu is the set of all derivations from initial to terminal graphs that apply rules in P and satisfy the control condition, i.e. $DER(gtu) = \{G \underset{P,C}{\overset{*}{\Longrightarrow}} H \mid (G, H) \in SEM(I) \times SEM(T)\}$. In figures, the components I, P, C, and T of a graph transformation unit are preceded by the keywords *initial*, *rules*, *control*, and *terminal*, respectively.

3 Reversible Circuits

Reversible logic can be used for realizing reversible functions. Reversible functions are special multi-output functions and defined as follows.

Let $\mathbb{B} = \{0,1\}$ be the set of truth values with the negations $\overline{0} = 1$ and $\overline{1} = 0$ and *ID* be a set of identifiers serving as a reservoir of Boolean variables. Let $\mathbb{B}^X$ be the set of all mappings $a\colon X \to \mathbb{B}$ for some $X \subseteq ID$ where the elements of $\mathbb{B}^X$ are called *assignments*. Then a bijective Boolean (multi-output) function $f\colon \mathbb{B}^X \to \mathbb{B}^X$ is called *reversible*. A completely or incompletely specified irreversible function can be embedded into a reversible function. Usually these embedded functions require the addition of constant inputs and garbage outputs [11], which are inputs that are set to fixed Boolean values, or don't cares for all possible input values, respectively.

Reversible circuits are used for representing reversible functions because a reversible function can be realized by a cascade of reversible gates. Reversible circuits differ from conventional circuits: while conventional circuits rely on the basic binary operations and also fanouts are applied in order to use signal values on several gate inputs, in reversible logic fanouts and feedback are not directly allowed because they would destroy the reversibility of the computation [13]. Also the logic operators AND and OR cannot be used since they are irreversible. Instead a reversible gate library is applied. Several reversible gates have been introduced including the Toffoli gate [18], the Peres gate [14] and the Fredkin gate [6]. Toffoli gates are universal gates, i.e. all reversible functions can be realized by cascades of this gate type alone. Hence, Fredkin gates, Peres and inverse Peres gates can be substituted by a sequence of Toffoli gates, respectively.

A *(multiple-control) Toffoli gate* consists of a *target line* $t \in ID$ and a set $C \subseteq ID - \{t\}$ of *control lines* and is denoted by $TG(t, C)$. The gate defines the function $f_{t,C}\colon \mathbb{B}^X \to \mathbb{B}^X$ for each $X \subseteq ID$ with $\{t\} \cup C \subseteq X$ which maps an assignment $a\colon X \to \mathbb{B}$ to $f_{t,C}(a)\colon X \to \mathbb{B}$ given by $f_{t,C}(a)(t) = \overline{a(t)}$ if $a(c) = 1$ for all $c \in C$. In all other cases, $f_{t,C}(a)$ is equal to a. Hence, $f_{t,C}(a)$ inverts the value of the target line if and only if all control lines are set to 1. Otherwise the value of the target line is passed through unchanged. The values of all other lines always pass through a gate unchanged. Consequently, $f_{t,C}$ is a mapping on $\mathbb{B}^X$ which is inverse to itself and, therefore, reversible in particular. In the original definition of Toffoli [18], a Toffoli gate has two control lines. For drawing circuits, in this work the established convention is utilized. A Toffoli gate is depicted by using solid black circles to indicate control connections for the gate and the symbol $\oplus$ to denote the target line. See Fig. 1(a) for a typical example. Fig. 1(b) shows a Toffoli circuit.

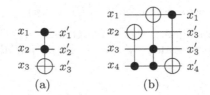

(a)   (b)

Fig. 1. A Toffoli gate with two control lines (a) as well as a Toffoli circuit (b)

4   Lines Graphs

We introduce the notion of lines graphs as a key concept to represent reversible gates and circuits. Let ID be a finite set of identifiers and $X \subseteq ID$. Then X induces the following graphs.

1. $X^{id} = (X, X, id_X, id_X, id_X)$ where id_X is the identity on X.
2. $X^{\emptyset} = (X, \emptyset, \emptyset, \emptyset, \emptyset)$.
3. $out(X^{id})$ contains X^{id} as a subgraph and carries an extra out_X-loop for each $x \in X$.
4. For $x \in X$, an x-line is a simple path from a node $v(x)$ to x where x may carry an extra loop labeled by out_x and $v(x)$ carries an extra loop labeled by x or in_x subject to the condition: In case of an in_x-loop, the path is not empty and the first edge may be labeled by 1 or 0 or $*$ while all other edges are unlabeled.
5. Let $line_x$ be an x-line for each $x \in X$ and $lines_X$ be the disjoint union of these lines, i.e, $lines_X = \sum_{x \in X} line_x$. Then a *lines graph* over X is every graph with $lines_X$ as subgraph that has no further y-loops or in_y-loops for $y \in ID$ and no further unlabeled edges.

Fig. 2 shows the special lines graphs X^{id}, $X^{\emptyset}$ and $out(X^{id})$ as well as a typical more general lines graph for $X = \{1, 2, 3, 4\}$.

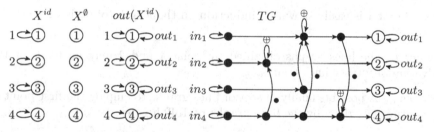

Fig. 2. Some lines graphs with $X = \{1, 2, 3, 4\}$

Example 1. Toffoli gates can be represented as lines graphs. Let $TG(t, C)$ be a Toffoli gate with $U = C \cup \{t\}$. Let $v(U) = \{v(x) \mid x \in U\}$ and $w(U) = \{w(x) \mid x \in U\}$ be two disjoint copies of U. Then $TG(t, C)$ induces the lines graph $lg(TG(t, C))$ over ID defined as follows:

$$lg(TG(t, C)) = (U + v(U), U + v(U) + w(U), \hat{s}, \hat{t}, \hat{l})$$

where (1) each edge $x \in U$ points from $v(x)$ to x and is unlabeled, i.e. $\hat{s}(x) = v(x)$, $\hat{t}(x) = x$, $\hat{l}(x) = *$, (2) each edge $v(x) \in v(U)$ is an x-labeled loop at the node $v(x)$, i.e. $\hat{s}(v(x)) = \hat{t}(v(x)) = v(x)$, $\hat{l}(v(x)) = x$, and (3) each edge $w(x) \in w(U)$ points from $v(x)$ to the node $v(t)$ (representing the target line of the gate) and is labeled by $\oplus$ if $x = t$ and by $\bullet$ otherwise, i.e. $\hat{s}(w(x)) = v(x)$, $\hat{t}(w(x)) = v(t)$, $\hat{l}(w(x)) = $ if $x = t$ then $\oplus$ else $\bullet$ for all $x \in U$. (We assume that

$*, \oplus, \bullet \notin ID$.) While the U-edges establish the lines consisting of single edges, the $v(U)$-edges allow access to the lines and the $w(U)$-edges identify the gate by separating the target line from the control lines.

Fig. 3 shows the lines graphs of the Toffoli gates $TG(2, \{4\})$, $TG(1, \{3, 4\})$ and $TG(4, \{1\})$.

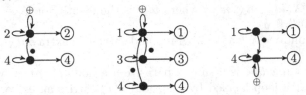

Fig. 3. The lines graphs of three Toffoli gates

Proposition 1.   *1. Let $LG = (LG_X)_{X \subseteq ID}$ be a family of sets of lines graphs LG_X for each $X \subseteq ID$. Let $R(LG)$ be the set of rules of the form*

$$r(lg) = (U^{id} \supseteq U^{\emptyset} \subseteq lg) \text{ for } lg \in LG_U, \ U \subseteq ID.$$

Let $X \subseteq ID$. Then all graphs derived from X^{id} by applying rules in $R(LG)$ are lines graphs over X.

2. Let for $i = 1, 2$, lg_i be lines graphs over $X_i \subseteq ID$ and $X_1 \cap X_2 = \emptyset$. Then the disjoint union $lg_1 + lg_2$ is a lines graph over $X_1 \cup X_2$.

Proof. Point 1 is easily shown by induction on the lengths of derivations. Point 2 is obvious.

Remark 1. The lines graphs obtained by point 1 and 2 may be used for the derivation of further lines graphs.

Example 2. A possible family of sets of lines graphs to employ the first point of Proposition 1 is given by the lines graphs of Toffoli gates:

$$TGLG_U = \{lg(TG(t, C)) \mid U = C \cup \{t\}\} \text{ for } U \subseteq ID.$$

In particular, each $TGLG_U$ contains $\#U$ elements.

5   Graph Representation of Toffoli Circuits

The corresponding set of rules $R(TGLG)$ can be used to generate graph representations of Toffoli circuits with the set X of lines. To achieve this, $out(X^{id})$ is used as initial graph and the x-loops are replaced by in_x-loops at the end. The corresponding transformation unit is given in Fig. 4.

According to the first point of Proposition 1, each terminal graph derived by *Toffoli graphs*(X) has, for each $x \in X$, a non-empty simple path from the node with the in_x-loop to the node with the out_x-loop, and these paths are pairwise disjoint. There are no further nodes, and all further edges are either $\oplus$-loops or $\bullet$-edges. If we assume an order on X, i.e. $X = \{x_1, \ldots, x_m\}$, then the terminal graphs can be drawn in the usual way of Toffoli circuits.

Toffoli graphs(X)
initial: $out(X^{id})$
rules: $R(TGLG)$
  $term(x)$: $x \hookrightarrow\!\bullet 1 \quad \longrightarrow \quad in_x \hookrightarrow\!\bullet\!\longrightarrow\!\bullet 1 \qquad$ for $x \in X$
terminal: $forbidden(\ \{\ x \hookrightarrow\!\bullet \mid x \in X\ \}\ )$

Fig. 4. The graph transformation unit *Toffoli graphs*(X)

Example 3. Starting from the initial graph $out(\{1,2,3,4\}^{id})$, one can apply the rules $r(lg(TG(4,\{1\}))), r(lg(TG(1,\{3,4\})))$ and $r(lg(TG(2,\{4\})))$ in this order followed by the four possible termination steps (see Fig. 5). This derives the Toffoli graph TG in Fig. 2.

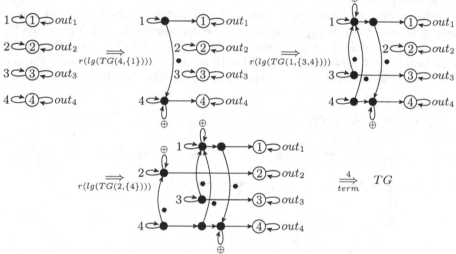

Fig. 5. A derivation of a Toffoli graph

The generation of graph representations of Toffoli circuits requires that the set of lines is fixed at the beginning. To become more flexible in this respect, one may use the empty graph $\emptyset$ as initial graph, add the rules $line(x) : \emptyset \longrightarrow x \hookrightarrow\!\widehat{x}$ for $x \in ID$ and forbid x-loops for arbitrary $x \in ID$ as terminal graph class expression. The resulting unit is called *Toffoli graphs* and allows one to create lines whenever they are needed.

Let der be a derivation $out(X^{id}) \overset{*}{\Longrightarrow} G$ in *Toffoli graphs*(X) with a terminal graph G, and let $r(lg(TG(t_1,C_1))) \cdots r(lg(TG(t_n,C_n)))$ be the sequence of rules in $R(TGLG)$ which are applied in der in this order. Then $TG(t_n,C_n) \cdots TG(t_1,C_1)$ is the Toffoli circuit represented by G and denoted by $TC(der)$. Conversely, let $TC = TG(t_n,C_n) \cdots TG(t_1,C_1)$ be a Toffoli circuit with the set of lines $X \subseteq ID$. Then starting with $out(X^{id})$ and applying the rules $r(lg(TG(t_j,C_j)))$ for $j = 1,\ldots,n$, and $term(x)$ for all $x \in X$, forms a derivation der_{TC} with $TC(der_{TC}) = TC$. The semantic relation between Toffoli circuits and their graph representation is studied in Section 6.

Sometimes it is helpful and convenient to allow constant inputs. This is easily obtained if we add the terminating rules $true(x)$: $x \leftharpoondown\!\!\bullet 1 \longrightarrow in_x \leftharpoondown\!\!\bullet \underset{1}{\longrightarrow}\bullet 1$ and $false(x)$: $x \leftharpoondown\!\!\bullet 1 \longrightarrow in_x \leftharpoondown\!\!\bullet \underset{0}{\longrightarrow}\bullet 1$ for $x \in ID$ to the unit *Toffoli graphs*. The resulting unit is called *Toffoli graphs with constants*.

6   Evaluation

As each Toffoli gate $TG(t, C)$ with $t \in X$ and $C \subseteq X - \{t\}$ defines a reversible function $f_{t,C}$ on $\mathbb{B}^X$, a Toffoli circuit $TC = TG(t_n, C_n) \cdots TG(t_1, C_1)$ computes the sequential composition $f_{TC} = f_{t_1,C_1} \circ \cdots \circ f_{t_n,C_n}$. As $f_{t,C}$ for each Toffoli gate $TG(t, C)$ is inverse to itself, $f_{t_n,C_n} \circ \cdots \circ f_{t_1,C_1}$ is inverse to f_{TC} so that f_{TC} is reversible.

In this section, we introduce the evaluation of Toffoli graphs. Let the set of terminal graphs derived in *Toffoli graphs*(X) be denoted by $L(Toffoli\ graphs(X))$. Let an assignment $a: X \to \mathbb{B}$ be represented by the graph $a^\bullet$ being X^{id} with an extra loop at each node $x \in X$ labeled with $a(x)$. Let $assign(X)$ denote the set of all these graphs. Then the evaluation can be specified by the unit in Fig. 6.

evaluation
initial: $L(Toffoli\ graphs(X)) + assign(X)$
rules:

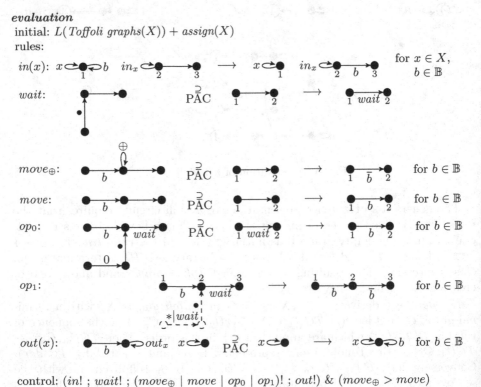

control: $(in!\ ;\ wait!\ ;\ (move_\oplus\ |\ move\ |\ op_0\ |\ op_1)!\ ;\ out!)\ \&\ (move_\oplus > move)$
$\&\ (op_0 > op_1)$

Fig. 6. The graph transformation unit *evaluation*

Initially, one has a Toffoli graph G and an assignment graph $a^{\bullet}$ for $a\colon X \to \mathbb{B}$. The rule $in(x)$ transfers $a(x)$ to the first edge of the x-line for all $x \in X$. The rule $wait$ labels all line edges leaving a controlled node with $wait$. If a line edge is unlabeled, but its predecessor is labeled with a truth value, then the truth value is moved. It is negated in case of a $\oplus$-loop at its source due to the priority of $move_{\oplus}$ over $move$. If a line edge is labeled with $wait$ and its predecessor with a truth value, then it depends on the control lines whether op_0 or op_1 is applied. If one of the control lines carries 0, then the truth value is moved forward without change. As this rule application has higher priority, all control lines must carry 1 if op_1 is applicable. In this case, the truth value is moved forward and negated. Eventually, all line edges carry truth values. At the end, the truth values on the last line edges are transferred to the the x-looped nodes defining a resulting assignment $\hat{a}\colon X \to \mathbb{B}$. These derivations define a function $f_G\colon \mathbb{B}^X \to \mathbb{B}^X$ by $f_G(a) = \hat{a}$ for all $a\colon X \to \mathbb{B}$.

The evaluation of a Toffoli graph coincides with the reversible function computed by the corresponding Toffoli circuit as the following correctness result shows. The proof is omitted because of lack of space.

Theorem 1. Let $G \in L(\text{Toffoli graphs}(X))$ and der be a derivation of G. Let $TC(der)$ be the induced Toffoli circuit. Then $f_G = f_{TC(der)}$.

Example 4. To illustrate the evaluation of Toffoli graphs, Fig. 7 shows the beginning of an evaluation of the Toffoli graph TG from Example 3 with respect to the assignment $a(1) = a(3) = a(4) = 1$ and $a(2) = 0$. We show the states of evaluation after applying in four times and $wait$ three times, after applying op_1 twice and $move$ three times and after applying op_0 and $move$ once and out four times where only the resulting assignment is drawn explicitly while TG' refers to the final Toffoli graph with 0- and 1-line edges only.

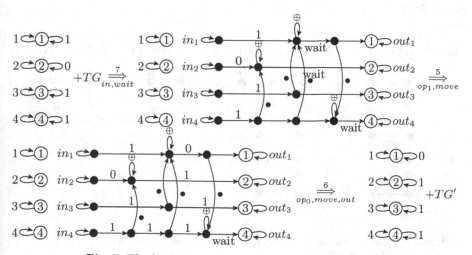

Fig. 7. The beginning of an evaluation of a Toffoli graph

7   From Reversible Functions to Toffoli Graphs

In this section, we transform reversible functions into Toffoli graphs and provide a graph-transformational analogy to one of the synthesis procedures for Toffoli circuits one encounters in the literature (cf., e.g., [12]).

Let $f\colon \mathbb{B}^X \to \mathbb{B}^X$ for some $X \subseteq ID$ be a reversible function. One may consider the elements of $\mathbb{B}^X$ as nodes and the pairs $a \mapsto f(a)$ as edges. Due to the bijectivity of f, each of these nodes has one incoming and one outgoing edge. Therefore, the graph consists of disjoint cycles. To be able to identify the nodes explicitly, each node $a \in \mathbb{B}^X$ gets a loop labeled with a. Moreover, the other edges get an auxiliary label for technical reasons.

Definition 1. Let $f\colon \mathbb{B}^X \to \mathbb{B}^X$ be a reversible function. Then f induces the graph $gr(f, \alpha) = (\mathbb{B}^X, \mathbb{B}^X \times \{rep, \alpha\}, s_f, t_f, l_f)$ with $s_f((a, rep)) = t_f((a, rep)) = l_f((a, rep)) = a$ and $s_f((a, \alpha)) = a$, $t_f((a, \alpha)) = f(a)$, $l_f((a, \alpha)) = \alpha$ for all $a \in \mathbb{B}^X$. The subgraph with the nodes $\mathbb{B}^X$ and the edges $\mathbb{B}^X \times \{rep\}$ is denoted by $gr(\mathbb{B}^X)$. The auxiliary graph $aux(\mathbb{B}^X)$ contains $gr(\mathbb{B}^X)$ as subgraph and, additionally, an unlabeled loop at each node. The set of graphs $gr(f, \alpha)$ for all reversible functions f is denoted by $\mathcal{G}(RF(X), \alpha)$.

To construct a Toffoli circuit that computes a reversible function $f\colon \mathbb{B}^X \to \mathbb{B}^X$, the basic idea in [12] is the following. Choose Toffoli gates $TC(t_1, C_1)$, $TC(t_2, C_2), \cdots, TC(t_i, C_i), \cdots$ starting with $i = 1$ until, for some $i = n$, $f_{t_1,C_1} \circ \cdots \circ f_{t_n,C_n} = f$. This is equivalent to $f_{t_{i+1},C_{i+1}} \circ \cdots \circ f_{t_n,C_n} = f_{t_i,C_i} \circ \cdots \circ f_{t_1,C_1} \circ f$ because $f_{t,C} \circ f_{t,C} = id_{\mathbb{B}^X}$ for all Toffoli gates $TG(t, C)$. Eventually, one may reach a circuit with $f_{t_n,C_n} \circ \cdots \circ f_{t_1,C_1} \circ f = id_{\mathbb{B}^X}$ such that $f_{t_1,C_1} \circ \cdots \circ f_{t_n,C_n} = f$. The same idea can be employed on the graph representation (see Fig. 8).

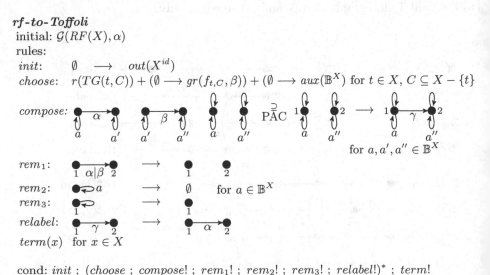

rf-to-Toffoli
initial: $\mathcal{G}(RF(X), \alpha)$
rules:

init: $\emptyset \longrightarrow out(X^{id})$

choose: $r(TG(t, C)) + (\emptyset \longrightarrow gr(f_{t,C}, \beta)) + (\emptyset \longrightarrow aux(\mathbb{B}^X))$ for $t \in X,\ C \subseteq X - \{t\}$

term(x) for $x \in X$

cond: *init* ; (*choose* ; *compose*! ; *rem$_1$*! ; *rem$_2$*! ; *rem$_3$*! ; *relabel*!)* ; *term*!
terminal: *forbidden(* ●——α——● *)*

Fig. 8. The graph transformation unit *rf-to-Toffoli*

A derivation in *rf-to-Toffoli* starts in $gr(f, \alpha)$ for some reversible function $f\colon \mathbb{B}^X \to \mathbb{B}^X$. The application of *init* adds the initial graph of *Toffoli graphs*(X) disjointly. Then the following sequence of rule aplications is done: First, *choose* applies some $r(TG(t, C))$ to the Toffoli graph component and adds $gr(f_{t,C}, \beta)$ and $aux(\mathbb{B}^X)$ disjointly. Secondly, *compose* takes all α-edges from a to $f(a)$ and all β-edges from $f(a)$ to $f_{t,C}(f(a))$ for all $a \in \mathbb{B}^X$ and adds γ-edges from a to $f_{t,C}(f(a))$ to the auxiliary component. Thirdly, $gr(f, \alpha), gr(f_{t,C}, \beta)$ and the unlabeled loops of the auxiliary component are removed. Finally, all γ-edges are relabeled with α resulting in the graph $gr(f_{t,C} \circ f, \alpha)$. Therefore, this sequence of derivation steps can be iterated as long as one likes until the terminating rules are applied. The result is a disjoint union of a Toffoli graph G with lines X and some graph $gr(\hat{f}, \alpha)$. It is terminal if $\hat{f} = id_{\mathbb{B}^X}$ because only in this case all α-edges are loops and hence not forbidden. The transformation of reversible functions into Toffoli graphs is correct in the following sense.

Theorem 2. Let $f\colon \mathbb{B}^X \to \mathbb{B}^X$ be a reversible function. Let $gr(f, \alpha) \overset{*}{\Longrightarrow} G + gr(id_{\mathbb{B}^X}, \alpha)$ be a derivation in *rf-to-Toffoli*. Then $f = f_G$.

Proof. As discussed above, the derivation decomposes into a derivation $der - (out(X^{id})) \overset{*}{\Longrightarrow} G$ in *Toffoli graphs*(X) and a derivation $gr(f, \alpha) \overset{*}{\Longrightarrow} gr(id_{\mathbb{B}^X}, \alpha)$. In the *choose*-steps, the chosen gate rules $r(TG(t_1, C_1)), \cdots, r(TG(t_n, C_n))$ are accompanied by the rules $\emptyset \to gr(f_{t_1, C_1}, \beta), \cdots, \emptyset \to gr(f_{t_n, C_n}, \beta)$. Due to Theorem 1, we get $f_G = f_{TC(der)}$ for $TC(der) = TG(t_n, C_n) \cdots TG(t_1, C_1)$. Moreover, the above reasoning yields $gr(id_{\mathbb{B}^X}, \alpha) = gr(f_{t_n, C_n} \circ \cdots \circ f_{t_1, C_1} \circ f, \alpha)$ such that $f = f_{t_1, C_1} \circ \cdots \circ f_{t_n, C_n} = f_{TC(der)} = f_G$.

8   From Binary Decision Diagrams to Toffoli Circuits

Binary decision diagrams (BDDs) represent formulas in propositional logic as acyclic graphs, in which nodes represent variables or constants (cf. [3,20,5]). The left graph in Fig. 9 is a BDD for the formula $x_1 \wedge x_2$.

Due to the fact, that AND is irreversible, it needs to be embedded into a reversible function by adding new constant input lines where in case of AND one yields the result and the other can be considered as garbage. In Fig. 9 *in* and *out* indicate constant inputs and garbage outputs, respectively. The following synthesis models this kind of embedding in a general setting.

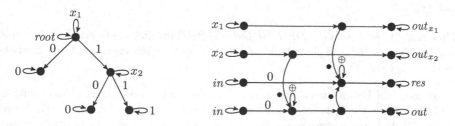

Fig. 9. A binary decision diagram and a Toffoli graph for $x_1 \wedge x_2$

For each variable assignment, there is a unique path in the corresponding BDD from the root to a leaf, which represents the result. For example, if $x_1 = 0$ and $x_2 = 1$, we get the path x_1 ⌐•———^{0}—→•⌐○0, so that the result is 0.

Since BDDs as well as Toffoli circuits can be seen as graphs, it is natural to transform them via graph transformation rules. In this first approach we do not allow shared nodes, i.e., all considered BDDs are binary trees.

In the following, the set of all BDDs is denoted by $\mathcal{B}$ and the set of all variables occurring in a BDD bdd by $Var(bdd)$. Each node with a variable loop is called a *var node* and each node with a b-loop ($b \in \mathbb{B}$) is called a *constant node*. In the standard definition of BDDs only leaves contain constants. The variable at a var node v is denoted by $var(v)$ and the value at a constant node v by $val(v)$. Moreover, for each var node v, let $low(v)$ be the child of v, to which a 0-edge points and let $high(v)$ be the other child of v.

We assume that the variables are taken from an underlying set $\mathcal{V}$ and that there is a bijection $\hat{\ }: \mathcal{V} \rightarrow \hat{\mathcal{V}}$ such that $\mathcal{V}$ and $\hat{\mathcal{V}}$ are disjoint. For technical simplicity, we allow lines of the form in ⌐•———$^{1|0}$—→• ··· •———→•⌐○out and in_y ⌐•———→• ··· •———→•⌐○w where $w \in \{out, res\}$ and y is some variable not occurring in any other line of the Toffoli graph. (Please note that this variant can be easily transformed into the Toffoli graphs in Section 5 by an adequate renaming of labels.)

The graph transformation unit in Fig. 10 specifies the conversion of BDDs to Toffoli graphs and is based on [21]. Each initial graph consists of a BDD, say bdd. According to the control condition, the rule $start$ is applied at first. It generates the output node res which delivers the result of evaluating the function $f_{bdd}: \mathbb{B}^{Var(bdd)} \rightarrow \mathbb{B}$ represented by bdd. Afterwards, the rule out is applied as long as possible. This generates a node x for each variable x in bdd such that x has an out_x-loop as well as an $\hat{x}$-loop. These nodes represent a subset of the output nodes of the generated Toffoli graph. Then the rules $r_1, \ldots, r_7$ are applied as long as possible. Since every rule application of $r_1, \ldots, r_7$ consumes one var node of bdd, these rules are applied a linear number of times. Finally, the rule in replaces each $\hat{x}$-loop by an in_x-loop. In each derived graph the number of nodes with an $\hat{x}$-loop is bounded by the number of variables in bdd. Hence, the unit terminates after a linear number of rule applications.

The graph on the right of Fig. 9 is the result of transforming the BDD on the left of the figure. In the following correctness theorem, we ignore the constant inputs and the garbage outputs, i.e., every Toffoli graph G induces a function $f: \mathbb{B}^Y \rightarrow \mathbb{B}^X$ where Y is the set of input variables and X is the set of output variables.

Theorem 3. $bdd \stackrel{*}{\Longrightarrow} G \in DER(BDD\text{-}to\text{-}Toffoli)$ implies (1) G is a Toffoli graph and (2) $f_{bdd} = pr_{res} \circ f_G$ where $pr_{res}(a) = a(res)$ for each $a: X \rightarrow \mathbb{B}$ with $\{res\} \cup Var(bdd) \subseteq X$.

Proof. Let bdd be an initial graph and let $bdd \stackrel{n}{\Longrightarrow} G_n$ be a derivation of length n which follows the control condition. Let $\{x_1, \ldots, x_m\} = Var(bdd)$. Then the first $m+1$ derivation steps comprise one application of $start$ and m applications of out, and for each $n > m + 1$ the graph G_n can be sketched as in Fig. 11(a),

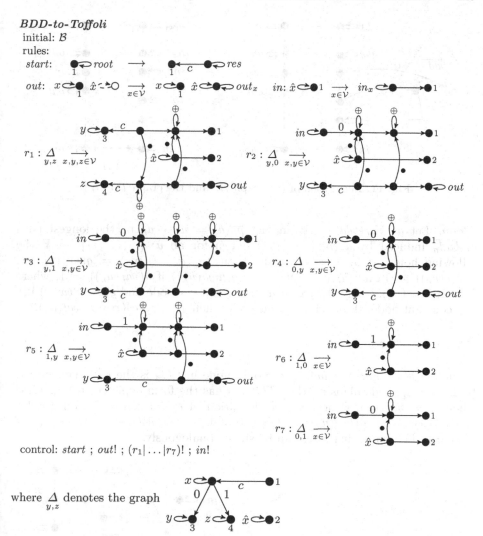

Fig. 10. The graph transformation unit *BDD-to-Toffoli*

where for $i = 1, \ldots, m$, $y_i \in \{\hat{x}_i, in_{x_i}\}$, $\overline{bdd_n}$ is the consumed part of *bdd*, and $v_1, \ldots, v_l$ are the roots of the remaining subtrees.

Let $TG(G_n)$ be the graph obtained from G_n by (1) applying rule *in* as long as possible, (2) deleting every item of *bdd*, apart from $v_1, \ldots, v_l$, and (3) replacing every *c*-edge e by an unlabeled opposite edge and an in_{v_i}-loop at the source v_i, for $i = 1, \ldots, l$ (see Fig. 11(b)).

Then by induction we get that $TG(G_n)$ is a Toffoli graph. Hence, every terminal graph is a Toffoli graph, because eventually, all *bdd*-items are removed and no $\hat{x}_i$-loop is left. This proves point 1 of the theorem. Point 2 follows from the following claim.

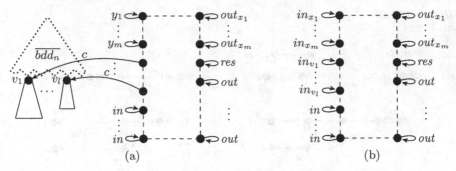

Fig. 11. Illustration of the graph G_n and the graph $TG(G_n)$

Claim. Let $a: Var(bdd) \to \mathbb{B}$ and let $\overline{p_n}(a) = u_1 \cdots u_k$ be the longest path in $\overline{bdd_n}$ induced by a $(n > m + 1)$. Then for all $\hat{a}: \{v_1, \ldots, v_l\} \to \mathbb{B}$ the following holds: If u_k is a var node, then $f_{G_n}(a \cup \hat{a})(res) = g(low(u_k))$ if $a(var(u_k)) = 0$ and $f_{G_n}(a \cup \hat{a})(res) = g(high(u_k))$ if $a(var(u_k)) = 1$, where for $v \in \{low(u_k), high(u_k)\}$ $g(v) = \hat{a}(v)$ if v is a var node and $g(v) = val(v)$ if v is a constant node. If u_k is a constant node, then $f_{G_n}(a \cup \hat{a})(res) = val(u_k)$.[1]

Proof of the claim.

Induction base. For $n = m + 2$, the only node in $\overline{bdd_n}$ is the root, i.e., $k = 1$. If the last applied rule is r_1, then $TG(G_n)$ has the form depicted in Fig. 12(a) where $x = var(u_k)$. Hence, by the definition of evaluation of Toffoli graphs, $f_{G_n}(a \cup \hat{a})(res) = \hat{a}(low(u_k))$ if $a(x) = 0$ and $f_{G_n}(a \cup \hat{a})(res) = \hat{a}(high(u_k))$ if $a(x) = 1$. The remaining cases can be shown analogously.

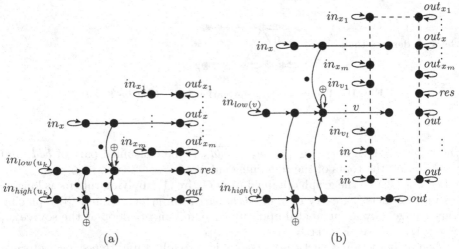

Fig. 12. Illustration of $TG(G_{m+2})$ and $TG(G_{n+1})$ after applying rule r_1

---

[1] $a \cup \hat{a} \in \mathbb{B}^{Var(bdd) \cup \{v_1, \ldots, v_l\}}$ with $(a \cup \hat{a})|_{Var(bdd)} = a$ and $(a \cup \hat{a})|_{\{v_1, \ldots, v_l\}} = \hat{a}$.

Induction step. Let r be the last rule applied in the derivation of G_{n+1}. If $r = in$, then the claim holds by induction hypothesis. Otherwise $r \in \{r_1, \ldots, r_7\}$. Let v be the node of bdd removed in the last step of the derivation and let $var(v) = x$. Then we can evaluate the new gates inserted by r obtaining an input a' for $TG(G_n)$ defined by $a'(y) = a(y)$ for $y \in Var(bdd)$ and $a'(y) = \hat{a}(y)$ for $y \in \{v_1, \ldots, v_l\} \setminus \{v\}$, and $a'(v)$ depends on r. Moreover, $\overline{p_n}(a'|_{Var(bdd)}) = u_1 \cdots u_{k-1}$ and $f_{G_{n+1}}(a \cup \hat{a})(res) = f_{G_n}(a')(res)$. If $v \neq u_k$, then by induction hypothesis we are done. Otherwise, let $r = r_1$. Then $TG(G_{n+1})$ can be depicted as in Fig. 12(b). If $v = low(u_{k-1})$, then $a'(var(u_{k-1})) = 0$ and by definition and induction hypothesis $f_{G_{n+1}}(a \cup \hat{a})(res) = f_{G_n}(a')(res) = a'(low(u_{k-1})) = a'(v) = \hat{a}(low(v))$ if $a(x) = 0$, and $f_{G_{n+1}}(a \cup \hat{a})(res) = \hat{a}(high(v))$, otherwise. The remaining cases can be shown analogously.

9 Conclusion

We have started to develop a graph-transformational framework for the modeling and analysis of reversible circuits. As the first steps, we have introduced lines graph representations of reversible circuits, generated, evaluated and synthesized them by graph transformation units and proved the correctness of evaluation and two synthesis variants. Further investigations will emphasize the significance of the approach:

1. Instead of multi-controlled Toffoli gates with single control lines, one may consider other kinds of gates like Peres gates, Fredkin gates, swap gates or Toffoli gates with multi target lines and multi polarity.
2. Another quite interesting topic will be to apply the graph-transformational approach to quantum circuits which are reversible circuits on vectors of qbits rather than bits.
3. The study of synthesis can be continued by including other variants and by considering efficiency issues in detail.
4. In the literature, one encounters quite a variety of optimization procedures with respect to various measurements. Graph transformation rules are expected to be suitable tools for local optimization.

Acknowledgment. We are greatful to the anonymous reviewers for their valuable comments.

References

1. Athas, W., Svensson, L.: Reversible logic issues in adiabatic CMOS. In: Proc. Workshop on Physics and Computation, PhysComp 1994, pp. 111–118 (1994)
2. Bennett, C.H.: Logical reversibility of computation. IBM Journal of Research and Development 17(6), 525–532 (1973)
3. Bryant, R.E.: Graph-based algorithms for boolean function manipulation. IEEE Trans. Computers 35(8), 677–691 (1986)
4. Cuykendall, R., Andersen, D.R.: Reversible optical computing circuits (1987)

5. Drechsler, R., Sieling, D.: Binary decision diagrams in theory and practice. International Journal on Software Tools for Technology Transfer 3(2), 112–136 (2001)
6. Fredkin, E.F., Toffoli, T.: Conservative logic. International Journal of Theoretical Physics 21(3/4), 219–253 (1982)
7. Houri, S., Valentian, A., Fanet, H.: Comparing CMOS-based and NEMS-based adiabatic logic circuits. In: Dueck, G.W., Miller, D.M. (eds.) RC 2013. LNCS, vol. 7948, pp. 36–45. Springer, Heidelberg (2013)
8. Knill, E., Laflamme, R., Milburn, G.: A scheme for efficient quantum computation with linear optics. Nature 409, 46–52 (2001)
9. Kreowski, H.-J., Kuske, S., Rozenberg, G.: Graph transformation units – an overview. In: Degano, P., De Nicola, R., Meseguer, J. (eds.) Montanari Festschrift. LNCS, vol. 5065, pp. 57–75. Springer, Heidelberg (2008)
10. Merkle, R.C.: Reversible electronic logic using switches. Nanotechnology 4(1), 21 (1993)
11. Miller, D.M., Wille, R., Dueck, G.: Synthesizing reversible circuits for irreversible functions. In: Núñez, A., Carballo, P.P. (eds.) Proc. 12th Euromicro Conference on Digital System Design, Architectures, Methods and Tools, DSD 2009, pp. 749–756. IEEE (2009)
12. Miller, D.M., Maslov, D., Dueck, G.W.: A transformation based algorithm for reversible logic synthesis. In: Proc. of the 40th Design Automation Conference, DAC 2003, pp. 318–323. ACM (2003)
13. Nielsen, M., Chuang, I.: Quantum Computation and Quantum Information. Cambridge Univ. Press (2000)
14. Peres, A.: Reversible logic and quantum computers. Phys. Rev. A 32, 3266–3276 (1985)
15. Ren, J., Semenov, V., Polyakov, Y., Averin, D., Tsai, J.-S.: Progress towards reversible computing with nSQUID arrays. IEEE Transactions on Applied Superconductivity 19(3), 961–967 (2009)
16. Rozenberg, G. (ed.): Handbook of Graph Grammars and Computing by Graph Transformation. Foundations, vol. 1. World Scientific, Singapore (1997)
17. Thapliyal, H., Srinivas, M.B.: The need of DNA computing: reversible designs of adders and multipliers using Fredkin gate (2005)
18. Toffoli, T.: Reversible computing. In: de Bakker, J., van Leeuwen, J. (eds.) ICALP 1980. LNCS, vol. 85, pp. 632–644. Springer, Heidelberg (1980)
19. Vos, A.D.: Reversible Computing - Fundamentals, Quantum Computing, and Applications. Wiley (2010)
20. Wegener, I.: Branching Programs and Binary Decision Diagrams. SIAM (2000)
21. Wille, R., Drechsler, R.: BDD-based synthesis of reversible logic for large functions. In: Design Automation Conference, pp. 270–275. ACM (2009)

Towards Process Mining
with Graph Transformation Systems

H.J. Sander Bruggink

Universität Duisburg-Essen, Germany
sander.bruggink@uni-due.de

Abstract. This paper is about process mining with graph transformation systems (GTSs). Given a set of observed transition sequences, the goal is to find a GTS – that is a finite set of graph transformation rules – that models these transition sequences as well as possible. In this paper the focus is on real-word processes such as business processes or (human) problem solving strategies, with the aim of better understanding such processes. The observed behaviour is not assumed to be either complete or error-free and the given model is expected to generalize the observed behaviour and be robust to erroneous input. The paper presents some basic algorithms that obtain GTSs from observed transition sequences and gives a method to compare the resulting GTSs.

1 Introduction

Process mining [14] is the automatic retrieval of process models from data that is obtained from observing real-world processes. Conventionally, the input data consists of process logs of traces consisting of named events and the output model is given in the form of a flow-centric model, such as a Petri-net, which models which events causally depend on each other and which events can occur concurrently. The state of the system at various times within a process is considered a black box: it is not modelled what components the system consist of, what attributes these components have and what relations they have to each other.

In many applications, however, we are not only interested in the events and their causal relationships, but also in the state of the system and how it changes in the course of the process. In order to accurately model such processes, a modelling formalism is required which can model the state of a system as well as the state changes. Graph transformation is such a formalism.

This paper is about process mining with graph transformation systems (GTSs). The goal is as follows: A transition sequence (trace) consists of a sequence of transitions. Given a set of such traces as input (actually, the traces are preprocessed so that isomorphic graphs are identified, so the input is really a transition system of graphs), we want to find a GTS (a set of graph transformation rules) that models the given input. A model is considered good if it fits the provided input relatively well and is sufficiently simple.

As is in general the case in process mining, there is usually not a single "best model" for a given input. Some models might be very good in one aspect

H. Giese and B. König (Eds.): ICGT 2014, LNCS 8571, pp. 253–268, 2014.

(for example, size) and less good in another (for example, fitting the input). Also, the input does not in general contain every allowed transition, so the returned models are required to generalize from the input. Some models might fit the input too tightly (overfitting), while others generalize too much (underfitting). Also, the input might contain errors, so process mining algorithms should be robust to noise. See also [3].

The contribution of this paper is that it defines the above goal of process mining for GTSs, gives a method to compare the quality of the generated models, and finally, presents a number of algorithms for process mining with GTSs. The first two items are heavily influenced by related work in the process mining community. Three of the algorithms are basic ones which are mainly used as starting points for the others. The latter two algorithms are a genetic algorithm and an algorithm which, starting from given GTS, greedily mutates the system to a better one. Interestingly, although the input consists of only *positive* information (transitions observed in practice), the latter two algorithms are also able to induce *negative* application conditions from it (this is also the case in [15]).

Applications. Applications of this work fall in two areas. First, process mining in general is often used to find models of business processes. The input then consists of (automatically generated) process logs of traces that are actually observed in practice. The found models can be used to gain insight into how the processes are carried out in practice, to find bottlenecks and suggest operational improvement, or to compare actual to prescribed behaviour. Process mining with GTSs is useful to find models of business processes where state information is recorded rather than named events.

Second, process mining for GTS can be used to obtain models of (human) behaviour, for example of problem solving strategies. This is useful for, among others, man-machine interaction, decision support systems and interactive learning software. A model of expert behaviour in a problem-solving exercise, for example, can be used to generate useful, online hints for novices. For these kinds of models, named events often do not suffice, because actions depend more on the current state of the solution rather than on the way the state was reached.

Related work. Apart from related work about process mining, related work exists in the model transformation community, where *model transformation by example* was proposed as a method of specifying model transformations [15,16,2,13,5].

The focus in model transformation by example is quite different from process mining, however. In model transformation by example, the focus is on engineering: there is a specific target model that the user of the system wants to generate. In process mining the target model cannot be uniquely determined. Generating a model is supposed to provide insight in observed processes.

Also, in model transformation, the meta-models of the source and target models are typically disjoint. For example, a UML class diagram is translated to a relational database specification. This means that nodes and edges that are created by a rule application are not part of the match in later rule applications. In contrast, in this paper we consider general graph transformation.

In [1] models of software systems are generated by observing the behaviour of the system. Like with model transformation by example, the resulting model is very specific: it must model all observed behaviour. Negative application conditions are only obtained from *negative* observations: a set of transitions that were prohibited to occur is given as input to the algorithm.

2   Preliminaries

In order to fix terminology and notation, we start by defining some well-known notions. When $f: A \to B$ is partial function we write $f(x) = \bot$ if there is no $y \in B$ such that $f(x) = y$. We call A the *domain* of f (written $Dom(f)$) and B its *codomain* (written $Cod(f)$). The subset of A for which f is defined is called the *domain of definition* of f, while the *image* of f is the set $\{y \in B \mid \exists x \in A : f(x) = y\}$. For a set A, A^* is the set of finite sequences of elements of A. For a (total) function $f: A \to B$, we define $f^*: A^* \to B^*$ as $f^*(a_1 \cdots a_n) = f(a_1) \cdots f(a_n)$.

In this paper we work with hypergraphs. Hypergraph are a generalization of directed (multi)graphs where each edge can be incident to any number of nodes.

Definition 1 (Hypergraph). *Let a set of labels Σ be given. A hypergraph over Σ is a structure $G = \langle V, E, att, lab \rangle$, where V is a finite set of vertices or nodes, E is a finite set of edges such that $V \cap E = \varnothing$, $att: E \to V^*$ is a function which assigns a sequence of incident nodes to each edge, and $lab: E \to \Sigma$ is a function which assigns a label to each edge.*

In the following, we will refer to hypergraphs simply by the word *graph*. As a notational convention, we will denote, for a given graph G, its components by V_G, E_G, att_G and lab_G, unless otherwise indicated.

Graphically, graphs are represented as follows. Black dots (•) denote nodes. Optionally, the name of the node is written next to it. Edges are represented by rectangles with the label written inside, connected to the incident nodes. If necessary, the position of the connected node within the $att(e)$ sequence is written on the connecting line. The most common kind of edges are binary edges, that is edges e such that $|att(e)| = 2$. For clarity, these edges will be denoted by arrows.

Definition 2 (Graph Morphism). *Let G, H be two graphs over some label set Σ. A (partial) graph morphism $\varphi: G \to H$ from G to H consists of two partial functions $\varphi_V: V_G \to V_H$ and $\varphi_E: E_G \to E_H$, such that for each edge $e \in E_G$ it holds that $att_H(\varphi_E(e)) = \varphi_V^*(att_G(e))$ and $lab_H(\varphi_E(e)) = lab_G(e)$.*

If φ is a graph morphism from G to H, G is called the domain *of φ (written $Dom(\varphi)$) and H is called the* codomain *of φ (written $Cod(\varphi)$).*

A graph morphism φ is total (injective, surjective) *when both φ_V and φ_E are total (injective, surjective) functions. It is an* isomorphism *if it is total, injective and surjective. Graphs G and H are isomorphic, written $G \simeq H$, if there exists an isomorphism $\varphi: G \to H$.*

If $\varphi: F \to G$ and $\psi: G \to H$ are graph morphisms, the composition of φ and ψ is denoted $\psi \circ \varphi$.

We will often drop the subscripts V, E and simply write φ instead of φ_V, φ_E. We will often use injective (not necessarily total) morphisms of the following form. An injective morphism $\varphi \colon G \to H$ is a *partial inclusion* if

$$\varphi_V(x) = \begin{cases} x & \text{if } x \in V_G \cap V_H \\ \bot & \text{otherwise} \end{cases} \quad \text{and} \quad \varphi_E(x) = \begin{cases} x & \text{if } x \in E_H \cap E_H \\ \bot & \text{otherwise.} \end{cases}$$

As the graph transformation formalism we employ the single-pushout approach to graph transformation with negative application conditions [7].

Definition 3 (Graph Transformation System).

(i) *A rule is a pair $\rho = \langle \varphi, N \rangle$, where φ is an injective (but not necessarily total) morphism $\varphi \colon L \to R$ (called the rule morphism) and $N = \{\nu_1, \ldots, \nu_p\}$ is a finite set of negative application conditions, where each negative application condition is a total injective morphism $\nu_i \colon L \to N_i$.*
Here, L is called the left-hand side of ρ, and R its right-hand side. Given a rule ρ, $Lhs(\rho)$ and $Rhs(\rho)$ denote the left-hand side and right-hand side of ρ, respectively.

(ii) *A graph transformation system (GTS) is a set of rules.*

Convention 1. As a convention, we will assume that the rule morphisms and negative application conditions of all rules are partial inclusions.

The context of a rule consists of those nodes and edges of the left-hand side for which the rule morphism is defined. Given the convention above, the context of rule $\rho \colon L \to R$ has $E_L \setminus E_R$ as edges and $V_L \setminus V_R$ as nodes.

Graphically, rules are represented as follows. The left-hand side and the right-hand side of the rule as drawn next to each other, with a $\Rightarrow$ symbol between them. Nodes of the left-hand side are assumed to be mapped to nodes of the right-hand side with the same name (nodes which do not have a name, or have a name that does not occur in the right-hand side, are assumed not be matched to anything). For clarity, edges are not given names. Unless otherwise indicated, it is assumed that edges of the left-hand side are mapped to edges of the right-hand side whenever possible.

Definition 4 (Transition). *A transition is an injective (but not necessarily total) morphism $t \colon G \to H$. Here, G is called the source of t, and H its target. Given a transition t, $Src(t)$ and $Tgt(t)$ denote its source and target, respectively.*

Transitions t and t' are called *isomorphic* if there exist isomorphisms $1_{\mathrm{src}} \colon Src(t) \to Src(t')$ and $1_{\mathrm{tgt}} \colon Tgt(t) \to Tgt(t')$ such that $t' \circ 1_{\mathrm{src}} = 1_{\mathrm{tgt}} \circ t$.

GTSs generate transitions in the following way:

Definition 5 (Graph Transformation). *Let $\rho = \langle \varphi, N \rangle$ be a rule and let $m \colon Lhs(\rho) \to G$ be total injective morphism from the left-hand side of ρ into some graph G (m is called the match).*

The rule ρ is applicable to the match m if there is no negative application condition $(\nu\colon Lhs(\rho) \to G') \in N$ such that there exists a total injective morphism $\nu'\colon G' \to G$ with $\nu' \circ \nu = m$.

If ρ is applicable to the match m, then the rule ρ and match m generate the transition t from G to H, written $t\colon G \to_{\rho,m} H$, if the diagram on the right is a pushout in the category of graphs and graph morphisms.

$$
\begin{array}{ccc}
L & \xrightarrow{\;\rho\;} & R \\
{\scriptstyle m}\big\downarrow & & \big\downarrow \\
G & \xrightarrow{\;t\;} & H
\end{array}
$$

Given a GTS $\mathcal{R}$ and a graph G, we denote by $\mathcal{R}(G)$ the set of transitions generated by $\mathcal{R}$ with source G, that is:

$$\mathcal{R}(G) = \{t \mid t\colon G \to_{\rho,m} H \text{ for some } \rho \in \mathcal{R} \text{ and match } m\colon Lhs(\rho) \to G\}.$$

The above definition of a graph transformation step $G \to_{\rho,m} H$ can be algorithmically described as follows. Let a rule $\rho = \langle \varphi\colon L \to R, N \rangle$ and a match $m\colon L \to G$ be given. The rule can be applied to the match if none of the negative application conditions is breached. If the rule is applicable, for all nodes v of L such that $\varphi(v)$ is undefined, the image $m(v)$ is removed from G, along with all edges incident with it. Similarly, for all edges e of L such $\varphi(e)$ is undefined, the image $m(e)$ is removed from G. Then the elements y of R that have no pre-image over φ are added and connected with the remaining elements, as specified by φ. This results in the graph H.

The input of the process mining algorithms will be a transition system, that is, a collection of transitions.

Definition 6 (Transition System). *A transition system is a tuple $\langle Q, Tr \rangle$, where Q is a set of graphs and Tr is a multiset[1] of transitions, such that for each $t \in Tr$, $Src(t) \in Q$ and $Tgt(t) \in Q$. Given a transition system $\mathcal{S} = \langle Q, Tr \rangle$ and a graph G, we denote by $\mathcal{S}(G)$ the multiset of all transitions with a source isomorphic to G, that is: $\mathcal{S}(G) = \{t \in Tr \mid Src(t) \simeq G\}$.*

3   Process Mining with Graph Transformation Systems

In this section we give a formal definition of the problem we are considering in this paper, and present a function which can measure how well graph transformation systems model the given input.

3.1   From Traces to Graph Transformation Systems

Informally, process mining is finding a model – in our case a graph transformation system – which models a set of input traces as well as possible. The traces are sequences of graphs plus information about what nodes and edges of a graph correspond to what nodes and edges in the next graph. In other words, in our setting a trace is a sequence of transitions. Note that the traces are already

---

[1] We take a multiset of transitions so that we can count how often a certain transition occurs in the input of a process mining algorithm.

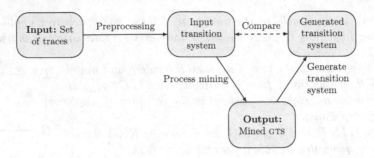

Fig. 1. Process mining with graph transformation systems

the result of a modelling step: from a collection of real-world courses of events, single steps have been identified, relevant information has been selected, and the system states between the steps have been modelled as a graph. Although an important part of process mining, this paper focuses on the automatable parts of process mining.

In a pre-processing step we construct from the traces a transition system by identifying isomorphic graphs. This transition system is given as input to the process mining algorithm, which produces a GTS as output. This GTS is a candidate model for the process under investigation.

Since the process mining algorithms do not possess domain-specific knowledge, a human observer is required to analyze the output model and decide if it is good. To do this, we generate a transition system corresponding to the output GTS. Typically, this transition system is infinite, so we have to cut off its generation at some point. We make sure, however, that for all graphs occurring in the input transition system, all outgoing transitions have been generated. The quality of the mining algorithm partially depends on how similar the generated transition system is to the input transition system.

Schematically, the process of process mining is depicted in Fig. 1.

3.2 Analyzing Models

The quality of a mined GTS as a model consists of two components [3]:

- *Conformance.* The GTS must conform to the given input. That is, the number of transitions that occur in the input transition system but are not generated by the GTS (*false negatives*) and the number of transitions that are generated by the GTS but are not in the input transition system (*false positives*) should be as small as possible. However, since the input can be incomplete or contain errors, the model should generalize the input, so the best solution models generally contain some false negatives and false positives. This component corresponds to the quality measures *replay fitness*, *precision* and *generalization* of [3].
- *Simplicity.* The GTS should provide insight in the process it is modelling, and so should not be too complex to be understood by a human reader.

The complexity of a GTS is mainly determined by its size (number of rules, number of nodes and edges and number of negative application conditions). This component corresponds to the quality measure *simplicity* of [3].

As written above, evaluating the output of the process mining algorithms is still a job for the human user because for a good analysis domain knowledge is required. However, for use within the algorithms we also need an automatic evaluation procedure for assigning a *fitness* to each mined GTS. For this, we define two quality measures: one which measures conformance and another which measures simplicity. In the end, the quality of the system is a weighted average of these two.

Measuring conformance. For measuring conformance, we compare the input transition system and the mined GTS. Since the mined GTS in general generates an infinite transition system, we restrict to the graphs occurring in the input. For each of those graphs, we compute a fitness value that depends on the number of outgoing transitions in the input system that are generated by the GTS, the number of outgoing transitions not generated by the GTS (the false negatives), and conversely, the transitions generated by the GTS which are not among the outgoing transitions in the input transition system (the false negatives).

The function which measures conformance has two parameters: p^+, the penalty for false positives, and p^-, the penalty for false negatives. Let $\mathcal{R}$ be the mined GTS and $\mathcal{S} = \langle Q, Tr \rangle$ be the input transition system. For each graph G in the input transition system, we calculate the conformance of G with respect to $\mathcal{R}$ as follows:

$$conf_\mathcal{S}(\mathcal{R}, G) = \frac{p^+ \cdot |FPos| + p^- \cdot |FNeg|}{|\mathcal{S}(G)|},$$

where

$$FPos = \{t \in \mathcal{R}(G) \mid \text{there is no } t' \in \mathcal{S}(G) \text{ such that } t \simeq t'\}$$
$$FNeg = \{t \in \mathcal{S}(G) \mid \text{there is no } t' \in \mathcal{R}(G) \text{ such that } t \simeq t'\}$$

Finally, the conformance of the entire GTS $\mathcal{R}$ is the mean of the conformances for all graphs in $\mathcal{S}$, that is

$$conf_\mathcal{S}(\mathcal{R}) = \frac{\sum_{G \in Q} conf_\mathcal{S}(\mathcal{R}, G)}{|Q|}.$$

Here, a lower number means a better conformance.

Measuring simplicity. For measuring how simple a GTS is, we take the (weighted) sum of the nodes and edges in the rules. Nodes in the right-hand side and the negative application conditions which correspond to nodes of the left-hand side are not counted. Of this number, we take a polynomial function (so that very large rules are overproportionally bad), and then take the sum of these values for all the rules.

The simplicity function has the following parameters:

p_E : edges weight d : degree of the polynomial

p_V : weight for number of nodes c : multiplier of the polynomial

Taking into account Convention 1, the node and edge set of a rule $\rho = \langle \varphi, N \rangle$, with $L = Lhs(\rho)$ and $R = Rhs(\rho)$, are defined by

$$V_\rho = V_L \cup V_R \cup \bigcup_{\nu \in N} V_{Cod(\nu)} \qquad E_\rho = E_L \cup E_R \cup \bigcup_{\nu \in N} E_{Cod(\nu)}$$

Now, the simplicity for a rule ρ is defined as $simp(\rho) = c \cdot (p_V \cdot |V_\rho| + p_E \cdot |E_\rho|)^d$. Finally, for a GTS $\mathcal{R}$, the simplicity is: $simp(\mathcal{R}) = \sum_{\rho \in \mathcal{R}} simp(\rho)$. Again, lower values are better than higher values.

Combined fitness. Given an input transition system $\mathcal{S}$ and a mined GTS $\mathcal{R}$, the combined fitness of $\mathcal{R}$ with respect to $\mathcal{S}$ is the weighted average of these two values. That is, given a parameter $0 < w < 1$: $fitness_\mathcal{S}(\mathcal{R}) = w \cdot conf_\mathcal{S}(\mathcal{R}) + (1 - w) \cdot simp(\mathcal{R})$.

Choosing the right values for the parameter of the conformance and simplicity functions is hard. They depend on personal preference and the application area. For the running example in this paper I chose the following defaults: $p^+ = 1$, $p^- = 20$, $p_E = 1$, $p_V = 1$, $d = 2.5$, $m = 0.01$ and $w = 0.7$.

4   Algorithms for Process Mining

In this section I present some basic algorithms which can be used for process mining. All of the algorithms were implemented in a prototype graph transformation mining tool. Before describing the algorithms themselves, I present a running example which is used as a first demonstration of each algorithm.

Running example. To demonstrate the algorithms, we consider a maze game as running example. (This is an "expert modelling" example rather than a business process example.) Each maze has a start and a finish; additionally, the current position of the player is marked. An example of such a maze is presented in Fig. 2. The player may move to any adjacent position which is not blocked by a wall. The goal is to move from start to finish. If the player arrived at the finish position, he or she can exit the maze (this also counts as a move).

A maze is modelled as a graph as follows. Each position in the maze is represented by node. Each node is connected by a binary, C-labelled edge to all nodes which represent adjacent positions (this means that there will be a C-edge in two directions); if there is wall between the two positions, additionally, there is a binary, B-labelled edge between the corresponding nodes. Finally, the nodes corresponding to the starting location, finish location and the current player location, are marked with unary edges labelled by S, F and P, respectively. If the player has exited the maze, the P-labelled edge is not present. The graph model of the maze of Fig. 2 is given in Fig. 3.

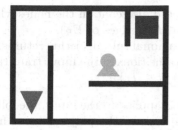

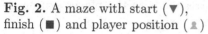

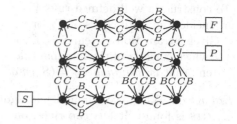

Fig. 2. A maze with start (▼), finish (■) and player position (♟)

Fig. 3. Graph representation of the maze of Fig. 2

The input given to the various algorithms will be a single trace corresponding to the player moving from start to finish in the shortest possible way. Note that in practical applications the input will consist of more traces; for demonstration purposes this single trace suffices because it contains several "move"-steps.

4.1 The Maximal Rule Algorithm

The maximal rule of a transition is the rule which has the transition itself as rule morphism and no negative application conditions. Such rule were previously considered in [6]. The maximal rule algorithm consists of the maximal rules of all transitions in the input transition system.

This algorithm leads to a large number of rules (one for each transition in the input). Also, each rule is typically quite large, because all of the context is present. On the other hand, the produced system models the input very accurately – in many cases the transition system of the mined graph transformation system is exactly equal to the input transition system. In fact, for practical purposes the maximal rule algorithm usually models the input system *too* accurately: the input is over-fitted. For instance, in the running example the resulting GTS only models the single trace available in the input. In the model it is not possible to divert from the optimal path by taking a wrong turn.

Still, the maximal rule algorithm is useful for the comparison with other algorithms, and for providing input to algorithms which optimize existing GTS (see Sect. 4.4 and Sect. 4.5).

4.2 The Minimal Rule Algorithm

The minimal rule of a transition consists of only those parts which are modified (deleted or added) by the transition. Such rules were previously considered in [1] and [6]. Given a transition $t: G \to H$, the minimal rule corresponding to t is constructed as follows (taking into account Convention 1): $minimal(t): L \to R$ is the smallest rule such that:

- $E_L = E_G \setminus E_H$ and $E_R = E_H \setminus E_G$;
- $V_L \supseteq (V_G \setminus V_H)$ and $V_R \supseteq (V_H \setminus V_G)$;
- att_L, att_R, lab_L and lab_R are the corresponding components of G and H restricted to their new domains.

(To construct a well-defined rule, V_L and V_R need also contain the nodes which are incident in G to edges of E_L or incident in H to edges of E_R).

Given a transition system $\mathcal{S} = \langle Q, Tr \rangle$, the minimal rule algorithm returns the GTS which consists of the minimal rules of all transitions in the input transitions system, that is: $minimal(\mathcal{S}) = \langle S, \{minimal(t) \mid t \in Tr\} \rangle$.

Example 1. When the minimal rule algorithm is applied to the input, the following GTS is found. It has two rules; one corresponds to the player moving from one position to another, while the other corresponds to the player exiting the maze.

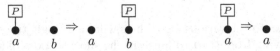

Since there is no context given, the model allows the player to move from any location to any other location in the maze, and to exit the maze from an arbitrary location. Needless to say, this model is not very good.

4.3 The Context Algorithm

The *context algorithm* tries to augment the minimal rule with context edges, that is, edges which occur both in the left-hand side and in the right-hand side of the rule, and thus act as conditions on the context which must hold before the rule can be applied. A similar algorithm was previously considered in [1] (there the resulting rule was called *maximal rule*). The algorithm works as follows:

1. Partition the set of transitions of the input into subsets of which all partitions have the same minimal rule.
2. For each of the partitions, add context to the minimal rule which is present in all transitions of the partition.

To find the common context of transitions, consider two transitions t_1 and t_2 which have the same minimal rule ρ (if there are more than two transitions, the procedure for two transitions is repeated to find context that is common to all transitions). The situation is as depicted on the 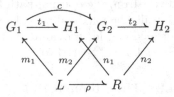 right. Common context is represented by a partial morphism $c\colon G_1 \to G_2$ such that the diagram commutes. The rule with additional context is now constructed by taking the subgraph of G_2 which contains nodes and edges in the image of c as left-hand side L', and the subgraph of H_2 which contains nodes and edges in the image of $c\,; t_2$ ("the context") or in the image of n_2 ("nodes and edges created by the rule") as the right-hand side R'.

It remains to describe how the context morphism c is found. For performance reasons, a greedy matching algorithm was implemented. The algorithm will not find a "maximal" context morphism in all cases, but yields very satisfying results in practice.

Algorithm 1. Get a context morphism

```
 1: input: transitions t₁, t₂, where tᵢ: Gᵢ → Hᵢ, with the same minimal rule ρ: L → R
 2:          injective total morphisms m₁: L → G₁ and m₂: L → G₂
 3: output: partial morphism c: G₁ → G₂ such that c ∘ m₁ = m₂
 4: begin
 5:     c ← empty morphism from G₁ to G₂
 6:     for x ∈ nodes and edges of L do
 7:         extend c by m₁(x) ↦ m₂(x)
 8:     end for
 9:     E₁ ← list of edges of G₁ that are incident to a node in the image of m₁
10:     E₂ ← list of edges of G₂ that are incident to a node in the image of m₂
11:     for e₁ ∈ E₁ do
12:         select an edge e₂ ∈ E₂ such that e₁ can be mapped to e₂ in c
13:         if such an e₂ exists then
14:             extend c by e₁ ↦ e₂ (also mapping incident nodes if necessary)
15:             append to E₁ all edges adjacent to e₁ not already in E₁
16:             append to E₂ all edges adjacent to e₂ not already in E₂
17:         end if
18:     end for
19:     return c
20: end
```

The algorithm is presented in Algorithm 1. In Lines 5–8 the context morphism c is initialized in such a way that $i_2 = i_1\,;c$. When we extend c by adding mappings, this property will still hold. After the initialization, we try to extend c. We keep track of two lists of edges, E_1 and E_2, which are the edges of G_1 and G_2 we try to add to the context. For each of the edges e_1 of E_1 we try to find an edge of E_2 that e_1 can be mapped to. The edge e_1 can be mapped to an edge e_2 if e_1 and e_2 have the same label and arity, no other edge has been mapped to e_2 yet, and all nodes incident to e_1 are either not mapped to any node yet, or mapped to the corresponding node of e_2. If such an e_2 can be found, we perform the mapping; otherwise $c(e_1)$ is left undefined.

Since it is a greedy algorithm, it will not always find the "best" context morphism. In particular, if some edge e_1 can be mapped to more than one candidate edge e_2, the algorithm may pick a sub-optimal one. In practice, however, the algorithm seems to perform quite well.

Example 2. For the running example, the context algorithm returns a graph transformation system which consists of two rules. The rule corresponding to the player moving is displayed on the right. The algorithm has discovered that the player can only move to adjacent squares. The nodes c and d and their

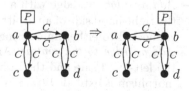

incident edges seem to be superfluous, but they do represent context which is there in every transition of the input. Still, the model allows the player to move through a wall.

The other rule corresponds to the player exiting the maze. Since there is only a single exiting transition in the input, this rule will be isomorphic to that transition (in other words, it is the maximal rule).

In the example, the second rule is very large because it models a transition which occurs only once, so the entire transition consists of "common context". This rule ruins the GTS's size and therefore its fitness. To prevent this from happening, I also implemented a variant of Algorithm 1 in which lines 15 and 16 are omitted. Because of this change, the context will only contain edges that are incident to nodes in the minimal rule.

4.4   The Genetic Algorithm

Genetic algorithms are a class of algorithms that mimic the process of biological evolution. In the latter half of the 20th century genetic algorithms evolved from mere computer simulations of biological systems to general optimization procedures used in a wide variety of application domains. An introduction to genetic algorithms can be found in [12].

A genetic algorithm is an iterative algorithm. In each iteration (called *generation*) there is a set of possible problem solutions (called the *population*). A *fitness function* which measures how well the individuals perform at solving the problem is supplied. The next generation is generated from individuals which are either taken from the current generation unmodified, are individuals from the current generation which are mutated in some way, or are formed by combining two individuals from the current generation (crossover). Individuals with high fitness have a better chance of producing offspring for the next generation (*survival of the fittest*). A genetic algorithm terminates after a fixed number of generations, or when the fitness does not improve anymore.

Genetic algorithms were previously used for process mining in [4] and for model transformation by example in [5]. I implemented a genetic algorithm to mine a GTS from an input transition system. In this case, the solutions are GTSs (in a sense, we mutate programs, which is also called *genetic programming* [10]). The fitness function is as described in Sect. 3.2. The following mutations are implemented:

- *Add a rule.* A randomly generated rule is added to the GTS.
- *Remove a rule.* A rule of the GTS is randomly chosen and removed from the system.
- *Add an edge.* An edge with a random label is added to the left-hand side or the right-hand side of an arbitrary rule. Each tentacle of the edge is randomly connected to either an existing node of the graph or a newly created one.
- *Add an edge to the context.* An edge with a random label is added to both the left-hand side and the right-hand side of an arbitrary rule, and the rule morphism is extended to map the new edge in the left-hand side to the new edge in the right-hand side.
- *Remove an edge.* A random rule from the GTS is chosen and in that rule a random edge is removed from either the left-hand side or the right-hand side.

Algorithm 2. Genetic process mining algorithm

1: **parameters:** $N_m =$ number of mutations (default: 100)
2:                          $N_c =$ number of crossovers (default: 20)
3:                          $Cap =$ maximum population size (default: 100)
4: **input:** transition system S
5: **begin**
6:     *population* ← list of systems obtained from other algorithms
7:     **while** best fitness changed during the last 100 generations **do**
8:         remove duplicates from *population*
9:         throw away systems with lowest fitness until $|population| \leq Cap$
10:        **for** $k \in \{1 \ldots N_m\}$ **do**
11:            choose random $\mathcal{R}$ from *population*
12:            choose random mutation
13:            mutate $\mathcal{R}$ according to mutation
14:            add mutated $\mathcal{R}$ to *population*
15:        **end for**
16:        **for** $k \in \{1 \ldots N_c\}$ **do**
17:            choose random $\mathcal{R}_1, \mathcal{R}_2$ from *population*
18:            add crossover of $\mathcal{R}_1$ and $\mathcal{R}_2$ to population
19:        **end for**
20:     **end while**
21:     **return** best of *population*
22: **end**

- *Remove an edge from the context.* A random rule $\rho\colon L \to R$ from the GTS is chosen. From L a random edge, for which $\rho(x)$ is defined, is chosen. Finally, x is removed from L and $\rho(x)$ is removed from R.
- *Add a negative application condition.* A random rule $\rho\colon L \to R$ is chosen. A negative application condition, which is equal to L with a random edge added to it, is added to ρ.
- *Remove a negative application condition.* An arbitrary negative application condition from some rule of the graph transformation system is removed.

The following crossover is implemented:

- Given GTSs $\mathcal{R}$ and $\mathcal{S}$, partition the rules of both systems by minimal rule. The crossover generates a new GTS which consists, for all minimal rules ρ, either all rules $\mathcal{R}$ that have ρ as minimal rule, or all rules of $\mathcal{S}$ that have ρ as minimal rule.

The initial population consists of the GTSs which were generated by the maximal rule algorithm, the minimal rule algorithm and the context algorithm, together with a number of randomly generated systems. The specific variant of genetic algorithm that we used is presented in Algorithm 2.

Example 3. Since genetic algorithms are inherently non-deterministic, running the algorithm multiple times will not always generate the same mined GTS. A typical found solution is the following GTS, which consists of the following rules:

$$\rho_1 = \quad a \overset{\substack{P \\ C}}{\underset{B}{\bullet\longleftrightarrow\bullet}} b \quad \neq \quad a \overset{\substack{P \\ C}}{\bullet\dashrightarrow\bullet} b \quad \Rightarrow \quad a \overset{\substack{P \\ C}}{\bullet\dashrightarrow\bullet} b \qquad \rho_2 = \quad a \overset{P}{\underset{F}{\bullet}} \Rightarrow a \underset{F}{\bullet}$$

The first rule corresponds to the player moving from one location to another. The genetic algorithm has discovered that the player can only move to adjacent locations. The superfluous context is not present (since removing it yields a simpler GTS which generates the same transitions). Unlike the algorithms previously considered, the genetic algorithm has also generated a negative application condition which represents the fact that the player cannot go through walls.

The second rule corresponds to the player exiting the maze. As for the moving rule, superfluous context has been removed.

4.5   The Greedy Mutation Algorithm

The GTS found by the genetic algorithm for our running example was quite good. However, the algorithm has quite bad running times. Therefore, I also tried a greedy algorithm, which only stores the graph transformation system which currently has the best fitness, then tries to mutate this system with an random mutation from the genetic algorithm, and discards the result if the fitness did not increase because of the mutation. If the fitness didn't improve for a certain amount of iterations, the algorithm terminates and returns the current best system.

For the running example, the results of this algorithm are as good as the results of the genetic algorithm. The algorithm is much faster, however (see Table 1). Still, as expected, it does sometimes get stuck in local optima with very low fitness.

4.6   Experimental Results

The running example used in this paper is of course quite simple. In order to test how well the algorithms scale to larger input I generated a number of random traces (from some given GTS) and ran the algorithm on them. Each trace consists of 20 transitions. The results are listed in Table 1. From the table it is clear that

Table 1. Run times in seconds of the algorithms with various examples; algorithms were terminated after 5 minutes ("–" means time-out)

Example	Run time (s)				
	Max. rule	Min. rule	Context	Gen. alg.	Greedy
Maze example	0.015	0.010	0.021	50	2
Random (10 traces)	0.031	0.017	0.031	201	8
Random (50 traces)	0.091	0.082	0.083	–	27
Random (100 traces)	0.202	0.175	0.223	–	66
Random (200 traces)	0.426	0.285	0.398	–	77
Random (500 traces)	1	0.370	0.575	–	184

all algorithms, except the genetic algorithm, scale well to larger input sizes. The genetic algorithm times out in all but trivial examples and requires more study to asses scalability (I conjecture its running time is linear with respect to the total number of transitions in the input, albeit with a large constant).

5 Conclusion and Further Research

This paper introduces process mining with GTSs. Five algorithms were presented, three basic ones and two more elaborate ones: a genetic algorithm and a greedy algorithm which mutates candidate graph transformation systems to improve their fitness; the latter two can induce negative application conditions from positive information only. The algorithms were tested on some small examples.

For future research, it is required to test the approach on case studies of larger size from various application areas (business processes, expert modelling, decision support, etc). To be able to support larger inputs, the existing algorithms must be optimized and other algorithms, such as *particle swarm optimization* (previously used in the related field of model transformation by example [9]) studied.

Currently, the genetic algorithm and the greedy mutation algorithm, have big problems with disjunctive application conditions.[2] For example, suppose that the running example would not have a single label C, but instead four labels R, D, L, U specifying the direction of the adjacent location, and the player can move along edges with all four of those labels. The minimal rule of all such transitions is equal, but there is no meaningful common context. Experiments with a mutation which duplicates a rule and modifies all copies in a different way were unsuccessful. This problem will be addressed in future research.

For modelling behaviour, stochastic [8] or probabilistic graph transformation systems [11] might be more suitable. For this, it is needed to generalize the algorithms to such models.

The applicability of graph transformation rules depends only on the graph on which they are applied. This means that all relevant information must be explicitly modelled. For example, when rule ρ_2 can only be applied after rule ρ_1 has been applied at least once, it must be explicitly modelled in the graph that ρ_1 has been applied. For process mining, this is undesirable, because one needs to consciously decide to include this information beforehand. This could be solved by hybrid algorithms, where conventional approaches are combined with the approach of this paper, which is also future research.

Acknowledgements. The author whishes to thank the anonymous reviewers for their valuable comments and remarks.

References

1. Alshanqiti, A., Heckel, R., Khan, T.A.: Learning minimal and maximal rules from observations of graph transformations. In: Proceedings of GT-VMT 2013 , vol. 58. ECEASST (2013)

---

[2] The other algorithms also have this problem, but "by design".

 2. Balogh, Z., Varró, D.: Model transformation by example using inductive logic programming. Software and System Modeling 8(3), 347–364 (2009)
 3. Buijs, J.C.A.M., van Dongen, B.F., van der Aalst, W.M.P.: On the role of fitness, precision, generalization and simplicity in process discovery. In: Meersman, R., et al. (eds.) OTM 2012, Part I. LNCS, vol. 7565, pp. 305–322. Springer, Heidelberg (2012)
 4. Alves de Medeiros, A.K., Weijters, A.J.M.M., van der Aalst, W.M.P.: Genetic process mining: An experimental evaluation. Data Mining and Knowledge Discovery 14(2), 245–304 (2007)
 5. Faunes, M., Sahraoui, H., Boukadoum, M.: Genetic-programming approach to learn model transformation rules from examples. In: Duddy, K., Kappel, G. (eds.) ICMT 2013. LNCS, vol. 7909, pp. 17–32. Springer, Heidelberg (2013)
 6. Große, S.: Process Mining mit Graph transformations systemen. Bachelor's thesis. Universität Duisburg-Essen (2012)
 7. Habel, A., Heckel, R., Taentzer, G.: Graph grammars with negative application conditions. Fundamenta Informaticae 26, 287–313 (1996)
 8. Heckel, R., Lajios, G., Menge, S.: Stochastic graph transformation systems. Fundamenta Informaticae 74 (2006)
 9. Kessentini, M., Sahraoui, H., Boukadoum, M.: Model transformation as an optimization problem. In: Czarnecki, K., Ober, I., Bruel, J.-M., Uhl, A., Völter, M. (eds.) MODELS 2008. LNCS, vol. 5301, pp. 159–173. Springer, Heidelberg (2008)
10. Koza, J.R.: Genetic Programming: On the Programming of Computers by Means of Natural Selection. MIT Press (1992)
11. Krause, C., Giese, H.: Probabilistic graph transformation systems. In: Ehrig, H., Engels, G., Kreowski, H.-J., Rozenberg, G. (eds.) ICGT 2012. LNCS, vol. 7562, pp. 311–325. Springer, Heidelberg (2012)
12. Mitchell, M.: An Introduction to Genetic Algorithms. MIT Press (1996)
13. Saada, H., Dolques, X., Huchard, M., Nebut, C., Sahraoui, H.: Generation of operational transformation rules from examples of model transformations. In: France, R.B., Kazmeier, J., Breu, R., Atkinson, C. (eds.) MODELS 2012. LNCS, vol. 7590, pp. 546–561. Springer, Heidelberg (2012)
14. van der Aalst, W.: Process Mining: Discovery, Conformance and Enhancement of Business Processes. Springer (2011)
15. Varró, D.: Model transformation by example. In: Nierstrasz, O., Whittle, J., Harel, D., Reggio, G. (eds.) MoDELS 2006. LNCS, vol. 4199, pp. 410–424. Springer, Heidelberg (2006)
16. Wimmer, M., Strommer, M., Kargl, H., Kramler, G.: Towards model transformation generation by example. In: Proceedings of the HICSS 2007. IEEE (2007)

Jerboa: A Graph Transformation Library for Topology-Based Geometric Modeling

Hakim Belhaouari[1], Agnès Arnould[1], Pascale Le Gall[2], and Thomas Bellet[1]

[1] University of Poitiers, Laboratory Xlim-SIC UMR CNRS 7262,
Bd Marie et Pierre Curie, BP 30179, 86962 Futuroscope Cedex, France
{hakim.belhaouari,agnes.arnould,thomas.bellet}@univ-poitiers.fr
[2] Laboratoire MAS, Ecole Centrale Paris
Grande Voie des Vignes, 92295 Chatenay-Malabry, France
pascale.legall@ecp.fr
https://sicforge.sp2mi.univ-poitiers.fr/jerboa

Abstract. Many software systems have to deal with the representation and the manipulation of geometric objects: video games, CGI movie effects, computer-aided design, computer simulations... All these softwares are usually implemented with ad-hoc geometric modelers. In the paper, we present a library, called Jerboa, that allows to generate new modelers dedicated to any application domains. Jerboa is a topological-based modeler: geometric objects are defined by a graph-based topological data structure and by an embedding that associates each topological element (vertex, edge, face, etc.) with relevant data as their geometric shape. Unlike other modelers, modeling operations are not implemented in a low-level programming language, but implemented as particular graph transformation rules so they can be graphically edited as simple and concise rules. Moreover, Jerboa's modeler editor is equipped with many static verification mechanisms that ensure that the generated modelers only handle consistent geometric objects.

Keywords: Topology-based geometric modeling, labelled graph transformation, rule-based modeler tool-set, static rule verification, generalized map.

1 Introduction

Context. Geometric modeling is the branch of computer science that focuses on modeling, manipulation and visualization of physical and virtual objects. Over the past decade, numerous generic tools have been developed to assist the conception of dedicated modelers (3D modeler for game design, CAD for mechanical design, and so on). Even if such tools usually offer a wide set of modeling operations ready to use, they may be not sufficient to answer new requirements that are outside their scope or involve complex transformations. Fig. 1 gives two examples of such specific modelers. Fig. 1(a) shows a modeler dedicated to architecture which provides a specific operation of extrusion. From a 2D map, the operation constructs a floor by distinguishing walls, doors and

H. Giese and B. König (Eds.): ICGT 2014, LNCS 8571, pp. 269–284, 2014.
© Springer International Publishing Switzerland 2014

windows. The right figure (Fig. 1(b)) presents a modeler devoted to the modeling
of plant growth (here a pear) and based on L-systems [TGM⁺09]. Both modelers
have in common that the consistency of manipulated objects is ensured by the use
of an underlying structure acting as a skeleton, called the topological structure.
Other pieces of informations attached to objects, as position, color, density, etc.
are of geometric or physical nature, and called *embedding*.

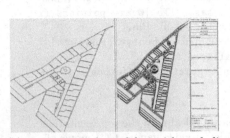 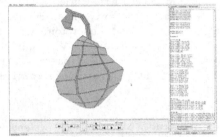

(a) Architectural modeler with a dedi-   (b) L-System modeler simulating plant
cated extrusion operation   growth

Fig. 1. Two geometric modelers with different application domains

Contribution. Graph transformations are already often used as key ingredients
of dedicated software applications [BH02]. We present a java library, called Jerboa, based on graph transformations and designed to assist the development of
new modelers whose objects and operations are specific to their application domains. All applications developed by means of the Jerboa library share the same
topological model, the one of *generalized maps* (*G-map* for short) [Lie91], that
can be viewed as a particular class of labelled graphs. Operations on geometric
objects are specified by the developer as *rules*. These rules fall within the general framework of graph transformations, more precisely of the DPO approach
on labelled graphs [EEPT06, HP02] and of rules with variables [Hof05]. The two
modelers glimpsed in Fig. 1 have been developed using the Jerboa library. While
classically data structures and operations of modelers are hand-coded in a low-
level programming language, modelers built over the Jerboa library inherit from
a predefined data structure implementing G-maps embedded (and being generic
with respect to object dimensions). Once data structures associated to the embedding are given, operations are then defined as simple and concise rules. The
workflow of a Jerboa-aided modeler development can be briefly described as follows: first, the user develops the data structures needed to represent embedding
data, then, using the JerboaModelerEditor, he/she implements operations that
will be used in the final application by writing one rule per operation. The rule
editor comes with some static analysis mechanisms that verify both topological and geometric consistencies of objects. Finally, the Jerboa rule application
engine ensures that rules are correctly applied. Note that rules manipulated in
the Jerboa library have been studied in some of our previous works. We first introduced in [PACLG08] special variables to generically denote topological cells.

As a proof of concept, [BPA+10] presented a first prototype of a rule-based modeler. But unlike Jerboa, this prototype was not generic with respect to embedding data, but designed with a single predefined embedding (position of 2D or 3D points). We introduced in [BALG11] rules built over embedding variables and provided with syntactic conditions related to geometric constraints. The new Jerboa library combines all these previous contributions with some additional efficiency concerns.

Related work. Rule-based languages have previously been used for twenty years in the context of geometric modeling. In particular, L-systems [PLH90] have been introduced to model plant growth. As L-systems are based on iterated applications of a set of rules until a stop condition is satisfied, they are suited to represent arborescent patterns, like flowers or trees [MP96, KBHK07]. Moreover, L-systems have already been used in a topological-based context in [TGM+09] to model internal structure of wood, or to model leaves growth. Inspired by L-systems, grammars were introduced to model buildings or to generate them from aerial pictures [VAB10] in order to be displayed in navigation applications. All these applications built over L-systems or grammars are defined by a limited set of specialized high level operations. To our knowledge, while the latter are often abstractly specified by means of some kinds of rules, they are mostly hand-coded in a classical way. On the contrary, our rule-based approach remains independent from the application domain and avoids any hand-coding, except the step of rule writing.

Outline of the article. In Section 2, we briefly present the topological model of generalized maps, and the way geometric elements are attached. We then introduce the first elements of JerboaModelerEditor. In Section 3, we explain by means of examples how an operation is defined as a rule in Jerboa and focus on graph transformation techniques involved in the rule application engine. We then present the different verification mechanisms that assist the design of correct rules. Lastly, before concluding the paper in Section 5, we discuss about the efficiency of the Jerboa library in Section 4.

2   Object Data Structure: Embedded Generalized Maps

2.1   Generalized Maps

As already stated in the introduction, we choose the topological model of generalized maps (or G-maps) [Lie91] because they provide an homogeneous way to represent objects of any dimension. This allows us to use rules for denoting operations defined on G-maps in an uniform way [PACLG08]. Moreover, the G-map model comes with consistency constraints characterising topological structures. Obviously, these constraints have to be maintained when operations are applied to build new objects from existing objects.

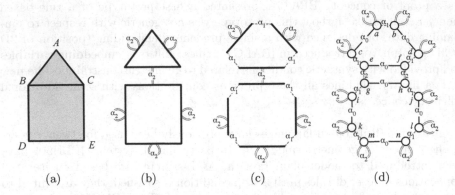

Fig. 2. Cell decomposition of a geometric 2D object

The representation of an object as a G-map comes intuitively from its decomposition into topological cells (vertices, edges, faces, volumes, etc.). For example, the 2D topological object of Fig. 2(a) can be decomposed into a 2-dimensional G-map. The object is first decomposed into faces on Fig. 2(b). These faces are *linked* along their common edge with an α_2 relation: the index 2 indicates that two cells of dimension 2 (faces) share an edge. In the same way, faces are split into edges connected with the α_1 relation on Fig 2(c). At last, edges are split into vertices by the α_0 relation to obtain the 2-G-map of Fig 2(d). Vertices obtained at the end of the process are the nodes of the G-map graph and the α_i relations become labelled arcs: for a 2-dimensional G-map, i belongs to $\{0, 1, 2\}$. According to the notation commonly used in geometric modeling, the labelling function is denoted α and an arc will be qualified as α_i-labelled. However, for simplicity purpose, when representing G-maps as particular graphs, arcs will be directly labelled by integer. Hence, for a dimension n, n-G-maps are particular labelled graphs such that arcs are labelled in the $[0, n]$ interval of integers.

In fact, G-maps are non-oriented graphs as illustrated in Fig. 2(d) : labelled non-oriented arcs represent a pair of reversed oriented arcs that are identically labelled. Notice that in order to be more readable, in all figures given in the sequel, we will use the α_i graphical codes introduced in Fig. 2(d) (simple line for α_0, dashed line for α_1 and double line for α_2) instead of placing a label name (α_i or i) near the corresponding arc. So, in the following, the way non-oriented arcs are drawn will implicitly indicate the arc label values.

Topological cells are not explicitly represented in G-maps but only implicitly defined as subgraphs. They can be computed using graph traversals defined by an originating node and by a given set of arc labels. For example, on Fig. 3(a), the 0-cell adjacent to e (or object vertex attached to the node e) is the subgraph which contains e, nodes reachable from the node e using arcs labelled by α_1 or α_2 (nodes c, e, g and i) and the arcs themselves. This subgraph is denoted by $G\langle\alpha_1\alpha_2\rangle(e)$, or simply $\langle\alpha_1\alpha_2\rangle(e)$ if the context (graph G) is obvious, and

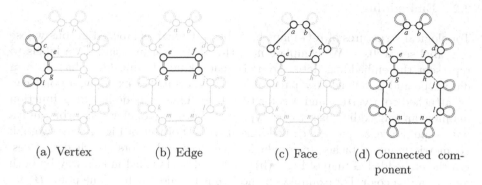

(a) Vertex   (b) Edge   (c) Face   (d) Connected com-
                                                              ponent

Fig. 3. Reconstruction of orbits adjacent to e

models the vertex B of Fig. 2(a). On Fig. 3(b), the 1-cell adjacent to e (or object edge attached to the node e) is the subgraph $G\langle\alpha_0\alpha_2\rangle(e)$ that contains the node e and all nodes that are reachable by using arcs labelled by α_0 or α_2 (nodes e, f, g and h) and the corresponding arcs. It represents the topological edge BC. Finally, on Fig. 3(c), the 2-cell adjacent to e (or object face attached to e) is the subgraph denoted by $\langle\alpha_0\alpha_1\rangle(e)$ and built from the node e and arcs labelled by α_0 or α_1 and represents the face ABC. In fact, topological cells (face, edge or vertex) are particular cases of *orbits* denoting subgraphs built from an originating node and a set of labels. We will use an ordered sequence of labels, encoded as a simple word o and placed in brackets $\langle\rangle$, to denote an orbit type $\langle o\rangle$. In addition to the orbit types already mentioned, we can mention the orbit $\langle\alpha_0\alpha_1\alpha_2\rangle(e)$ on Fig. 3(d) representing the whole connected component.

For a graph G with arc labels on $[0, n]$ to represent an n-G-map, it has to satisfy the following topological consistency constraints:

- *Non-orientation constraint*: G is non-oriented, i.e. for each arc e of G, there exists a reversed arc e' of G, such that the source of e' is target of e, target of e' is source of e, and e and e' have the same α label,
- *Adjacent arc constraint*: each node is the source of exactly $n + 1$ arcs respectively α_0 to α_n-labelled,
- *Cycle constraint*: for every i and j verifying $0 \leq i \leq i + 2 \leq j \leq n$, there exists a cycle labelled by $ijij$ starting from each node.

These constraints ensure that objects represented by G-maps are consistent manifolds [Lie91]. In particular, the cycle constraint ensures that in G-maps, two i-cells can only be adjacent along $(i - 1)$-cells. For instance, in the 2-G-map of Fig. 2(d), the $\alpha_0\alpha_2\alpha_0\alpha_2$ cycle constraint implies that faces are stuck along edges. Let us notice that thanks to loops (see α_2-loops in Fig. 2(d)), these three constraints also hold at the border of objects.

2.2   Embedding

Topological structures of n-G-maps have been defined as labelled graphs whose
arc label set is $[0, n]$. We complete now this definition by using node labels to
represent the embedding data. Actually, each kind of embedding has its own
data type and is defined on a particular type of orbit. For example, a point can
be attached to a vertex, and a color to a face. Thus, a node labelling function
defining an embedding has to be typed by both a (topological) orbit and a
data type. For example, the embedding of the 2D object of Fig. 4(a) is twofold:
geometric points attached to topological vertices and colors attached to faces.
On the embedded G-map of Fig. 4(b), each node is labelled in this way by both
a point and a color. For example, the node g is labelled by both the point B and
the light color.

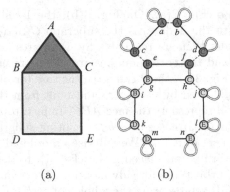

(a)   (b)

Fig. 4. Representation of 2D object with multiple embedding functions

Actually, for embedded G-maps, we characterize a node labelling function
as an *embedding operation* $\pi : \langle o \rangle \to \tau$ where π is an operation name, τ is
a data type and $\langle o \rangle$ is a domain given as an n-dimensional orbit type. For an
embedded G-map G equipped with an embedding $\pi : \langle o \rangle \to \tau$, we generically
denote $\lfloor \tau \rfloor$ the set of values associated to the data type τ and $g.\pi$ the value
associated to a node v by π. For the object of Fig. 4, the embedding operation
$point : \langle \alpha_1 \alpha_2 \rangle \to point\_2D$ associates the values A, B, C, D, E of type $point\_2D$,
suggesting some 2-dimensional coordinates, to the topological vertices. Similarly,
an embedding operation $color : \langle \alpha_0 \alpha_1 \rangle \to color\_RGB$ associates RGB coordinates to the topological faces.

For an embedding operation $\pi : \langle o \rangle \to \tau$, it is expected that an embedded
G-map G verifies that all nodes of any $\langle o \rangle$-orbit share the same value by π,
also called π-label or π-embedding. This defines the *embedding constraint* that
an embedded G-map G defined on a family $\Pi = (\pi : \langle o \rangle \to \tau)$ of embedding
operations has to satisfy [BALG11]: for each π in Π, each node is π-labelled on
$\lfloor \tau \rfloor$ and for all nodes v and w of G if v and w belong to the same orbit of type
$\langle o \rangle$, then v and w are labelled with the same π-label (i.e. $v.\pi = w.\pi$).

2.3   Creation of a New Modeler

The previous subsections highlight that a modeler is statically defined by its topological dimension and by the profile of considered embedding operations. These static data will intuitively be the first inputs that should be entered by using our editor, named *JerboaModelerEditor*, whose main function is the creation of a new modeler. The first step for creating a new modeler is thus to give its name and its dimension. The second step is the specification of all embeddings through a *pop-up*, that requires informations such as the embedding name, the associated orbit type and the data type. For execution issues, data types are given in terms of built-in or user-defined data types of the underlying programming language (in our case, Java).

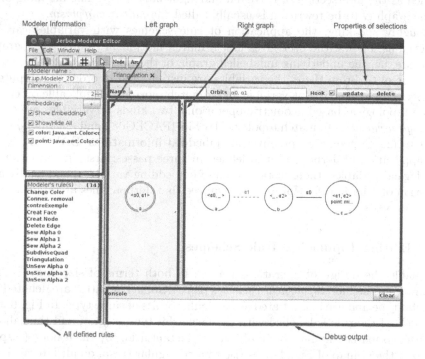

Fig. 5. Main interface of JerboaModelerEditor

The main *JerboaModelerEditor* window is organized in several parts (see Fig. 5). The upper leftmost box presents the previously described core informations as the modeler name, its dimension and the list of embeddings. The other parts of the window are dedicated to the description of operations. The lower leftmost box contains the current list of available operations (specified as rules) for the modeler under construction. The central boxes are used for the edition of a rule. More details are given in the next section devoted to the treatment (edition, verification, application) of rules.

3   Operations on G-Maps Defined as DPO Rule Schemes

3.1   Formal Framework

The formal background of Jerboa rules is the DPO approach [EEPT06], more precisely, the DPO approach of [HP02] devoted to labelled graphs, extended with variables in the style of [Hof05]. Roughly speaking, a DPO rule is in the form of a span of inclusions $L \hookleftarrow I \hookrightarrow R$, where L is the pattern to be matched in a graph G under modification and R is the new pattern to be inserted in the transformed graph, I being the rule interface, the common part of L and R. Our concrete syntax (see Fig. 6) contains only two graphs: the left-hand (resp. right-hand) side corresponds to the graph L (resp. R), the graph interface I being implicitly defined as the intersection of left-hand and right-hand sides. The mapping of L in the graph G to be rewritten is usually called the *match morphism*.

Roughly speaking, the application of rule schemes with variables can be sketched as follows: the user gives first a match morphism from the left graph structure, i.e. the underlying unlabelled graph, of the rule towards to the graph structure of G. From there, the variables are instantiated in order to compute a classical DPO rule without variable and a match morphism applicable on G. To define topological-based geometric operations, two kinds of variables are used: *topological variables* to match topological orbits [PACLG08], and *embedding variables* to match geometric or physical embedded informations [BALG11]. Thus the application of Jerboa rules is defined in three passes: first, instantiation of topological variables, then, instantiation of embedding variables, and finally application of a basic rule. However, we will see that Jerboa rules are applied in a one-pass process.

3.2   Editing Topological Rule Schemes

To enable the design of operations generic in both terms of size and of orbit nature, rules include *topological variables* [PACLG08]. The latter are denoted by an orbit type and are instantiated as particular orbits of same type. In Fig. 6, the left node of the rule is labelled with the face orbit type $\langle \alpha_0 \alpha_1 \rangle$, and thus allows the user to match any face of an object. Its instantiation by the a node (resp. g node) of the G-map of Fig. 4 gives rise to a triangular (resp. quadrilateral) face.

All nodes of the right-hand side carry an orbit type of the same length, but that can differ by deleting or relabelling arcs. Thus, all considered patterns are isomorphic up to some orbit correspondence. More precisely, by using the special

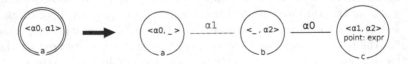

Fig. 6. Jerboa Rule of the triangulation for 2G-map

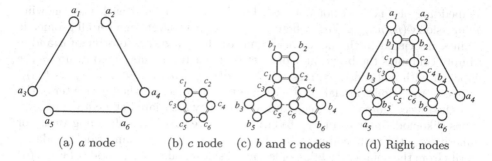

(a) a node   (b) c node   (c) b and c nodes   (d) Right nodes

Fig. 7. Topological instantiation

character '_', topological variables allow us to delete arcs. The left hand-side node $a : \langle \alpha_0 \_ \rangle$ combined with the face attached to the node a of the G-map of Fig. 4 allows us to build the G-map on Fig. 7(a): α_0-labelled arcs are preserved while α_1 ones are deleted. Thus, edges of the matched face are disconnected. Similarly, arcs can be relabelled. The instantiation of the node $c : \langle \alpha_1 \alpha_2 \rangle$ with the face attached to the node a of the G-map of Fig. 4 leads to the G-map of Fig. 7(b), where α_0-labelled arcs are relabelled to α_1 ones, and α_1 arcs to α_2 ones. Thus, a dual vertex of the matched face is added. Afterward, once each node is instantiated by an orbit, all these orbits are linked together. More precisely, each arc of the right-hand side graph is duplicated in several arcs linking instantiated nodes sharing the same index. For example, the instantiation of the α_0-arc linking nodes b and c of the right-hand side graph leads to 6 α_0-arcs linking b_i and c_i nodes in G-map on Fig. 7(c). Thus, edges are added around the dual vertex. Finally, the instantiation of the right-hand side of the rule of Fig. 6 on a triangle face gives rise to the G-map of Fig. 7(d).

Some left nodes of Jerboa rules are denoted with a double circle, and called *hook* nodes. Thus, in Fig. 6, the left node a is an hook. To apply a Jerboa rule, each hook node of the rule scheme must be map to a node of the target object. From an association between hook nodes and target graph nodes, the Jerboa library automatically computes the match morphism (if it exists).

3.3   Editing Geometrical Rule Schemes

The second rule edition step concerns the embedding counterpart of operations. In Fig. 6, since *point* is an embedding operation, the left node a of the rule is implicitly labelled with the *a.point* embedding variable. When instantiating the topological variable by a particular orbit, embedding variables are duplicated as many times as the size of the considered orbit. Thus for example on Fig. 7, all instantiated nodes a_1 to a_6 are implicitly labelled with embedded variables $a_1.point$ to $a_6.point$.

On the right-hand side of rules, embedding variables are put together in expressions built upon user-defined operations on embedding data types and some predefined iterators on the orbits. In Fig. 6, the full expression of point embedding of c node is not detailed, but it can be defined with the following expression $\Phi(point_{\langle \alpha_0, \alpha 1 \rangle}(a))$ where $point_{\langle \alpha_0, \alpha 1 \rangle}(a)$ collects in a set all geometric points associated to the nodes belonging to the $\langle \alpha_0, \alpha 1 \rangle$ face orbit of a and Φ simply computes the barycenter of a set of geometric points. c_1 to c_6 nodes are labelled with embedded expressions $\Phi(point_{\langle \alpha_0, \alpha 1 \rangle}(a_1))$ to $\Phi(point_{\langle \alpha_0, \alpha 1 \rangle}(a_6))$. Thus, c_1 to c_6 nodes are labelled with a common value, i.e. the barycenter of the matched face. Moreover, right nodes without any associated embedding expressions (like nodes a and b of Fig. 6) either preserve matched embedding values or inherit from their embedding orbit. For instance, the right hand side a node inherits from the point embedding of left hand side a node: each a_i node of Fig. 7(d) keeps its initial point embedding (as collected by the match morphism). Since the b node is α_1-linked with the a node in Fig. 6, it belongs to the same $\langle \alpha_1 \alpha_2 \rangle$ vertex orbit and inherits from the point embedding of a node. More precisely, each b_i node of Fig. 7(d) inherits from the value of the point embedding of the corresponding a_i node. Thus, as a result, point embeddings of the matched face are preserved and the point embedding of the new vertex added by c is set to the barycenter of the face.

As previously explained, the Jerboa editor illustrated in Fig. 5, allows the user to graphically edit left and right rule scheme patterns. In addition, an informative toolbar summarizes the current selection and offers buttons to create/modify the rule scheme, especially topological and embedding labels. Nonetheless, it allows to change many settings like hide/draw the alpha link name, color convention and so on. Finally, the editor can generate an image from the current rule (Jerboa rules of this article were created with the SVG export).

3.4   Syntax Checking

The main advantage offered by this editor is an on-the-fly verification of the rule sheme, avoiding compilation errors and debugging. At each rule modification, the editor checks simple syntactic properties numbered from (i) to (v) thereafter. (i) usual lexical analyses are applied to identifiers and expressions annotating rules. (ii) all labels carried by arcs are of form α_i with i an integer less than or equal to the modeler dimension. (iii) all orbit types labelling nodes are of the same size, that is, contain exactly the same number of (α or _)-elements; thus, all nodes instantiated patterns are isomorphic only up to relabelling and deletion of arcs. (iv) each hook node should be labelled by a full orbit type, that is, by $\langle \alpha_i \ldots \alpha_j \rangle$ without any '_' occurring in it. Indeed, at the instantiation step, the targeted orbit is built by performing a graph traversal from a selected node, and guided by the arc labels given in the full orbit type. (v) lastly, each connected component of the left hand side of the rule should contain exactly one single hook. This last condition allows us to compute an unique match morphism (if it exists) or to trigger a warning message.

Beyond these basic syntactic properties, we showed in [PACLG08, BALG11] that some additional syntactic properties on rules can be considered to ensure object consistency preservation through rule application, that is, to ensure that

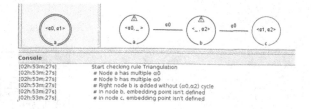

Fig. 8. Topology verification by JerboaModelerEditor

resulting objects satisfy both the topological consistency and embedding constraints given in Sections 2.1 and 2.2. For example, the preservation of the adjacent arc constraint can be ensured by checking for each node how many arcs are linked to. First, all removed and added nodes should have exactly $n + 1$ arcs respectively α_0 to α_n-labelled, including arcs that are implicitly handled by orbit types attached to them. In Fig. 6, the b node has two explicit arcs linked to it and labelled resp. by α_0 and α_1 and one implicit arc linked to it and labelled by α_2, provided by its topological type ($\langle \_\alpha_2 \rangle$). Secondly, preserved nodes must have the same links on both sides. In Fig.6, on the left, the a node, typed by $\langle \alpha_0 \alpha_1 \rangle$, is given with two implicit arcs, resp. labelled by α_0 and α_1, and on the right, the a node, being typed by $\langle \alpha_0 \_ \rangle$, keeps an implicit α_0-arc and is linked to a new α_1-arc.

To preserve the embedding constraint, Jerboa checks that each embedding orbit of Jerboa rule carries one and only one embedding expression. But the instantiation of topological variables produces several embedding expressions which do not necessarily have the same value. Jerboa computes the first one and ignores the following ones: by default, the computed value is attached to all nodes belonging to the embedding orbit.

By lack of space, other advanced syntactic conditions are not more detailed. All syntactic conditions are on-the-fly checked by the JerboaModelerEditor. The editor shows the encountered errors directly in the graph and identifies the incriminated nodes. In Fig. 8, nodes a and b are decorated by warning pictograms. All errors are detailed in the console: the right-hand side a node has two α_0 links and loses an α_1 link, and so on.

3.5   Jerboa Rule Application

The Jerboa rule application has been sketched by means of the three following steps: the instantiation of topological and embedding variables, and the application itself. In practice, these steps are done at once by the rule application engine encompassed in the Jerboa library. Let us emphasize that this rule application engine is unique and is able to apply any rule of any generated modeler.

First the engine instantiates the topological variables and computes the match morphism if possible. For example the application of the Jerboa rule of Fig. 6

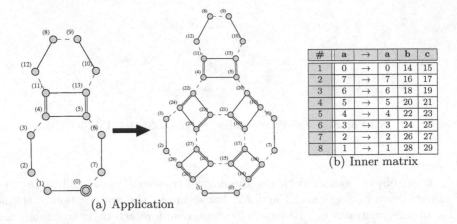

(a) Application

#	a	→	a	b	c
1	0	→	0	14	15
2	7	→	7	16	17
3	6	→	6	18	19
4	5	→	5	20	21
5	4	→	4	22	23
6	3	→	3	24	25
7	2	→	2	26	27
8	1	→	1	28	29

(b) Inner matrix

Fig. 9. Rule application and inner structure of the triangulation

on the left object[1] of Fig. 9(a) allows to complete the first column of the inner matrix given on Fig. 9(b). For that, once the user has mapped the hook node a of the rule scheme to the node (0) of the object, one line is created in the matrix for each node of the orbit $\langle \alpha_0 \alpha_1 \rangle$ of (0). For a rule with multiple hooks, this step can fails and the engine triggers an exception.

Secondly, the right part of the inner matrix is completed in accordance with the right-hand side of the rule. As shown on Fig. 9(b), names of preserved nodes are copied, and names of added nodes are created. Then, the created part of the rewritten object is computed. This part corresponds to the two last columns of nodes b and c on Fig. 9(b). Each arc of the rule is duplicated for each inner matrix line. Thus, the α_0-arc between b and c nodes produces α_0-arcs between (14) and (15) nodes, ..., between (28) and (29) nodes. First, each arc of the instantiated orbit of the hook is translated for each inner matrix columns up to relabelling or removing on added nodes. The α_1-arc between (0) and (7) is α_2 relabelled between (14) and (16), and between (15) and (17), the α_0-arc between (7) and (6) is not added between (16) and (18), and α_1 relabelled between (17) and (19), ..., the α_0-arc between (1) and (0) is not added between (28) and (14), and α_1 relabelled between (29) and (15).

Before further topological modification, the new embedding values are computed. Indeed, evaluations of embedded expressions depend on initial embedding values, but also on the topological links in the initial object. We use the fact that all orbits of same embedding type necessarily contain nodes sharing the same embedding value (with respect to this considered embedding function). The engine computes new embedding values but does not yet replace them in the nodes to avoid any corruption of the next embedding value computations. Instead, those values are memorized in a private buffer of the engine. The treatment of

---

[1] Note that *graphviz* exportation is used to generate from our generic viewer that uses barycentric coordinates as exploded view.

topological variables is completed by relabelling the preserved nodes with new embedding values and connecting them to the added nodes with appropriate embedding values.

This way of processing Jerboa rule application calls for a comment on the management of the embedding values. The case of the merge of two embedding orbits could lead to a non-deterministic choice between the two original embedding values. To avoid such a situation, the editor asks for a unique embedding expression, that is sufficient to ensure the embedding expression constraint.

4   Discussion

Examples. We briefly introduce some non-trivial operations. First, we propose the Catmull-Clark operation used to smooth face by performing some face subdivision mechanisms. The rule for a 3D modeler is described on[2] Fig. 10(a) and Fig. 10(b) illustrates the successive applications of this rule on a mesh.

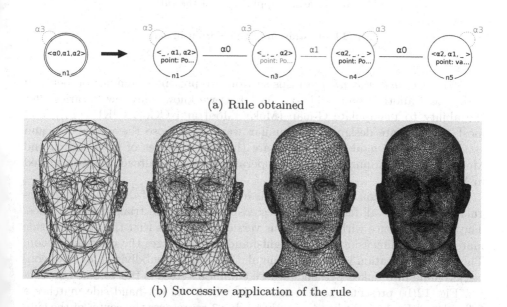

(a) Rule obtained

(b) Successive application of the rule

Fig. 10. Illustration of the Catmull-Clark smoothing operation

Second, the *Sierpinski carpet* is a fractal on a 2D surface (in higher dimension, this fractal is called Menger sponge). The rule, presented on Fig. 11(a), matches a face and produces eight faces from it. Fig. 11(b) shows successive applications of this rule on a simple 2D face.

---

[2] By lack of space, complete embedding expressions are omitted.

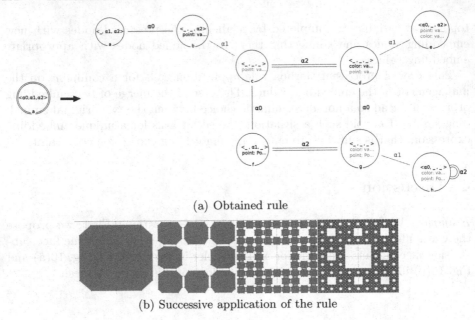

(a) Obtained rule

(b) Successive application of the rule

Fig. 11. Illustration of the Sierpinski carpet

Performance and evaluation. In this section, we present a comparison between Jerboa and another similar library. As far as we know, only few libraries offer the ability to manipulate G-map (Moka [Mod] and CGoGN [KUJ+14]), and most of them are designed in a similar way. We choose the well-known and established Moka modeler and compare the performances of both Jerboa and Moka by benchmarking them on two operations that are already fully optimized in Moka.

The first operation (see Fig. 12(a)) is a generalization of the basic face triangulation in which all faces of a connected component are triangulated at once. The main difference with the previous version (Fig. 6) is that the left-hand side matches more dimension and the right-hand side manages the third dimension. The second operation is volume triangulation, i.e. the subdivision of any (convex) volume into linked pyramids whose common vertex is the volume barycenter. Fig. 12(b) presents the Jerboa rule in which the left-hand side matches a full volume. In the right-hand side, the node **n3** represents the center of the subdivision while nodes **n2** and **n3** represent sides of inserted faces. Let us notice that the border faces remain unmodified.

Tests have been executed on a Core i7-2600 – 3.4GHz with 8GB of RAM, under Linux/Ubuntu 12.04 LTS system with JDK 7 of Oracle. We computed the average time of operation execution in both modelers with a single embedding of geometric points[3] to represent 3D objects. We used the same objects for both libraries, with various sizes (from 4 nodes to 3 millions nodes). To summarize,

---

[3] By default, a Moka modeler is only defined by this unique embedding

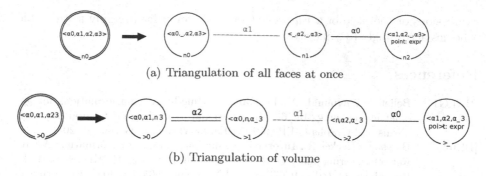

(a) Triangulation of all faces at once

(b) Triangulation of volume

Fig. 12. Two Jerboa rules of triangulation operations

Jerboa shows better performances than Moka for the triangulation of all faces at once whereas Moka is better considering the volume triangulation. There is no better optimization than tuned optimization carefully performed by a developer, and Jerboa is implemented in Java language that is recognized to be slower than C++: this can explained why generally, Moka has still better performances than Jerboa. The better performances of Jerboa for the face triangulation operation is due to the fact that the Moka developer has intensively reused existing codes.

To conclude this section, we wish to emphasize the ease of developing new operations using Jerboa. Classically, the development of a modeler's operation is done by hand-coding with all induced problems: debugging, verification steps on customized objects, and so on. While developers usually mix inside the same code topological and embedding considerations, developing with Jerboa imposes a clear separation of topological and embedding aspects, and even more, requires that static parts are declared before designing operations. Thus, as regards the code development, using Jerboa brings two clear advantages: a significant gain in time and a high level of code quality. For instance, for the two operations discussed above and given in Fig. 12, we only took half a day.

5   Conclusion and Future Works

This article introduces a novel tool set dedicated to rule-based geometric modeling based on G-maps. This tool set includes the JerboaModelerEditor, that allows a fast characterization of any new dedicated modeler and a fast design of its operations in a graphical manner, assisted by static verification steps. When the design is over, the Jerboa library produces a full featured modeler kernel that can be used in a final application. Moreover, the produced modeler is highly reliable as generated rules take benefit from graph transformation techniques ensuring key consistency properties.

Jerboa has been successfully used in other works, especially for the adaptation of L-Systems with G-map or in an architecture context (see Fig. 1). These experimentations allowed us to identify some required features for the next version of Jerboa, as the need of stronger verification mechanisms of the embedding expressions or the need of a *rule script* language in order to apply several rule schemes accordingly to some strategies.

References

[BALG11]   Bellet, T., Arnould, A., Le Gall, P.: Rule-based transformations for geometric modeling. In: 6th International Workshop on Computing with Terms and Graphs (TERMGRAPH), Saarbrucken, Germany (2011)

[BH02]   Baresi, L., Heckel, R.: Tutorial introduction to graph transformation: A software engineering perspective. In: Corradini, A., Ehrig, H., Kreowski, H.-J., Rozenberg, G. (eds.) ICGT 2002. LNCS, vol. 2505, pp. 402–429. Springer, Heidelberg (2002)

[BPA+10]   Bellet, T., Poudret, M., Arnould, A., Fuchs, L., Le Gall, P.: Designing a topological modeler kernel: A rule-based approach. In: Shape Modeling International (SMI 2010). IEEE Computer Society (2010)

[EEPT06]   Ehrig, H., Ehrig, K., Prange, U., Taentzer, G.: Fundamentals of Algebraic Graph Transformation. Monographs in Theoretical Computer Science. An EATCS Series. Springer (2006)

[Hof05]   Hoffmann, B.: Graph transformation with variables. Formal Methods in Software and System Modeling 3393, 101–115 (2005)

[HP02]   Habel, A., Plump, D.: Relabelling in graph transformation. In: Corradini, A., Ehrig, H., Kreowski, H.-J., Rozenberg, G. (eds.) ICGT 2002. LNCS, vol. 2505, pp. 135–147. Springer, Heidelberg (2002)

[KBHK07]   Kniemeyer, O., Barczik, G., Hemmerling, R., Kurth, W.: Relational growth grammars - a parallel graph transformation approach with applications in biology and architecture. In: Schürr, A., Nagl, M., Zündorf, A. (eds.) AGTIVE 2007. LNCS, vol. 5088, pp. 152–167. Springer, Heidelberg (2008)

[KUJ+14]   Kraemer, P., Untereiner, L., Jund, T., Thery, S., Cazier, D.: CGoGN: n-dimensional meshes with combinatorial maps. In: 22nd International Meshing Roundtable. Springer (2014)

[Lie91]   Lienhardt, P.: Topological models for boundary representation: A comparison with n-dimensional generalized maps. Computer-Aided Design 23(1) (1991)

[Mod]   Moka Modeler. XLim-SIC, http://moka-modeller.sourceforge.net/

[MP96]   Měch, R., Prusinkiewicz, P.: Visual models of plants interacting with their environment. In: 23rd Conference on Computer Graphics and Interactive Techniques, SIGGRAPH. ACM (1996)

[PACLG08]   Poudret, M., Arnould, A., Comet, J.-P., Le Gall, P.: Graph transformation for topology modelling. In: Ehrig, H., Heckel, R., Rozenberg, G., Taentzer, G. (eds.) ICGT 2008. LNCS, vol. 5214, pp. 147–161. Springer, Heidelberg (2008)

[PLH90]   Prusinkiewicz, P., Lindenmayer, A., Hanan, J.: The algorithmic beauty of plants. Virtual laboratory. Springer (1990)

[TGM+09]   Terraz, O., Guimberteau, G., Merillou, S., Plemenos, D., Ghazanfarpour, D.: 3Gmap L-systems: An application to the modelling of wood. The Visual Computer 25(2) (2009)

[VAB10]   Vanegas, C.-A., Aliaga, D.-G., Benes, B.: Building reconstruction using manhattan-world grammars. In: Computer Vision and Pattern Recognition (CVPR). IEEE (2010)

Author Index